SCHOOL OF ORIENTAL AND AFRICAN STUDIES
University of London

Please return this book on or before the last date shown

Long loans and One Week loans may be renewed up to 10 times
Short loans & CD's cannot be renewed
Fines are charged on all overdue items

Online: http://lib.soas.ac.uk/patroninfo
Phone: 020 7898 4197 (answerphone)

_ 8 AUG 2014

INDIAN INK

INDIAN INK

Script and Print in the Making of the English East India Company

MILES OGBORN

THE UNIVERSITY OF CHICAGO PRESS

CHICAGO AND LONDON

MILES OGBORN is professor of geography at Queen Mary, University of London.

The University of Chicago Press, Chicago 60637
The University of Chicago Press, Ltd., London
© 2007 by The University of Chicago
All rights reserved. Published 2007
Printed in the United States of America

16 15 14 13 12 11 10 09 08 07   1 2 3 4 5

ISBN-13: 978-0-226-62041-1 (cloth)
ISBN-10: 0-226-62041-7 (cloth)

Library of Congress Cataloging-in-Publication Data

Ogborn, Miles.
    Indian ink : script and print in the making of the English East India Company /
Miles Ogborn.
        p. cm.
    Includes bibliographical references and index.
    ISBN-13: 978-0-226-62041-1 (cloth : alk. paper)
    ISBN-10: 0-226-62041-7 (cloth : alk. paper)
        1. East India Company—History.   2. Printing—Political aspects—India—Bengal—
History.   3. Bengal (India)—Colonization—History.   4. England—Commerce—
History.   I. Title.
    DS465.O43 2007
    954'.14031—dc22

                                                                        2006032457

TO MAURICE OGBORN (1907–2003)
AND OLIVE OGBORN (1911–)

CONTENTS

BL	British Library
BMS	Baptist Missionary Society Archives, Angus Library, Regent's Park College, Oxford
Cal. S.P., *1513–1616*	W. N. Sainsbury (ed.) (1862) *Calendar of State Papers. Colonial Series: East Indies, China and Japan, 1513–1616* (Longman, Green, Longman & Roberts, London). Citations are to entry numbers.
Cal. S.P., *1622–1624*	W. N. Sainsbury (ed.) (1878) *Calendar of State Papers. Colonial Series: East Indies, China and Japan, 1622–1624* (Longman & Co., London). Citations are to entry numbers.
Cal. S.P., *1625–1629*	W. N. Sainsbury (ed.) (1884) *Calendar of State Papers. Colonial Series: East Indies, China and Persia, 1625–1629* (Longman & Co., London). Citations are to entry numbers.
Eur Mss.	European Manuscripts, Oriental and India Office Collection, British Library
IOR	India Office Records, Oriental and India Office Collection, British Library
OIOC	Oriental and India Office Collection, British Library
RFSG, *Despatches,* *1670–1677*	Records of Fort St. George (1911) *Despatches from England, 1670–1677* (Government Press, Madras).
RFSG, *Despatches,* *1680–1682*	Records of Fort St. George (1914) *Despatches from England, 1680-1682* (Government Press, Madras).
RFSG, *Diary,* *1672–1678*	Records of Fort St. George (1910) *Diary and Consultation Book, 1672–1678* (Government Press, Madras).
RFSG, *Diary,* *1678–1679*	Records of Fort St. George (1911) *Diary and Consultation Book, 1678–1679* (Government Press, Madras).

RFSG,   Records of Fort St. George (1913) *Diary and Consultation*
Diary, 1681   *Book, 1681* (Government Press, Madras).

༺༻

ACKNOWLEDGMENTS

M any people have helped me in the long process of researching and writing this book. They have done so through conversations both extended and brief; by sharing unpublished work, references, ideas, and images; by deploying linguistic skills; and by engaging in critique (gentle or not) when earlier versions of this work were presented at seminars and conferences. I owe particular debts to Jayani Bonnerjee, Richard Bourke, Huw Bowen, Stephen Caffey (who gave me the title for chapter 6), Dan Clayton, Andrew S. Cook, Hal Cook, Phil Crang, Michael Dodson, Florence D'Souza, Markman Ellis, Jim Epstein, David Featherstone, Anindita Ghosh, Andrew Grout, Darab Haghipour, Santhi Hejeebu, Amin Jaffer, Shompa Lahiri, David Lambert, Javed Majeed, Peter Marshall, Robert Mayhew, Kapil Raj, Fiona Ross, Jim Secord, Álvaro Sequeira Pinto, Sujit Sivasundaram, Philip Stern, and John Styles. I know most of them will have their doubts, but I accept my responsibilities for what I have done with their wise counsel.

This work would certainly not have been possible without the support and inspiration of other historical geographers, particularly Mona Domosh, Felix Driver, Michael Heffernan, David Livingstone, Chris Philo, and especially Charles Withers, a fellow geographer of the book. At the Department of Geography, Queen Mary, University of London, I am very lucky to have colleagues and friends who are willing to tolerate someone so irretrievably stuck in the past, especially Alison Blunt, Ray Hall, Roger Lee, Jon May, Philip Ogden, Alastair Owens, Bronwyn Parry, David Pinder, Simon Reid-Henry, Adrian Smith, and Nigel Spence. I am very grateful to Edward Oliver for drawing the maps.

This work has been supported by a Philip Leverhulme Prize from the Leverhulme Trust (2001–2003) and a small grant from the British Academy. It has benefited from the assistance of librarians in the Oriental and India

Office Collection and the Rare Books and Manuscript rooms of the British Library, the Special Collections room and Goldsmiths' Library of the University of London's Senate House Library, the William Andrews Clark Memorial Library (Los Angeles), the Huntington Library (San Marino), and the New York Public Library. I would like to thank Blackwell Publishers for letting me reuse material originally published as "Writing travels: power, knowledge and ritual on the English East India Company's early voyages," in *Transactions of the Institute of British Geographers* (2002), and Hodder Arnold for letting me reproduce material first published in *Cultural Geographies*.

This book would not have been brought into being as an object in the world without the rare combination of enthusiasm and calm professional skill possessed by Christie Henry at the University of Chicago Press. I also thank the publisher's readers for their insightful comments and encouraging words, and Michael Koplow for his careful and flexible copyediting.

Others have helped by not asking too much. Nicholas Robson, Juliette Enser, John (J.P.) Horan, John Maher, Charlie Smith, Michael Pryke, Christian Frost, Scott McCracken, Michael Fitzgerald, and Patrick Fitzgerald; Jon and Joan; Kate, Nick, and Cora; Harriet, Andrew, Matthew, and Louis; Bridgit and David Nash; Gráinne and David Forrai; and, of course, Jane Ogborn (for whom books are not a thing but *the* thing) will all wonder how it took so long. This book is dedicated to my grandfather, an actuary and company historian who knew the power of words and numbers, and my grandmother, a woman who loves a good story.

In the end it is Catherine Nash without whom none of this would be worthwhile. She knows what it means that the monkey of the inkpot can now unfold its silken paws, drink its dark draft, and sit back on its haunches to rest.

PREFACE

Imagine the natural philosopher Robert Boyle in around 1689 (Figure 1). He sits confidently in his robes, a Christian and a gentleman in an ornate but decorous room. He gestures towards the book lying upon the table in front of him, and he turns up the corner of a page. Since Boyle was one of the founders of a new mode of experimental knowledge buttressed by Protestantism, printing, and the procedures of the Royal Society, it may be assumed that this single volume contained significant truths. Imagine, however, all the other books and papers beyond the frame that have had Boyle's hands or eyes upon them, or the hands and eyes of those he engaged to read and write for him. What might they have to say about the certainties and uncertainties of ways of knowing and communicating?[1] In particular, imagine the ways those handwritten and printed sheets, as loose leaves or bound into volumes, connected this virtuoso to the far-off East Indies, where he himself had never been. What might then be discovered about the making of knowledge, the determination of truth, and the exercise of power in an age of trade and empire?

Within his extensive correspondence we do indeed find that Boyle's world was also the world of the English East India Company. The Company had been established in the late sixteenth century by London merchants who were keen to exploit the direct sea route from Asia to bring valuable spices to European markets. In 1600 they were granted a royal charter by Queen

---

1. On Boyle, experimental knowledge, and books see Steven Shapin and Simon Schaffer (1985) *Leviathan and the Air-Pump: Hobbes, Boyle, and the Experimental Life* (Princeton University Press, Princeton, NJ); Steven Shapin (1994) *A Social History of Truth: Civility and Science in Seventeenth-Century England* (University of Chicago Press, Chicago); and Adrian Johns (1998) *The Nature of the Book: Print and Knowledge in the Making* (University of Chicago Press, Chicago). For Boyle's amanuenses, see Ann Blair (2004) "Focus: scientific readers—an early modernist's perspective," *Isis*, 95, pp. 420–30.

Figure 1 Robert Boyle, c. 1689. (National Portrait Gallery, London.)

Elizabeth I, and with it a potentially lucrative monopoly on trade beyond the Cape of Good Hope. At both the beginning and end of the 1670s Robert Boyle was a member of the Court of Committees that managed the Company's operations.[2] By then the English Company had been forced out of the

2. Robert Boyle (2001) *The Correspondence of Robert Boyle*, Michael Hunter, Antonio Clericuzio, and Lawrence M. Principe (eds.) (Chatto and Pickering, London), vol. 4: *1668–1677*, p. 458 n.

Indonesian archipelago and the spice islands by the more powerful Dutch East India Company (the Verenigde Oost-Indische Compagnie, or VOC). They had, however, found a very rich trade exchanging bullion for Indian cotton and silk cloth through coastal factories (as their trading posts were known) organized from Bombay (now Mumbai) and Fort St. George (later Madras, now Chennai). Boyle would therefore have been party to discussions over the organization of the annual fleet of East Indiamen, the management of the factories, and the prices of Indian cloth in London, as well as over the roles of king and Parliament in the renewal of the Company's charter and the crucial question of its continued monopoly position in those boom years. The Company was the centerpiece of late seventeenth- and early eighteenth-century English overseas trade. From different political and economic points of view, it was either vital to the nation's prosperity or a significant drain upon national wealth. Many took an interest in it.[3] Boyle himself owned East India stock, as did the Royal Society. He also used his position in the Company to find places for those who sought his patronage.[4] However, mindful perhaps of criticisms of the workings of the stock market, a late seventeenth-century novelty and concern, the natural philosopher was keen to stress that his involvement with the Company was out of "the desire of knowledge, not profit."[5]

The Company's world was one made on paper as well as on land and sea. Boyle's relationships to the Company were often worked out in script through networks of letter writers. Henry Oldenberg, with his correspondents all across Europe, kept Boyle informed of the shifting rivalries between the Dutch and English trading companies.[6] Boyle was also the recipient of letters from aspiring natural philosophers bound for India in the Company's service, or pondering their new and unfamiliar surroundings, who reported their findings on heat and cold, weather and terrestrial magnetism, plants

3. Between 1698 and 1708 there were two English East India companies, a new company being chartered on its promise to lend the king substantial sums. The old East India Company bought a substantial share of the new one and they were merged as the United Company of Merchants Trading to the East Indies in 1709.

4. *Correspondence of Robert Boyle*, vol. 4: *1668–1677*, Thomas Hyde to Boyle, 23 October 1671.

5. The Royal Society bought Company stocks from John Evelyn; see John Evelyn (1955) *The Diary of John Evelyn*, E. S. de Beer (ed.) (Clarendon Press, Oxford) vol. 4: *Kalendarium, 1673–1689*, pp. 297–98. Boyle is quoted in Michael Hunter (ed.) (1994) *Robert Boyle by Himself and His Friends with a Fragment of William Wotton's Lost* Life of Boyle (William Pickering, London) p. lxviii.

6. A. Rupert Hall and Marie Boas Hall (eds) (1966) *The Correspondence of Henry Oldenburg*, vol. 2: *1663–1665* (University of Wisconsin Press, Madison), Oldenburg to Boyle, 1 September 1664.

and animals, medicines and magical stones, and the customs and religions of strange peoples.[7] Their reports arrived in London with the valuable cargoes of cloth and spices, and with the Company's own letters and the accounts cast up for each of the factories. They were sometimes read at Royal Society meetings, and evaluated in ways that weighed the Company's own reputation in the scales of credit and credibility alongside questions of gentlemanly civility.[8] Some of these investigations were directed via lists of queries produced by the Society. In other cases Company ships were sent with particular "Philosophical commissions" from Boyle and his colleagues.[9] India and points east were a source of the combination of wonders and systematic observations that animated late seventeenth-century natural philosophy.[10]

Robert Boyle's Company life was also shaped by print. The mid-seventeenth century saw a massive increase in the production of printed materials in England. These were primarily concerned with religion and politics, but the power of the press was also harnessed in the realms of trade and natural philosophy. Boyle's correspondents gave their opinions, and sought his in return, on controversial tracts and periodicals relating to the politics and economics of the East India trade.[11] He certainly planned to send books to India, both his own philosophical texts and, later, religious works.[12] Indeed, Boyle's second spell as a Company director was largely taken up with the printing of religious texts in Indian languages with the aim of "Gospellizing" the natives. This would save both their souls and the Company's reputation,

---

7. *Correspondence of Robert Boyle*, vol. 2: *1662–1665*, Nathaniel Foxcroft to Boyle, 29 November 1664; vol. 3: *1666–1667*, Nathaniel Foxcroft to Boyle, 11 September 1666; and vol. 6: *1684–1691*, Thomas Mackrith to Boyle, 5 February 1686.

8. For example, see *Correspondence of Robert Boyle*, vol. 2: *1662–1665*, Robert Hooke to Boyle, 10 July 1663. See also Boyle's evaluation of a report, probably from Sir Edward Winter, the governor of Fort St. George from 1662 to 1665: "it is preferable to any I have been heitherto able to gain among the Marchants & Seamen: and comes from an ingenious Knight, whose Quality & station at the place where he is the Fort St George (intrusted to him by the famous East India Company of England) makes me conclude that the account he gives me, as imperfect as we may thinke it, is the best that can be easily procured upon the place...," Vol. 6: *Appendices*, [Boyle] to unknown correspondent, no date.

9. *Correspondence of Robert Boyle*, vol. 4: *1668–1677*, Oldenburg to Boyle, 28 January and 25 February 1668.

10. Lorraine Daston and Katherine Park (1998) *Wonders and the Order of Nature, 1150–1750* (Zone Books, New York). John Evelyn recorded his wonder at examples of Chinese paper and writing among items brought from the East. *Diary of John Evelyn*, vol. 3: *Kalendarium, 1650–1672*, pp. 373–74.

11. *Correspondence of Robert Boyle*, vol. 5: *1678–1683*, John Beale to Boyle, 1 and 8 July 1682.

12. *Correspondence of Robert Boyle*, vol. 4: *1668–1677*, Boyle to Samuel Clarke, 29 May 1669.

"remembering ourselves to be both Christians and Merchants."[13] However, Boyle's intentions in this direction came to little. Making such books was a difficult business. In London, there was his own ill health, the difficulties of organizing patronage, foot-dragging among other committees more concerned with profit than proselytizing, and the problems of obtaining "vocabularies for teaching the Malayan tongue to Merchants and Interpreters" that could be turned to more godly uses.[14] Without these word lists there was the danger that the faults of the available Dutch models would simply be reprinted, "and make the thing seem non-sense to the Natives for whom you intend it."[15] In addition, to ensure that what was written and printed might subsequently be spoken there was the need to secure appropriate positions in India for suitable chaplains. Securing these posts meant Boyle—judged "the only fit person to move in it," with his elevated status and East India Company connections—needed to approach the Company's powerful director, Sir Josiah Child. This controversial figure, whom Boyle's fellow Royal Society member John Evelyn had judged to be "most sordidly avaricious," was the target of allegations by the Company's critics of illegitimate private interests and the manipulation of stock prices, as well as being the author of widely read tracts on the public benefits of the India trade.[16] Money was an issue in other ways too. Boyle could hope to bankroll the printing project only when he was certain that his Irish affairs were in good shape after 1688. Other, similar publishing schemes were also hamstrung when dramatic falls in the value of East India Company stock had, as Boyle said, "so prejudicial an Influence upon my little Affairs, as forbids me to do what I should otherwise be able to keep the Printer from being quite despondent."[17]

---

13. *Correspondence of Robert Boyle*, vol. 4: *1668–1677*, Boyle to Robert Thompson, 5 March 1677. It might also serve the purposes of natural philosophers interested in Oriental languages. Vol. 6: *1684–1691*, Edward Bernard to Boyle, 1 August 1689.

14. *Correspondence of Robert Boyle*, vol. 4: *1668–1677*, Boyle to Robert Thompson, 5 March 1677, and Thomas Hyde to Boyle, 16 October 1677.

15. *Correspondence of Robert Boyle*, vol. 4: *1668–1677*, Thomas Hyde to Boyle, 16 October and 29 November 1677; and vol. 6: *1684–1691*, Hyde to Boyle, 23 February 1689. Hyde also sought, in the same letter, to profit from this hard-won specialist knowledge and the hard work of printing and correcting in "a strang Character." He wanted Boyle to recommend him for the position of minister at Stalbridge in Dorset, and putting his case noted that, in terms of printing in Malay, "I am sure there is no man now alive in England, who is able to serve you besides my self."

16. *Correspondence of Robert Boyle*, vol. 6: *1684–1691*, Hyde to Boyle, 11 June 1691, and *Diary of John Evelyn*, vol. 4: *Kalendarium, 1673–1689*, p. 306 [16 March 1683].

17. *Correspondence of Robert Boyle*, vol. 6: *1684–1691*, Hyde to Boyle, 23 February 1689; and vol. 5: *1678–83*, Boyle to Narcissus Marsh, 28 April 1683, on the proposal for an Irish grammar.

Further obstacles also remained to be overcome in India. Boyle learned from a letter from the Coromandel coast sent by the Company's agent at Fort St. George, Streynsham Master (with whom we will become very well acquainted in a later chapter), that "Gentue and Malabar" were the languages spoken there, and not Malay. Moreover, that there were "very few English who can speake the Languages," and no suitable "Vocabularies." Master also doubted that the chaplains currently in India had any enthusiasm for making converts or for understanding "the Manners, Coustumes, and Humours of the Natives," being "so well pleased with their own school-learning and manners that they undervalue all others," either English or Indian. There was, furthermore, no place for public worship, and there were great divisions between Anglican and Nonconformist chaplains and governors. Much to Master's frustration, but demonstrating an astute grasp of the dynamics of religion and politics in the seventeenth-century English confessional state, the local people refused to recognize the subtleties of Protestant theology involved in these differences. As he put it, "[we] can never make the Natives understand the differences between Conformists and Non-conformists to be any other than a Lover of the King and Lawes, and a Rebell of Cromwells party."[18] This reading of English religion as politics, together with the strength of Indian faith and the debating skills of the faithful, made it a vain hope that "Pagan worship" might be overcome by printing and shipping to India what Boyle envisioned as "a solid, but civilly pennd, Confutation of the Authentick Bookes wherein the Bramins religion is contain'd."[19]

Indeed, these problems with using print as part of European attempts to reshape Asian societies were encountered again long after Boyle's projects had failed, and within a new role for the Company. From the second half of the eighteenth century the English East India Company became the territorial ruler of a large part of India. Benefiting from the decline of the Mughal empire, and battling with other successor states, the Company was charged with the gathering of taxation and the administration of law in Bengal, Bihar, and Orissa. The Company introduced printing in Indian languages to northeast India in the service of imperial governance. However, the advent and administration of this company-state as part of the British empire was no simple matter. The principles, forms, and mechanisms of Company rule were all the subject of controversy both in India and in England. Once again,

18. *Correspondence of Robert Boyle*, vol. 6: *Appendices*, Streynsham Masters [sic] to Samuel Masters [sic], 9 December 1678.

19. *Correspondence of Robert Boyle*, vol. 4: *1668–1677*, Boyle to Robert Thompson, 5 March 1677.

those using print to enter or end these controversies faced the local difficulties of how to use print to make the books they wanted, and they had
problems with readers too (see chapter 6).

In Boyle's connections to the English East India Company, and in the letters, papers, and books through which they were pursued, can be found the
varied forms of "Indian ink" that are the substance of this book. There are
the politics of written charters giving royal authority and granting monopoly
privileges to groups of merchants. There are the necessities and vagaries of
long-distance communication by letter for which matters of veracity, authority, and civility needed to be established. There was the debating of
Company privileges in print between many different interests with unequal
access to money, patronage, and the publishing process. There was the importance of accurate information on the rise and fall of stock prices, and a
concern that stocks and shares were themselves troublesome things. And,
finally, there were the potentials and problems of printing in Eastern languages, and the uncertainty of readers' responses. This book considers Indian
ink that flowed from the reed, the quill, or the pen, or was imprinted by the
wooden hand press. It interprets Indian ink that was spilled for or against
the Company in Asia and Europe, and whether it shaped English, Persian,
or Bengali characters. All these varied uses of Indian ink demonstrate the
importance of different modes of writing to the English East India Company
in the seventeenth and eighteenth centuries. Robert Boyle was interested
and involved in them all because the Company, and those who sought to
either benefit from it or to challenge it, in Europe or in Asia, had to interest
and involve themselves in these modes of writing and in their potentially
powerful and far-reaching effects.

Robert Boyle's life and work signified then on matters of knowledge,
power, and practice, and it signifies now too. Historians of science have
used Boyle's engagement in the seventeenth-century scientific revolution
and the foundation of the Royal Society to argue for a conception of scientific
knowledge that understands it as an engagement with political concerns that
are inseparable from matters of practice. The forms of experimental life, the
rules that governed natural philosophers' relationships to one another, and
the matters that could or could not be discussed and adjudicated upon by
them have been used to demonstrate that "[s]olutions to the problem of
knowledge are solutions to the problem of social order."[20] Knowledge is
social because adjudication on truth is a matter of the social order that
guarantees how truths can be arrived at and secured. Knowledge-making is

20. Shapin and Schaffer, *Leviathan and the Air-Pump*, p. 332.

political because this attention to social order necessarily both requires and suggests modes of ordering that differ over the organization of power.[21] This, it is argued, is not merely about public pronouncements or the ideological positioning of science. The politics of knowledge works right down to the most basic level of practice. Which instruments and techniques are used, how they are used, and by whom they are used are also matters of political and social order and are vital to the making of scientific truth. Boyle's air pump has, therefore, now become known for its usefulness in separating contentious politics and credible science, and thereby making certain sorts of truth, as well as (indeed, at the same time as) its creation of a vacuum in which experiments might be done.[22]

In practice, therefore, Boyle's knowledge-making depended upon matters of geography and writing. His primary achievement in the conjoined realms of politics and natural philosophy was the creation of "an intellectual *space*."[23] At once abstract and concrete, public and of restricted access, this novel and enduring laboratory space permitted both the forms of experiment and the witnessing of these experiments by the philosophical community in a way that produced truth and provided a model of political order. Through the careful use of writing and printing, this space also supported modes of virtual witnessing that went far beyond its concrete confines to create a far-flung philosophical commonwealth and a notion of universal scientific truth.[24] As well as illuminating the forms of Indian ink at work in and around the English East India Company, Robert Boyle's life and work signal the importance in studies of the history of knowledge-making (within natural philosophy and beyond) of understanding the geographies, both locally situated and widely dispersed, of the material practices of writing, printing, and reading. This book argues that it matters what the geographies of these forms of Indian ink were: where and how they were produced, what journeys they took, and how and where they were consumed.

Overall, *Indian Ink* takes seriously the different forms of writing in and around the English East India Company in order to understand how changing relationships of knowledge and power shaped the encounter between Europe and Asia in the seventeenth and eighteenth centuries. It explores this complex terrain by examining the activities of the Company from its

21. Shapin, *A Social History of Truth*.

22. Bruno Latour (1993) *We Have Never Been Modern* (Harvester Wheatsheaf, Hemel Hempstead).

23. Shapin and Schaffer, *Leviathan and the Air-Pump*, p. 332.

24. Johns, *The Nature of the Book*; and David N. Livingstone (2003) *Putting Science in Its Place: Geographies of Scientific Knowledge* (University of Chicago Press, Chicago).

earliest tentative trading voyages in the first decades of the 1600s to the establishment of an extensive Indian empire after 1757. The book argues that the complexities of the exertion of power and the making of knowledge and profit in these mercantile and imperial worlds are revealed through the development and deployment of the forms of writing that were a crucial element of the Company's operation in both Europe and Asia. These range from royal letters in manuscript on parchment from Elizabeth I and James I addressing "princes" in the east to the printing of Asian-language texts in Bengal by British Orientalists as part of the forging of empire. Close examination of these forms of writing—reconstructing their material forms and the conditions of their production, the ways in which they were part of intricate administrative and political procedures, and the differences of interpretation that they provoked—opens a way into the intellectual and political histories of early modern trade and empire. Concentrating on these different forms of writing brings London and the "East Indies" into a single interpretative frame. Indian ink might be that used in London, Fort St. George or Bengal, or that which traveled between them, as script or print. Examining modes of writing and reading reveals power and knowledge in mercantile and imperial worlds as they were in the process of being made. Doing so can demonstrate the fragility of that power and the uncertainty of knowledge, as well as showing the ways in which both could become, through processes of inscription and reinscription, solidified into institutional routines and forms of rule. In order to do this *Indian Ink* brings together the concerns of imperial history with those of the history of the book to demonstrate how concentrating on the making and movement of writing can draw out geographies of knowledge and power that are both firmly located in particular places and stretched halfway across the globe. That is the tale that the geography of Boyle's books begins to tell.

CHAPTER ONE

The Written World

Writing Empire

Understanding the history of European engagements with the rest of the world through the written word is nothing new. There is, however, substantial debate over how that might be done. In the last twenty-five years some of the most highly charged disputes over the study of empire, and particularly the British empire in India, have related to questions of writing. The development of various forms of postcolonial theory, drawing most prominently on the work of Michel Foucault and Jacques Derrida, has provoked a strenuous debate across a broad interdisciplinary field about the relationships between imperial power and the written word.[1] It is questions of the relationship between writing and the written-about world that have engaged both those who have followed Edward Said in attempting to map out the terrain of colonial discourse and the critics who accuse them of mistaking rhetoric for reality.[2] And it is the relationship between power, resistance, and meaning that structures attempts in the wake of the Subaltern Studies group to deconstruct the documentary record of the imperial archive and has animated those who are suspicious of the power of theories of language to reveal the full force of colonialism.[3] Although it is important

---

1. For example, Chris Tiffin and Alan Lawson (eds.) (1994) *De-Scribing Empire: Post-colonialism and Textuality* (Routledge, London); Elleke Boehmer (1995) *Colonial and Postcolonial Literature: Migrant Metaphors* (Oxford University Press, Oxford); and Carol A. Breckenridge and Peter van der Veer (eds.) (1993) *Orientalism and the Postcolonial Predicament: Perspectives on South Asia* (University of Pennsylvania Press, Philadelphia).

2. Edward W. Said (1978) *Orientalism* (Routledge and Kegan Paul, London); Jyotsna G. Singh (1996) *Colonial Narratives/Cultural Dialogues: "Discoveries" of India in the Language of Colonialism* (Routledge, London); and John Mackenzie (1995) *Orientalism: History, Theory and the Arts* (Manchester University Press, Manchester).

3. Ranajit Guha and Gayatri Chakravorty Spivak (eds.) (1988) *Selected Subaltern Studies* (Oxford University Press, Oxford); Homi Bhabha (1994) *The Location of Culture* (Routledge,

not to minimize the significance of these debates, it is evident that they have frequently proceeded by oversimplifying the stance of their opponents instead of exploring the substantial and increasing degree of shared ground. While it might be stated in different ways, it is evident that there is a sense that the textual and the material are to be understood together in the exertion of imperial power over people and places. As Gayatri Chakravorty Spivak put it some time ago, "the concept-metaphor of the 'social text' is not the reduction of real life to the page of a book."[4] Also, while there are significant debates over both conceptualization and historical exemplars, there is, more often than not, an underlying consensus that formations of imperial power and knowledge are both potentially transformative and, at the same time, contested, fragmentary, contradictory, and anxious. These formations should be understood as "work in progress," involving the necessary engagement of Europeans with other peoples and their forms of power and knowledge, in an ongoing process of the construction and deconstruction of empires. The aim, therefore, should be to find ways of understanding imperial power and knowledge that begin from these premises and try to be adequate to them. What I want to argue here is that ways forward can be found by bringing together questions raised by the turn towards geographical interpretations in both the history of empire and the history of the book. This provides objects of study and modes of interpretation that can be used across the broad historical period covered in this book from the opening maneuvers in long-distance trade to the establishment of the formal structures of a territorial empire.

Historical accounts of empire are becoming more concerned with geography in two important and related ways: a reconsideration of empire as defined through a hierarchical geography of center and periphery; and an attention to the specific characteristics of imperial and colonial sites, territories, and networks. The first recognizes that the shape and politics of the growing and changing British empire from the seventeenth century onwards is not adequately captured by simple models based upon the assumed centrality of Britain within a hierarchical set of relationships to discrete overseas colonies, territories, and trading zones. Instead, alternative models of "networks" or "webs" seek to explore a different geography that allows a

London); and Robert J. C. Young (2001) *Postcolonialism: An Historical Introduction* (Blackwell, Oxford).

4. Gayatri Chakravorty Spivak (1988) "Subaltern studies: deconstructing historiography," in Guha and Spivak, *Selected Subaltern Studies*, p. 5. See also C. A. Bayly (1996) *Empire and Information: Intelligence Gathering and Social Communication in India, 1780–1870* (Cambridge University Press, Cambridge).

range of competing and contradictory relationships to come into view. These alternatives emphasize the vulnerability of empire as well as its dynamism. They stress the unity of empire, as a single network or web, as well as the multiple differentiation of sites within it, the many forms of connectivity between them, and the ever-changing nature and shape of those connections.[5] An important part of this has been to find ways to "treat metropole and colony in a single analytic field."[6] This has meant exploring the means by which imperial sites are rendered distinct and different from the metropolitan core in terms of what sorts of rights or forms of production and coercion are made possible in each. It has also meant insisting on the ways in which the history of the empire has shaped the making of nations, states, and identities in the metropolis. Political disputes over empire, changing patterns of consumption, and the agency of colonized and enslaved people demanding rights and recognition has meant that what was "out there" was also simultaneously "in here."[7] Investigations of this complex whole refuse, both as impossible and as undesirable, calls to provide totalizing accounts of this reconceived British empire, or the global geography of which it was a part. Instead the focus is on analyzing and tracking particular sites, connections, and movements.[8] As Kathleen Wilson puts it, "In one sense, empire as a unit was a phantasm of the metropole: all empire is local."[9]

This, then, is the second element of imperial history's renewed engagement with geography. Here the specific "local" geographies of imperial sites, territories, and networks are understood as a vital part of the exertion of imperial power, of the forging of new relationships between people and places,

5. Tony Ballantyne (2002) *Orientalism and Race: Aryanism in the British Empire* (Palgrave, Basingstoke); and David Lambert and Alan Lester (2006) "Introduction: imperial spaces, imperial subjects," in David Lambert and Alan Lester (eds.) *Colonial Lives across the British Empire: Imperial Careering in the Long Nineteenth Century* (Cambridge University Press, Cambridge) pp. 1–31.

6. Ann Laura Stoler and Frederick Cooper (1997) "Between metropole and colony: rethinking a research agenda," in Frederick Cooper and Ann Laura Stoler (eds.) *Tensions of Empire: Colonial Cultures in a Bourgeois World* (University of California Press, Berkeley) p. 4.

7. Catherine Hall (2002) *Civilising Subjects: Metropole and Colony in the English Imagination, 1830–1867* (Polity Press, Cambridge); Laurent Dubois (2004) *Avengers of the New World: The Story of the Haitian Revolution* (Belknap Press of Harvard University Press, Cambridge, Massachusetts); and Frederick Cooper (2005) *Colonialism in Question: Theory, Knowledge, History* (University of California Press, Berkeley).

8. Felicity A. Nussbaum (ed.) (2003) *The Global Eighteenth Century* (Johns Hopkins University Press, Baltimore); and Kathleen Wilson (ed.) (2004) *A New Imperial History: Culture, Identity and Modernity in Britain and the Empire, 1660–1840* (Cambridge University Press, Cambridge).

9. Kathleen Wilson (2003) *The Island Race: Englishness, Empire and Gender in the Eighteenth Century* (Routledge, London) p. 213 n. 74.

and of the making of the modes of resistance that challenged these imperial reconfigurations of the world. This is not the determining and supposedly invariant backdrop of oceans, mountains, and plains: the geography behind history. Instead, it recognizes that imperialism is constituted through its arrangements of spaces, places, landscapes, and networks of connection. In this vein, imperial history involves the investigation of the small-scale geographies of sites such as trading posts, mercantile offices, imperial and colonial cities, and plantations and slave gardens, or of encounters between Europeans and others on ship and shore.[10] Alternatively, it means investigating the survey and delineation of specific imperial and colonial spaces through exploration and mapping, or a variety of administrative schemes of property ownership, settlement, and cultivation.[11] Finally, it is pursued through demonstrating the ways in which specific global connections forged by natural philosophers, merchants, and political renegades to further projects of knowledge creation, profit making, and liberation from oppression transformed the worlds within which they were constructed.[12]

These new geographies of early modern empire are, therefore, attentive both to what happened in particular sites and settings and to what moved between them. In particular, there is a renewed attention to questions of knowledge that has involved a more careful tracing of the production, dissemination, and consumption of ideas and information through the circuits of empire. This can be seen in recent work drawing on the history of political thought that has paid careful attention to resituating arguments about imperial policy and practice in the intellectual traditions, administrative contexts, and ideological battles of which they were originally a part, as well

10. For example, Daniel W. Clayton (2000) *Islands of Truth: The Imperial Fashioning of Vancouver Island* (University of British Columbia Press, Vancouver); David Hancock (1995) *Citizens of the World: London Merchants and the Integration of the British Atlantic Community, 1735–1785* (Cambridge University Press, Cambridge); Felix Driver and David Gilbert (eds.) (1999) *Imperial Cities: Landscape, Display and Identity* (Manchester University Press, Manchester); Jill H. Casid (2005) *Sowing Empire: Landscape and Colonization* (University of Minnesota Press, Minneapolis); and Greg Dening (1992) *Mr Bligh's Bad Language: Passion, Power and Theatre on the Bounty* (Cambridge University Press, Cambridge).

11. For example, Matthew H. Edney (1997) *Mapping an Empire: The Geographical Construction of British India, 1765–1843* (University of Chicago Press, Chicago); and R. Cole Harris (2002) *Making Native Space: Colonialism, Resistance, and Reserves in British Columbia* (University of British Columbia Press, Vancouver).

12. For example, Richard Drayton (2000) *Nature's Government: Science, Imperial Britain, and the "Improvement" of the World* (Yale University Press, New Haven); Hancock, *Citizens of the World*; Joseph Roach (1996) *Cities of the Dead: Circum-Atlantic Performance* (Columbia University Press, New York); and Peter Linebaugh and Marcus Rediker (2000) *The Many-Headed Hydra: Sailors, Slaves, Commoners, and the Hidden History of the Revolutionary Atlantic* (Beacon Press, Boston).

as seeing how they moved between places.[13] These concerns are also evident in discussions of information gathering and knowledge creation in imperial settings that stress both the imperial purposes of surveillance, science, and scholarship and the importance of local contexts, practices, and relationships to what could be made known.[14] Finally, they are apparent in studies of imperial and mercantile connections and forms of integration that stress the importance of flows of knowledge and information in such diverse forms as the manuscript letters of merchants, the passing of news and ideas among enslaved people, the transportation of botanical specimens, and even the printed papers of scholars.[15] In all these cases, questions of imperial power can be investigated through relationships between knowledge and space that work through forms of communication in speech, script, and print.

The Geography of Writing

A productive way into these relationships between imperial spaces, knowledges, and powers is through recent work in the histories of reading, writing, and publishing that often goes under the catch-all title "the history of the book." This vibrant and growing area of scholarship has sought to transform the relationship between textual and literary studies on the one hand and social and cultural history on the other by emphasizing the importance of

13. For example, David Armitage (2000) *The Ideological Origins of the British Empire* (Cambridge University Press, Cambridge); and Anthony Pagden (1995) *Lords of All the World: Ideologies of Empire in Spain, Britain and France, c.1500–c.1800* (Yale University Press, New Haven). For India, see Robert Travers (2005) "Ideology and British expansion in Bengal," *Journal of Imperial and Commonwealth History*, 33:1, pp. 7–27; and Jon E. Wilson (2000) *Governing Property, Making Law: Land, Local Society and Colonial Discourse in Agrarian Bengal, c. 1785–1830*, unpublished Ph.D. thesis, University of Oxford.

14. For example, Bayly, *Empire and Information*; D. Graham Burnett (2000) *Masters of All They Surveyed: Exploration, Geography, and a British El Dorado* (University of Chicago Press, Chicago); Kapil Raj (2000) "Colonial encounters and the forging of new knowledge and national identities: Great Britain and India, 1760–1850," *Osiris*, 15, pp. 119–34; Sujit Sivasundaram (2005) "Trading knowledge: the East India Company's elephants in India and Britain," *Historical Journal*, 48:1, pp. 27–63; Felix Driver and Luciana Martins (eds.) (2005) *Tropical Visions in an Age of Empire* (University of Chicago Press, Chicago); and, more generally, David N. Livingstone (2003) *Putting Science in Its Place: Geographies of Scientific Knowledge* (University of Chicago Press, Chicago).

15. For example, David Hancock (2000) "'A world of business to do': William Freeman and the foundations of England's commercial empire, 1645–1707," *William and Mary Quarterly*, 3rd ser., 57:1, pp. 3–34; Laurent Dubois (2006) "An enslaved Enlightenment: rethinking the intellectual history of the French Atlantic," *Social History*, 31:1, pp. 1–14; Londa Schiebinger and Claudia Swan (eds.) (2005) *Colonial Botany: Science, Commerce, and Politics in the Early Modern World* (University of Pennsylvania Press, Philadelphia); and Ballantyne, *Orientalism and Race*.

the variety of material instantiations of the written word and the significance of readers' diverse appropriations of those artifacts within cultural
practices that involve and combine talk, handwriting, and the printed word
and image.[16] Doing so, its practitioners argue, makes possible a series of
histories of the production, distribution, and consumption of knowledge in
ways that can demonstrate how it was positioned and differentiated socially,
historically, and geographically, and how it was both put to work and contested.[17] These histories of the book locate the construction of meaning and
the workings of power in the concrete processes of the making, distribution,
and use of texts as material objects.[18]

As with the recent developments in imperial history it is striking how
concerned the histories of writing, of the book, and of reading have become
with questions of geography. It is, of course, an area that has been littered
with grand historical transitions: Walter Ong and Jack Goody's accounting
of the dramatic transformations in forms of society and consciousness in
the shift from orality to literacy; Marshall McLuhan and Elizabeth Eisenstein's elaborations of the revolutionary implications of moveable metal
type; and Rolf Engelsing's identification of a late eighteenth-century "reading revolution" in the move from the "intensive" reading of a few texts
to the "extensive" reading of many.[19] Yet, without denying the significant
changes brought by new forms of script and print and the increasing variety

16. Robert Darnton (1982) "What is the history of books?" *Daedalus*, 111:3, pp. 65–83; Robert
Darnton (1986) "First steps towards a history of reading," *Australian Journal of French Studies*,
23:1, pp. 5–30; Roger Chartier (1994) *The Order of Books: Readers, Authors, and Libraries in
Europe between the Fourteenth and Eighteenth Centuries* (Stanford University Press, Stanford);
David D. Hall (1996) *Cultures of Print: Essays in the History of the Book* (University of Mas-
sachusetts Press, Amherst); Kevin Sharpe (2000) *Reading Revolutions: The Politics of Reading in
Early Modern England* (Yale University Press, New Haven); and James Raven, Helen Small, and
Naomi Tadmor (eds.) (1996) *The Practice and Representation of Reading in England* (Cambridge
University Press, Cambridge).

17. Peter Burke (2000) *A Social History of Knowledge: From Gutenberg to Diderot* (Polity
Press, Cambridge).

18. Roger Chartier (ed.) (1989) *The Culture of Print: Power and the Uses of Print in Early
Modern Europe* (Polity Press, Cambridge); and Donald F. McKenzie (1986) *Bibliography and the
Sociology of Texts* (BL, London).

19. Jack Goody (1986) *The Logic of Writing and the Organization of Society* (Cambridge
University Press, Cambridge); Walter J. Ong (1988 [1982]) *Orality and Literacy: The Technolo-
gizing of the Word* (Routledge, London); Marshall McLuhan (1969 [1962]) *The Gutenberg Galaxy:
The Making of Typographic Man* (Signet Books, New York); Elizabeth L. Eisenstein (1979) *The
Printing Press as an Agent of Change: Communications and Cultural Transformations in Early
Modern Europe* (Cambridge University Press, Cambridge); and Ian Jackson (2004) "Approaches
to the history of readers and reading in eighteenth-century Britain," *Historical Journal*, 47:4,
pp. 1041–54.

of available texts, notions of singular, dramatic, and universal transformations have been dispersed into investigations of the simultaneous variety
of forms of the production, dissemination, and use of texts at any point
in time.[20] In each part of what Robert Darnton has called the "communications circuit" there is a focus on the local, concrete, and particular historical
geographies of writers, printers, manuscripts, books, booksellers, and many
different sorts of readers.[21]

In terms of the production of texts, questions of authorship remain important but have been redistributed across the range of people involved in
producing the written word. Thus, the "routine authorship" involved in
editing, compiling, correcting, annotating, and anthologizing, as well as the
practical processes of production by scribes, compositors, and pressmen, are
given their due in the making of manuscript and printed texts.[22] This ac-
knowledges that scribal copying and letterpress printing were not simply
processes of transparent replication, but active modes of production. Scribes
and printers were producers of new forms of the written word. First, they
changed the texts that came into their hands simply through the process
of changing their forms; in the choices of paper and binding, and of the
handwriting or typography that that involved. Second, they changed them
through the processes of editing that they saw as their duty as workers with
words. Only in very rare circumstances was the early modern text a sacro-
sanct one.[23] There were, of course, social rules and norms governing the
sorts of changes that were made, and what they might mean. These were
inscribed both in the norms of artisanal and professional communities of
scribes and printers and in the legal forms of regulation, licensing, and censorship that they negotiated in their working lives. Understanding these,

20. Adrian Johns (1998) "Science and the book in modern cultural historiography," *Studies in
History and Philosophy of Science*, 29A:2, pp. 167–94; and Marina Frasca-Spada and Nick Jardine
(eds.) (2000) *Books and the Sciences in History* (Cambridge University Press, Cambridge).

21. Darnton, "What is the history of books?"

22. Nick Jardine (2000) "Books, texts, and the making of knowledge," in Frasca-Spada and
Jardine (eds.), *Books and the Sciences in History*, p. 401; Robert Darnton (1979) *The Business of
Enlightenment: A Publishing History of the* Encyclopédie, *1775–1800* (Belknap Press of Harvard
University Press, Cambridge, Mass.); Robert Iliffe (1995) "Author-mongering: the 'editor' between
producer and consumer," in Anne Bermingham and John Brewer (eds.) *The Consumption of
Culture, 1600–1800: Image, Object, Text* (Routledge, London) pp. 166–92; and Adrian Johns (1998)
The Nature of the Book: Print and Knowledge in the Making (University of Chicago Press,
Chicago).

23. Darnton, *The Business of Enlightenment*; D. F. McKenzie (2002) *Making Meaning: "Printers of the Mind" and Other Essays*, edited by Peter D. McDonald and Michael F. Suarez (University
of Massachusetts Press, Amherst); and Harold Love (1993) *Scribal Publication in Seventeenth-
Century England* (Clarendon Press, Oxford).

for different forms of both script and print, means attending, then, to the specifics of places of production and their social organization. For example, for seventeenth-century London, Harold Love has situated the scribal publication of large numbers of political documents—parliamentary speeches and the manuscript equivalent of pamphlets—within the scriptoria inhabited by the armies of clerks necessary for the workings of the law, the church, and the government, particularly around the Inns of Court. He then differentiates this from the chain copying of single versions of politically contentious manuscripts by and for individual readers in what he calls "scribal communities" stretched across a network of country houses and other spaces.[24] For printing in the same period, Adrian Johns has argued that we need "a topography of the places of printed material in early modern London," and that it was this "social geography" of printing houses, bookshops, and the wider urban environment that "conditioned the knowledge that could be produced and encountered."[25] For both script and print, it is the detailed geographies of local processes of production that can explicate the process of the making of meaning in texts as they are produced as material objects.

These processes are also pursued via questions of distribution and dissemination, in order to render concrete the particularities and inequalities of access to information. In place of general formulations about print and the public sphere are more specific questions of the material geographies of texts. Where did letters, manuscripts, account books, pamphlets, newspapers, and books go, and into whose hands did they fall? That there is no simple relationship between publishing and the public sphere is evident in the variety of modes of publication from oral proclamation, through manuscript circulation, to licensed printing.[26] All constituted different relationships between the public and the private, and different modes of publication and distribution might be chosen for different sorts of texts and authors.[27] However, there always had to be journeys, carriers, destinations, and transactions as texts moved from writers to readers. Where these can be mapped out, there is the possibility of tracing the cultural geographies of the demand for particular forms of writing or particular books.[28] As Robert Darnton prosaically put

24. Love, *Scribal Publication*; and Peter Beal (1998) *In Praise of Scribes: Manuscripts and Their Makers in Seventeenth-Century England* (Clarendon Press, Oxford).

25. Johns, *The Nature of the Book*, pp. 61–62.

26. Love, *Scribal Publication*.

27. For example, on women authors choosing manuscript publication, see Jackson, "Approaches to the history of readers and reading."

28. Darnton, *The Business of Enlightenment*; Margaret Spufford (1981) *Small Books and Pleasant Histories: Popular Fiction and Its Readership in Seventeenth-Century England* (Methuen,

it, "The wagon, the canal barge, the merchant vessel, the post office, and the railroad may have influenced the history of literature more than one would suspect."[29] These geographies of distribution were, as Harold Love affirms, always political. The use of scribal publication by chain copying, with its non-metropolitan, dispersed, and serial form of reproduction beyond the purview of the Crown, meant that "the scribal text had a politics, and that this politics was also a geography." It was, he notes, a geography and politics of distribution and dissemination different from that characteristic of seventeenth-century print.[30]

Finally, accounts of the uses of written objects in script and print have also taken a material and geographical turn. In particular this involves recognizing reading itself as a differentiated and situated practice, taking different forms for different reading communities in different periods and places.[31] Here reading is understood as a significant cultural act, a process of appropriation that is shaped by the text and its materiality, but not governed by it.[32] A range of studies in the history of reading have therefore located readers in their specific contexts, elaborating not just where reading took place, but the array of written materials that they had at hand, what they did to them, what technologies they employed, and how reading fitted into readers' lives. Thus, the marginal annotations Dr. John Dee made in his library at Mortlake, Anna Larpent's combination of intensive reading and polite urban sociability in eighteenth-century London, and Gabriel Harvey's use of a book wheel to marshal texts in the service of Elizabethan political grandees in their country houses were all located readers making meaning in the process of reading specific texts in particular ways.[33] This suggests

---

London); John Feather (1985) *The Provincial Book Trade in Eighteenth-Century England* (Cambridge University Press, Cambridge); and Robin Myers and Michael Harris (eds.) (1990) *Spreading the Word: The Distribution Networks of Print, 1550–1850* (St Paul's Bibliographies, Winchester).

29. Darnton, "What is the history of books?" p. 77.

30. Harold Love (2002) "Oral and scribal texts in early modern England," in John Barnard and D. F. McKenzie (eds.) *The Cambridge History of the Book. Volume 4: 1557–1695* (Cambridge University Press, Cambridge) p. 109. It should be noted that scribal production could be rendered more or less centralized and hierarchical and print more or less dispersed by different forms of social organization such as the organization of clerks in offices or the practices of pirate printers.

31. Stephen Colclough (2000) *Reading Experience 1700–1840: An Annotated Register of Sources for the History of Reading in the British Isles* (University of Reading, Reading), p. iv; and Sharpe, *Reading Revolutions.*

32. Roger Chartier (1988) *Cultural History: Between Practices and Representations* (Polity Press, Cambridge).

33. William H. Sherman (1995) *John Dee: The Politics of Reading and Writing in the English Renaissance* (University of Massachusetts Press, Amherst); John Brewer (1997) *The Pleasures of*

the usefulness of reconstructing the "geography of reading" to emphasize this process of appropriation as a cultural practice located in place.[34] However, such histories also need to acknowledge the diverse uses to which the variety of written objects have been put. Harold Love, for example, draws attention to the need to understand the uses of words within the complex and changing relationships between voice, script, and print, which often required particular combinations of speech and writing. Thus, the authority of the king, the law, and Parliament in seventeenth-century England was a matter of the spoken word, but that might become a basis for action only via a written document. Equally, written forms of speech might act as authority and precedent in a court of law, but only when respoken in the right place at the right time.[35] Indeed, writings in script or print might serve their purposes in being venerated, brandished, or just stored away, rather than them actually needing to be read.

Posed in these ways, questions of the geography of the book—or, to avoid the concentration on a specific form, the geography of writing or communication—have become central to understanding the relationships between knowledge, power, and social change. Indeed, disagreements over matters as significant as the impact of the printing press in early modern Europe become disagreements over questions of geography. On the one hand, Elizabeth Eisenstein stresses the "products of a far-flung book trade" and Western Europe's "cosmopolitan commonwealth of learning" in order to argue for the revolutionary impact of printing on learning.[36] On the other hand, Adrian Johns proposes "that we explain the development and consequences of print in terms of how communities involved with the book as producers, distributors, regulators, and readers actually put the press and its products to use." As he continues, "[t]he best way to do *that*...is to

---

the Imagination: English Culture in the Eighteenth Century (HarperCollins, London) pp. 56–72; Lisa Jardine and Anthony Grafton (1990) "Studied for action': How Gabriel Harvey read his Livy," Past and Present, 129, pp. 30–78. See also Carlo Ginzburg (1980 [1976]) The Cheese and the Worms: The Cosmos of a Sixteenth-Century Miller (Johns Hopkins University Press, Baltimore); and Robert Darnton (1984) "Readers respond to Rousseau: the fabrication of romantic sensitivity," in The Great Cat Massacre and Other Episodes in French Cultural History (Penguin Books, Harmondsworth) pp. 209–49.

34. James A. Secord (2000) Victorian Sensation: The Extraordinary Publication, Reception, and Secret Authorship of Vestiges of the Natural History of Creation (University of Chicago Press, Chicago); and David N. Livingstone (2005) "Science, text and space: thoughts on the geography of reading," Transactions of the Institute of British Geographers, 30:4, pp. 391–401.

35. Love, Scribal Publication.

36. Elizabeth L. Eisenstein (2002) "An unacknowledged revolution revisited," American Historical Review, 107:1, p. 95.

adopt a local perspective, at least in the first instance."[37] Indeed, this dis-agreement signals the dual nature of the geographies of writing. Texts as material objects are made and find their uses in particular settings: settings that enter into their very constitution and modes of appropriation. At the same time, these textual objects are useful precisely because they move be-tween settings. We need ways, as Eisenstein and Johns both acknowledge, of recognizing simultaneously the groundedness and the mobility of written objects.[38] This means examining what might be meant by the term "local." In the "localist" turn in both the history of science and the history of the book there is always a danger of collapsing onto each other the sense of the local as the concrete and the particular and the local as a specific, fixed *place* with its own unique characteristics.[39] Doing so makes it easier to explain knowledge making as located forms of cultural practice, but harder to ex-plain how ideas, objects, and texts are made mobile.[40] In rethinking this it is right to question any easy use of Bruno Latour's notion of "immutable mobiles"—entities that can move without changing—and their use in the construction of knowledge within the "centers of calculation" where they are gathered together, particularly when this relies upon an unproblematic acceptance of the use of print in making mobile texts immutable.[41] How-ever, Latourian ideas of global networks that are local at every point, and of the multiple forms of situated work done on "inscriptions" in order to make them mobile, combinable, and able to represent the world within new forms of action, are consonant with the idea that the large-scale, the far-flung, and the cosmopolitan are all achieved through situated forms of work with their own "local" geographies.[42] Once again, this is a geography that involves both specific sites and the movements of people, knowledge, and things that join them together in particular configurations. These ideas can take their place within the elaboration of "localist" accounts of the

37. Adrian Johns (2002) "How to acknowledge a revolution," *American Historical Review*, 107:1, pp. 124–25.

38. See also Burke, *A Social History of Knowledge*, pp. 53–80; and James A. Secord (2004) "Knowledge in transit," *Isis*, 95, pp. 654–72.

39. On this dual turn to the "local", see Johns, "Science and the book." Roger Chartier, for example, often uses "local" simply to mean a particular context that might have no spatial referent; see, for example, Chartier, *The Culture of Print*, pp. 3–5.

40. Steven J. Harris (1998) "Long-distance corporations, big sciences, and the geography of knowledge," *Configurations*, 6, pp. 269–304; and Secord, "Knowledge in transit."

41. Bruno Latour (1987) *Science in Action: How to Follow Scientists and Engineers through Society* (Harvard University Press, Cambridge, Mass.); and Johns, "Science and the book."

42. Bruno Latour (1999) *Pandora's Hope: Essays on the Reality of Science Studies* (Harvard University Press, Cambridge, Mass.); and Johns, "How to acknowledge a revolution."

making of knowledge that argue that this making is also always a matter of communicating between contexts, and that this means giving due consideration to the "circulating practices" of "things-in-motion" in order to understand "knowledge in transit."[43] More simply, the same point can be made in the recognition that ships and other modes of transport can be, at one and the same time, both local and mobile sites for the work of making and communicating knowledge.[44] It is this dual sense of the geography of writing as simultaneously local and mobile that can help to rethink the nature of the connections between trade, empire, and the written word.

Writing, Trade, and Empire

It is not difficult to find work that connects the form of modes of communication, knowledge, and information—the material processes of writing and reading—with the making and decline of empires.[45] In 1949, as the drama of postwar decolonization was being played out across the world, the Canadian economist Harold Innis delivered a series of lectures in Oxford that were published the following year as *Empire and Communications*. Here Innis outlined a long-term history of empires structured by an insistence on the competing tendencies of lightweight portable media such as papyrus and paper to effect the control of space, and of more durable, heavy materials such as stone, clay, and parchment to permit the control of time.[46] Innis's focus was very much on the lineage of Western civilization, from Egypt, through Babylonia, Greece, and Rome to Western Europe. Chinese and Islamic developments were considered primarily in terms of how their initiation and spread of paper manufacture had shaped later European transformations. In tracing this history Innis was clear that any form of writing was a significant and power-laden transformation of the relationships between people. What he sought to emphasize was the importance of the nature of different media in those transformations.

43. Secord, "Knowledge in transit."

44. For example, see Richard Sorrenson (1996) "The ship as a scientific instrument in the eighteenth century," *Osiris*, 11, pp. 221–36, and the discussion of William Burchell's African collecting wagon in Felix Driver (2001) *Geography Militant: Cultures of Exploration and Empire* (Blackwell, Oxford) pp. 17–19.

45. Lucien Febvre and Henri-Jean Martin (1976 [1958]) "The geography of the book," in *The Coming of the Book: The Impact of Printing, 1450–1800* (Verso, London) pp. 167–215; and Benedict Anderson (1983) *Imagined Communities: Reflections on the Origin and Spread of Nationalism* (Verso, London).

46. Harold A. Innis (1986 [1950]) *Empire and Communications* (Press Porcépic, Victoria).

The subsequent influence of this often bizarre and elusive text is not entirely clear. Marshall McLuhan claimed in the 1960s that Innis "was the first person to hit upon the *process* of change as implicit in the *forms* of media technology," and that McLuhan's own book *The Gutenberg Galaxy* was merely "a footnote of explanation to his work."[47] Indeed, McLuhan, Walter Ong, and Jack Goody, in their explorations of the transformations wrought by script and print on oral and scribal cultures, all saw fit in varying degrees to present evidence from the recent colonial past alongside that of the classical world, ancient civilizations, and medieval Europe.[48] Doing so made the dual claim, like Innis, that these were universal transformations more or less invariant in their effects across time and space, and that they were significantly structured by relationships of power.

More recently, these issues have been debated in terms of the European conquest of the New World. Wary of Tzvetan Todorov's claim that the ability to manipulate signs, particularly through writing, was what gave Iberian *conquistadores* an objective and determining advantage over the unlettered peoples of the Americas, Stephen Greenblatt did, however, reaffirm that Europeans understood their representational technologies of alphabet and printing press as part of what made them superior to these barbarous savages.[49] What is emphasized, then, particularly by Walter Mignolo, is the will to power represented and enforced by European forms of literacy and book-knowledge in the Americas. Colonial domination gave the Spanish in Mexico and Peru the ability to make real and universal their culturally specific understanding that writing alphabetic script was the construction of signs, not the painting of pictograms by *tlacuilo* or the tying of quipu cords, and that reading words in book form was the interpretation of signs, not the telling of stories by looking at images on tree-bark (*amoxtli*). Mignolo, therefore, insists that the changes wrought in the forms of communication are matters of power, and sees this same process replicated time and again in the history of European colonization of other parts of the world.[50]

47. McLuhan, *The Gutenberg Galaxy*, p. 65, emphasis in original.

48. Goody, *The Logic of Writing*; and Ong, *Orality and Literacy*.

49. Stephen Greenblatt (1991) *Marvelous Possessions: The Wonder of the New World* (Clarendon Press, Oxford); and Tzvetan Todorov (1984) *The Conquest of America* (Harper and Row, New York).

50. Walter D. Mignolo (2003 [1995]) *The Darker Side of the Renaissance: Literacy, Territoriality, and Colonization*, 2nd ed. (University of Michigan Press, Ann Arbor); for a critique see Jorge Cañizares-Esguerra (2001) *How to Write the History of the New World: Histories, Epistemologies, and Identities in the Eighteenth-Century Atlantic World* (Stanford University Press, Stanford) pp. 63–69.

The interpretation of the form of modes of communication within the conflict-ridden history of imperialism and colonization is the purpose of a remarkable essay by the bibliographer Donald McKenzie on the Treaty of Waitangi in 1840. Here, McKenzie's expansion of bibliography to being a "sociology of texts" and his insistence that the formal properties of texts are essential to the interpretations that are made of them—ideas that have been fundamental to the revivification of the history of the book—are applied to the founding document of the British colonization of New Zealand.[51] McKenzie argues, like Greenblatt, that British missionaries in New Zealand in the early nineteenth century were convinced of the superiority of print literacy over oral modes of memory and communication, and that they fooled themselves into thinking that the fundamental transformation of an oral culture to a literate print culture had been achieved in only a few short years through Maori Bibles and mission schools. Significantly, it was this myth of literacy that underpinned the legitimation of the signing by forty-six Maori chiefs of a printed treaty in Maori that, the British claimed, handed sovereignty over the islands to Queen Victoria. Against this McKenzie argues that the act of signing the document meant quite different things to the Maori and the British. First, few of the chiefs could have read what they were signing. Second, the conception of passing over all rights of property and sovereignty by the act of making a mark on a piece of paper was not one that made sense within Maori notions of land and power. Finally, that a bibliographic study of the translations and variant forms of the treaties and other associated texts reveals that there was no final stable text in both Maori and English that could actually be said to have been agreed upon.

McKenzie has been criticized for overemphasizing the orality of Maori culture and failing to appreciate the productive appropriations that the Maori were very quick to make of both letter writing and printing in order to construct new identities that challenged British imperialism.[52] Be that as it may, McKenzie's argument and evidence can be used to demonstrate, in the spirit of this critique, the need to consider the contingencies of imperial power and the agency of colonized peoples, and the ways in which a close consideration of forms of writing, reading, and printing is well placed

51. D. F. McKenzie (1985) *Oral Culture, Literacy and Print in Early New Zealand: The Treaty of Waitangi* (Victoria University Press with the Alexander Turnbull Library Endowment Trust, Wellington). For an identification of this essay as central to McKenzie's work, see Roger Chartier (1997) *On the Edge of the Cliff: History, Language, and Practices* (Johns Hopkins University Press, Baltimore) p. 86.

52. Ballantyne, *Orientalism and Race*, pp. 147–48.

to reveal them. First, it is evident that the British required the signatures of Maori chiefs and were willing to print in Maori to get them. It is also apparent that the printer, William Colenso, was keenly aware of the local frailties of printing as a colonial technology. He had been expected to commence printing for the Church Missionary Society in 1834 with no galleys, cases, or printing paper, and was fortunate to have brought his own composing stick with him as none were provided. He also tried to argue at the time the treaty was signed that the chiefs did not know what they were signing, and that that put into question the legal force of the document. Indeed, it was Colenso who spotted the differences between the English and Maori versions, whereby the former gave the Crown the preemptive right of purchase and the latter seemed only to give Maori the right to sell their land. Second, McKenzie argues that the Maori chiefs understood the Treaty of Waitangi in the light of previous documents, particularly the Declaration of Independence. As he put it, "their sense that the independence (the *rangatiratanga*) and the sovereignty (the *mana*) they had affirmed in 1835 and reaffirmed by further subscriptions as late as 1839, was not nullified by the treaty. British Colonial Office attitudes may have changed in the meantime, but for the Maori one document did not supersede the other: they lived together, one complementing the other."[53] Rather than understanding the treaty signing as a clash between a mutually uncomprehending oral culture and print culture, it was McKenzie's intention to use it to stress that speech, manuscript, and print were "complementary modes," relating to each other in different ways for different people in different contexts.[54] The ways in which Maori and British interlocutors in New Zealand in 1840 understood the relationships between speaking, writing, and printing were clearly different, yet they both made use of and made interpretations of spoken, written, and printed words and signs. It was also evident that there was on both sides a need to try to understand what the other thought, and a differential ability to enforce any particular definition of the situation.

This sort of differentiated and power-laden, but not overdetermined, meeting of European colonists, traders, and travelers and other peoples across and within a varied ecology of speech, script, and print is evident from other colonial situations. In each case, particularly where people were learning of the power of alphabetic writing and printing with moveable type within the encounter, it is common to identify the use of literacy and print as the tools of a civilizing process backed by violence and also the local appropriation

53. McKenzie, *Oral Culture, Literacy and Print*, p. 41.
54. McKenzie, *Making Meaning*, p. 238; his intentions are set out in n. 5.

of written and printed objects in surprising ways. Alongside the violent destruction of imperial documentation by colonized peoples, texts also appear
as bodily adornments, as signals of rebellion sent by the gods, and among
grave goods, hinting at notions of their efficacy as symbols of power or instruments of healing.[55] Moreover, it is also common to find a substantial engagement of colonial powers with the languages and writing systems of colonized
peoples—often in the interests of making converts to Christianity—coupled
with a rapid and sophisticated appropriation of European forms of reading
and writing by indigenous peoples and others in order to establish rights and
identities within the new legal, religious, and administrative structures of
colony and empire.[56] These engagements through modes of communication
were unequal, but they were more than one-sided affairs. Indeed, a longer-
term view sees the construction of print as the preeminent technology of
anti-imperial nationalism.[57]

In situations where Europeans traded with and eventually sought to
incorporate into their empires people who already conducted religion, politics, and economic activity through forms of writing and printing that
were recognizable, if not immediately legible, to the newcomers, these re-
ciprocities were even more pronounced.[58] Early modern Europeans were well

55. For example, David D. Hall, "Introduction," in Hugh Amory and David D. Hall (eds.)
(2000) *A History of the Book in America*, vol. 1: *The Colonial Book in the Atlantic World* (Cambridge University Press, Cambridge) p. 25; Ranajit Guha (1983) *Elementary Aspects of Peasant
Insurgency in Colonial India* (Oxford University Press, Delhi) pp. 248–49; and Joanne Rappaport
(1994) "Object and alphabet: Andean Indians and documents in the colonial period," in Elizabeth Hill Boone and Walter D. Mignolo (eds.) *Writing without Words: Alternative Literacies in
Mesoamerica and the Andes* (Duke University Press, Durham) p. 284.

56. For example, Cañizares-Esguerra, *How to Write the History of the New World*; Jill Lepore
(1999) *The Name of War: King Philip's War and the Origins of American Identity* (Vintage Books,
New York); Ballantyne, *Orientalism and Race*; and E. Jennifer Monaghan (1993) "'Able and willing to read': the meaning of literacy to the Indians of colonial Martha's Vineyard," in Greg Brooks,
A. K. Pugh, and Nigel Hall (eds.) *Further Studies in the History of Reading* (United Kingdom Reading Association, Widnes) pp. 43–59. For the enslaved, see David Waldstreicher (1999) "Reading
the runaways: self-fashioning, print culture, and confidence in slavery in the eighteenth-century
mid-Atlantic," *William and Mary Quarterly*, 3rd ser., 56: 2, pp. 243–72; and Dubois, "An enslaved
Enlightenment."

57. Michael Warner (1990) *The Letters of the Republic: Publication and the Public Sphere in
Eighteenth-Century America* (Harvard University Press, Cambridge, Mass.); Anderson, *Imagined
Communities*; and Bayly, *Empire and Information*.

58. This should recognize both the colonial politics of print in Ireland and the continuities
between reading on the island and in other parts of the British Isles; see Robert Welch (2002) "The
book in Ireland from the Tudor re-conquest to the battle of the Boyne," in Barnard and McKenzie,
The Cambridge History of the Book in Britain, vol. 4 pp. 701–18; and Bernadette Cunningham
and Máire Kennedy (eds.) (1999) *The Experience of Reading: Irish Historical Perspectives* (Rare
Books Group of the Library Association of Ireland, Dublin).

aware that both gunpowder and printing, those twin engines of modernity, were Chinese inventions. Chinese, Japanese, and Korean states and empires were making extensive use of wood-block printing and lithography in the early modern period in the service of "surveillance and moral suasion."[59] In India, the Mughal emperors showed little interest in the printing press, introduced by Portuguese Jesuits to Goa in the 1550s, although Akbar collected a large number of European printed books for his library.[60] The rulers of successor states were even moved to suppress printing presses. Yet, as C. A. Bayly has argued, this was a sign of the strength of the scribal tradition and its handwritten instruments and techniques, not of its weakness. The Mughal empire had well-developed procedures for the collection, transmission, and storage of information in the service of Indo-Muslim ideals of kingship. Rulers valued their extensive libraries of books and manuscripts, often formed via gift giving and plunder. Their courts supported practitioners of the arts of calligraphy, miniature painting, and bookbinding.[61] Indeed, the arts of kingship were in large part scribal arts. The emperor's letters were revered for the beauty of their words and script, and Mughal royal chanceleries operated through elaborate scribal procedures. Indeed, as Bayly argues, "[t]he whole nobility had been brought up to revere the art of insha or letter-writing as a tool of literacy and as a form of regulating proper social relations" to the extent that "[t]he regular exchange of loving letters between commanders and nobles, not dry administrative correspondence, was the cement of the polity." The beautiful calligraphy, elaborate honorifics, and "flowery Persian" of official letters were a vital part of the management of rank and duty within and between these dynastic states.[62]

In the course of the seventeenth and eighteenth centuries the value given to official writing in Persian became just one part of an increasingly "literacy aware society" in north India.[63] Anindita Ghosh has described a vibrant world of writing and reading in preprint and precolonial Bengal. The smaller polities that grew up within the shell of declining Mughal power,

59. Bayly, *Empire and Information*, p. 19; Burke, *A Social History of Knowledge*; and Thomas Francis Carter (1925) *The Invention of Printing in China and Its Spread Westward* (Columbia University Press, New York).

60. John F. Richards (1995) *The Mughal Empire* (Cambridge University Press, Cambridge).

61. Bayly, *Empire and Information*; Shaikh Allaudin and R. K. Rout (1996) *Libraries and Librarianship during Muslim Rule in India* (Reliance Publishing House, Delhi); and Annemarie Schimmel (2004) *The Empire of the Great Mughals: History, Art and Culture* (Reaktion Books, London).

62. Bayly, *Empire and Information*, pp. 76–77.

63. Bayly, *Empire and Information*, p. 39.

the powerful landowners (*zamīndars*), and the thriving commercial economy all produced great demand for writers and writing. High caste Brahmins, *Kāyastha*s, and *Vaidya*s, from whom the literate professions had long
been drawn, occupied the upper reaches of government and worked in Persian, Sanskrit, and Bengali. Below them were "the vast numbers, often with
only an elementary knowledge of Bengali and Persian, who served in lesser
administrative positions in the government as clerks, news writers, subordinate military commanders, or managers and officials of princely house-
holds."[64] The many lowly clerks and accountants employed by revenue and
commercial establishments would have received their initial education in
reading, simple arithmetic, and the rules of correspondence and accounting in the *patthśālā*s, Bengali elementary schools. Indeed, as Ghosh argues,
even the illiterate were part of this world of manuscripts and of reading.
Their access to the written word was through *kathakatā*s, open-air "collective narrative sessions" where Brahmin narrators (*kathak*s) read, or rather
performed, religious texts based on the Hindu epics from manuscripts or
wooden tablets to active and involved audiences.[65]

Writing's power and value were therefore widely understood, used, and
valued in Indian civil society as well as by the state. Indian merchants kept
account books and used credit notes that were familiar to their European
counterparts. Religious sects produced and disseminated manuscript copies
of holy texts in large numbers. The resolution of conflicts at law increasingly depended upon written records. Those who could not read and write
found plenty who would do it for them, for pay, or through duty or goodwill.
Indeed, Bayly argues that a "self-confident set of scribal elites" who based
their status and independence on the power and utility of writing was "[t]he
most dynamic social formation of the period." Armed with an ethic of service to the state independent of any particular ruler, and therefore aware of
the need for vigilance to detect unjust rule, these people of the pen were a
vital part of what Bayly calls "the Indian ecumene," a public sphere of rational debate over interlocking concerns of politics, religion, and aesthetics
that existed across north India in the eighteenth and nineteenth centuries.[66]
This public sphere, without the printing press or the formal public meeting,
worked through oral and scribal modes of communication. Its mechanisms
were news, gossip, and opinion passed around and debated at druggists' stalls

64. Anindita Ghosh (2006) *Power in Print: Popular Publishing and the Politics of Language
and Culture in a Colonial Society* (Oxford University Press, New Delhi) p. 33.

65. Ghosh, *Power in Print*, p. 40.

66. Bayly, *Empire and Information*, p. 45.

and sweetshops, around mosques and temples, and in discussions of poetry. In written form it worked through personal letter writing, newsletters read aloud to crowds in the street, and placards posted at significant sites. Both before and after the British conquest of India, writing was a vital part of both the organization of state power and authority and the political processes through which that was questioned, limited, and contested. As Bayly has shown, the British empire's forms of knowledge were significantly shaped and limited by their engagement with India's existing geographies of information, knowledge, and writing. In the process of empire building what mattered were the different social and political forms through which information gathering, the uses of knowledge, and the deployment of writing were organized and the relationships between them.

Understanding writing and empire—or, even more broadly, writing and European overseas "expansion" through varied mercantile and imperial ventures—involves, therefore, consideration of a wide range of forms of script and print. These were produced, disseminated, and consumed in a variety of spaces and traveled through a range of networks. Some of these texts were directly produced through processes of cross-cultural encounter in interaction with speakers, readers, and writers of very different sorts. Others were administrative texts that facilitated long-distance trade, or that sought to secure its profits and privileges in the stock exchanges and parliaments of Europe.[67] Within the public spheres of print, texts debated the politics of trade and empire, reconfigured knowledge about the rest of the world, or were, quite simply, part of a late eighteenth-century reorientation of Britain's book export trade towards markets in India and North America.[68] There has certainly been an increased attention to the imperial, colonial, and postcolonial histories of the book.[69] What is argued here is that understanding the relationships between power and knowledge worked out through these histories requires attention to the geographies of these forms

67. For example, Hancock, *Citizens of the World*; and Hancock, "A world of business to do."

68. Kathleen Wilson (1995) *The Sense of the People: Politics, Culture and Imperialism in England, 1715–1785* (Cambridge University Press, Cambridge); Jeremy Osborn (2002) "India and the East India Company in the public sphere of eighteenth-century Britain," in H. V. Bowen, Margarette Lincoln, and Nigel Rigby (eds.) *The Worlds of the East India Company* (Boydell Press, Woodbridge) pp. 201–21; and Giles Barber (1982) "Book imports and exports in the eighteenth century," in Robin Myers and Michael Harris (eds.) *Sale and Distribution of Books from 1700* (Oxford Polytechnic Press, Oxford) pp. 77–105.

69. For example, the recent Colonial/Postcolonial History of the Book Project at the Open University in the United Kingdom and the long-standing interest in the colonial book evidenced in Amory and Hall (eds.), *The History of the Book in America*.

of writing. There are some productive examples in existing work that point in this direction. For example, Donna Merwick has reconstructed in astonishing detail the life, work, and death of a notary in seventeenth-century Dutch North America. She chronicles the eventual suicide of Adriaen Janse van Ilpendam as power shifted to the British and his scribal skills were displaced from the courtrooms of the new administration.[70] In many ways this microhistory of marginalization offers a counterweight to Angel Rama's more poetic evocation of the power and confidence of the Spanish American "lettered city," the "myriad of administrators, educators, professionals, notaries, religious personnel, and other wielders of pen and paper" that inhabited the viceregal capitals.[71] Turning to the printing press, Jill Lepore has shown how the production of printed accounts of King Philip's War in late seventeenth-century New England was a key part of the conflict between English settlers and Narragansett peoples. Despite the prior existence of Native American translators such as John Sassamon and the Nipmuck compositor and pressman James Printer, the printed page, the printing office, and the consumption of colonial printed material became arenas for the creation of oppositional settler and native identities.[72] Finally, various studies of reading have demonstrated its enganglements with empire. For example, the uses of John Dee's library as the campaign headquarters in the battle of the books that loomed large in the early modern imperial contest over new territories; the violent schemes of Irish plantation that were the occasion for Gabriel Harvey's readings of Livy with Thomas Smith Junior and Sir Humphrey Gilbert at Hill House and Theydon Mount in 1570 and 1571; and the variety of readings of English melodramatic novels that proliferated among nineteenth-century Indian readers in new sites such as the Calcutta Public Library or the Burra Bazar Family Literary Club.[73] In each of

---

70. Donna Merwick (1999) *Death of a Notary: Conquest and Change in Colonial New York* (Cornell University Press, Ithaca).

71. Angel Rama (1996) *The Lettered City* (Duke University Press, Durham) p. 18; see Cañizares-Esguerra, *How to Write the History of the New World*, for a more careful situating of these creole elites within what he calls "patriotic epistemology."

72. Lepore, *The Name of War*.

73. Sherman, *John Dee*; Patricia Seed (1995) *Ceremonies of Possession in Europe's Conquest of the New World, 1492–1640* (Cambridge University Press, Cambridge); Lisa Jardine (1990) "Mastering the uncouth: Gabriel Harvey, Edmund Spenser and the English experience in Ireland," in John Henry and Sarah Sutton (eds.) *New Perspectives on Renaissance Thought: Essays in the History of Science, Education and Philosophy in Memory of Charles B. Schmitt* (Duckworth, London) pp. 68–82; and Priya Joshi (2002) *In Another Country: Colonialism, Culture, and the English Novel in India* (Columbia University Press, New York). On resistant reading see also Srinivas Aravamudan (1999) "Equiano and the politics of literacy," in *Tropicopolitans: Colonialism and Agency, 1688–1804* (Duke University Press, Durham) pp. 233–88.

these cases there is an attempt to take seriously the claims of the material geographies of writing, printing, and reading within histories of European trade and empire.

The geography of the written word, therefore, offers ways of understanding the relationship between space, knowledge, and power in the practices of European trade and empire that show those practices in the process of their construction and operation. This can demonstrate the relationships between power and knowledge in the making, rather than simply assuming what they might be. A focus on the material practices of making, disseminating, and using documents in both script and print can decipher the entanglements of both a will to power and the many forms of agency that needed to be negotiated for trade and empire to be pursued and realized.[74] If one of the problems with imperial history is that "European agency too often remains undifferentiated, assumed, and unexplored," then the examination of writing, printing, and reading offers an opportunity to rethink and give empirical depth to this problem of agency in terms of the construction of differentiated relationships between power and knowledge.[75] Concentrating on the geographies of writing also requires the dual focus on small-scale spaces of production and consumption and on the further-flung networks of transmission, transportation, and dissemination. Following the written word through these spaces and journeys can, therefore, map out a geography that traces how trade and empire were done in place and in the relationships between places. Finally, conceiving of the geography of words and representations as a material geography of written objects that had to be constructed, transported, and acted upon can illuminate the potentially transformative effects of the technologies of information, but without those technologies determining everything. As careful accounts of the uses of script and print in different contexts, including the cross-cultural circumstances of trade and empire, can show, what matters is the social and political organization of the technologies of writing and the different meanings that are given to

---

74. A useful parallel here is with the practices and procedures of cartography. In the Indian case, as recent histories have shown, the processes of surveying and mapping involved both an investment in the power of the map and the need to use Indian labor and expertise in its construction. The instruments and practices of surveying were adapted to local conditions, local bodies, and local expertise, shaping the maps that were produced in the process. See Edney, *Mapping an Empire*; and Kapil Raj (2003) "Circulation and the emergence of modern mapping: Great Britain and early colonial India, 1764–1820," in Claude Markovits, Jacques Pouchepadass, and Sanjay Subrahmanyam (eds.) *Society and Circulation: Mobile People and Itinerant Cultures in South Asia, 1750–1950* (Permanent Black, Delhi) pp. 23–54.

75. Stoler and Cooper, "Between metropole and colony," p. 16.

these processes and their products. Writing has been fundamental to changing the world's geographies through its intimate implication in trading and empire building. The task is to show how that happened.

Indian Ink

This book pursues these concerns by studying the writing practices of one organization—the English East India Company—over a period of around two hundred years during which it was involved both in long-distance trade with Asia and in building an empire in India. The book considers, in broadly chronological order, significant moments in the development and use of new forms of writing in and around the Company. This means interpreting very different modes of writing that are primarily differentiated from each other by their forms and uses: heraldic manuscripts, merchants' accounts, political pamphlets, lists of stock prices, and official printed forms and regulations.[76] These are forms of writing that have to be considered as integral parts of practices—of trade, politics, economic policy, stock trading, and imperial government. The fundamental claim is that the Company, and those who sought to profit from it, depended on these forms of writing as a crucial part of constructing this new global geography. Royal letters were necessary to establish initial trading connections in Asia. A system of correspondence and accounting was a fundamental part of organizing and developing a trade in Asian commodities for European markets. The interests of the Company had to be defended, increasingly in print, or it could simply be made to disappear with the removal of its monopoly. Trading in stocks had to be constructed as a safe and legitimate part of everyday business through the provision of regular information on prices in print. Finally, the Company's empire building in late eighteenth-century Bengal was to be built upon the security and certainty that printing seemed to provide. In each instance attention is paid to the ways in which securing the written word aimed to secure the world of the Company itself. Doing so means careful attention to questions of production, dissemination, and consumption. However, as each chapter shows, that process was always a contested and problematic one. There could be no final security found in writing.

It needs to be recognized that this approach to the early modern engagement between Europe and Asia opens up particular lines of inquiry. These

76. Roger Chartier (1995) *Forms and Meanings: Texts, Performances, and Audiences from Codex to Computer* (University of Pennsylvania Press, Philadelphia).

accounts of writing and reading always need to be situated in terms of the
work of the Company of which they were a part. Putting pen to paper was not
the only thing that merchants or empire builders did, and writing practices
can only be understood in context. In doing this, studying a specific institution permits a close engagement with how space, power, and knowledge
were connected in the practices of trade and empire. This is, indeed, just
the sort of evidence required to reconstruct the histories and geographies
of writing as outlined above. The vast archive of the English East India
Company, and the Company's presence in other collections of material in
script and print, is both a testament to the power of writing in its operation
and a resource that allows the uses of that writing to be interrogated. Tracing the Company's operation through these archives across the seventeenth
and eighteenth centuries also allows attention both to the period when the
Company operated as merchants trading with Asia and to its founding of
a territorial empire. In general, historians have tended to concentrate on
one period or the other, although the importance of trade to the Company's
empire building, and vice versa, has been established.[77] However, following
the Company through its substantial transformations of purpose and practice allows different forms of the relationship between Company and state,
and between Europe and Asia, to come into view.[78] Considerable attention
is now being paid to the intersections of ideas and practices in the making
of a company-state in late eighteenth-century Bengal, and to the "colonial
transition" more generally. Doing so through nuanced accounts of the technologies and practices of Company governance permits careful assessment
of the patterning of the continuities with Indian political forms and processes and their intersection with the novel forms of power and governance
introduced by the colonial state.[79] *Indian Ink* contributes to those debates

---

77. See, for example, Kirti N. Chaudhuri (1978) *The Trading World of Asia and the English
East India Company, 1660–1760* (Cambridge University Press, Cambridge); and Peter J. Marshall
(1987) *Bengal: The British Bridgehead, Eastern India 1740–1828* (Cambridge University Press,
Cambridge). For contrasting perspectives on the relationships between trade and the passage to
empire, see C. A. Bayly (1983) *Rulers, Townsmen and Bazaars: North Indian Society in the Age of
British Expansion, 1770–1870* (Cambridge University Press, Cambridge); and Sudipta Sen (1998)
Empire of Free Trade: The East India Company and the Making of the Colonial Marketplace
(University of Pennsylvania Press, Philadelphia).

78. H. V. Bowen (2002) "'No longer mere traders': continuities and change in the metropolitan
development of the East India Company, 1600–1834," in Bowen, Lincoln, and Rigby (eds.) *The
Worlds of the East India Company*, pp. 19–32.

79. Bayly, *Empire and Information*; Wilson, *Governing Property, Making Law*; Sudipta Sen
(2002) *Distant Sovereignty: National Imperialism and the Origins of British India* (Routledge,

(particularly in chapter 6), but it also extends them to the period prior to territorial empire, and to the Company's workings in London as well as in Asia. Focusing on the material geographies of writing practices can therefore open up an intellectual and political history of trade as well as of empire.

As this makes clear, analyzing the English East India Company's writing practices requires attention to both Asia and Europe, and, more specifically, to the many different sites where the Company was at work. In part this is a matter of heeding the call to bring these different places into a single analytic frame. Yet, in very practical terms, considering trade as well as empire, and a focus on material objects within economic and political processes, already foregrounds the conjoined histories of different parts of the world.[80] The intention to trace the material geographies of the written word means this book is concerned as much with what was happening in London as with what was occurring in Aceh, Fort St. George, or Calcutta. Thus, chapters 2 and 3 reconstruct the processes of making trading connections through different forms of writing that passed between Europe and Asia. Chapter 2 discusses the uses of royal letters on the Company's early voyages to the Indian Ocean as ways of establishing trading agreements. Chapter 3 interprets the reform of the documentary procedures of the Company's Indian factories as trade expanded in the late seventeenth century. Both chapters necessarily consider both ends of the connections between Asia and England, as well as the forms of linkage between them. Chapters 4 and 5, by contrast, are almost wholly concerned with the metropolitan politics of print. Chapter 4 works through the Company's relationship with the public sphere of print and the politics of trade. Chapter 5 considers the specific contexts for the production of the first listings of stock prices in print for a general audience. Although each of these chapters is centered on London, and on the spaces of the production, consumption, and regulation of print, this inevitably soon extends to the Dutch Republic and the rest of England, as well as considering the representation in print of events in Asia. Finally, chapter 6 is primarily concerned with the local geographies of print in Bengal. However, once again, this cannot be understood without an appreciation of the interactions and flows of ideas, texts, and people between Bengal and Britain that were crucial to the making of a Company press, and were shaped by its products. Overall, therefore, *Indian Ink* traces a history of writing, trade, and empire

---

London); and the collection of papers introduced by Ian J. Barrow and Douglas E. Haynes (2004) "The colonial transition: South Asia, 1780–1840," *Modern Asian Studies*, 38:3, pp. 469–78.

80. The contrast is with the questions of identity that have primarily concerned historians of nineteenth-century imperialism; see Hall, *Civilising Subjects*.

that is located across a global geography of sites and connections both in and between Asia and Europe.

Interpreting the particular writing practices of the English East India Company also means avoiding the tendency in many studies of the written word and its material forms to concentrate on either script or print, often highlighting the differences between them in the process. Across the book as a whole the focus is diverted from questions of the differences that the shift to script or, especially, from script to print, might make. Instead the emphasis is placed on the ways in which modes of writing and printing were organized and the ways in which speech, script, and print were used together in particular situations. Within these situated histories there is, of course, attention to the explosion of printed material in England during the mid-seventeenth century and the implications of the new uses of letterpress printing in late eighteenth-century Bengal. However, the impact of print in each instance is seen to lie not in the properties of print technology itself, but in the social and political organization of printing and the many different uses of printed materials.

Finally, studying trade and empire through the workings of the East India Company has meant that the particular forms of writing that come under closest scrutiny are those—often written in English, but not exclusively so—that were produced in and around the Company itself. These make up a substantial archive that is patterned in particular ways.[81] Yet, understanding the uses of these forms of writing requires attention to contemporary Asian writing practices and Asian-language texts—for example, the production of the documents of political authority in the gunpowder empires of the east (chapter 2), the trading and accounting practices of Indian merchants (chapter 3), and the work of Indian linguists, chroniclers, and printers (chapter 6). I argue that the foregrounding of the Company's writing practices can demonstrate the complexities of power and agency involved in the production and use of writing, and that this serves the purpose of a geography of writing that is global in scope. However, I recognize that this is not the same as either foregrounding the writing practices of those that the Company traded with and sought to rule, or endeavoring to reconstruct in detail their experience of the forms of script and print produced by the Company.

Indian Ink offers an account of how a new world of knowledge and power was made between the beginning of the seventeenth century and the end

---

81. Betty Joseph (2004) *Reading the East India Company, 1720–1840: Colonial Currencies of Gender* (University of Chicago Press, Chicago); and Guha, *Elementary Aspects of Peasant Insurgency.*

of the eighteenth. It demonstrates that the geographies of writing, at once local and global, were a vital part of the reconfiguring of the relationships between Europe and Asia through trade and empire. Writing was not simply a commentary upon what happened, it was very much part of the action. This begins with the Company's earliest voyages.

CHAPTER TWO

Writing Travels: Royal Letters and the Mercantile Encounter

In March 1608, shortly before he sailed in the *Ascension*, a bundle of documents in different hands and distinct styles, on diverse kinds of paper and parchment, and appended with various seals and signatures, was delivered to Alexander Sharpeigh by the committees of the English East India Company. Carrying this set of writings was vital to the success of the trading venture he was leading, the Company's fourth voyage, to the Red Sea, India, and the Indonesian archipelago. It included Sharpeigh's commissions from king and Company, a dozen letters from James I to "Princes in the Indies," sailing directions, an invoice for his cargo, and a list of weights and scales.[1] These documents were part of a geography of early modern global connection and encounter that involved the tense compact between merchants and monarch in England, the passage of ships across the seas, and the relationships formed or not formed in the Indian Ocean between Europeans and Asians. This chapter argues that interpreting how these writings were made, carried, and exchanged can illuminate the forms of power, representation, and negotiation involved in the earliest encounters between the English company and the Asian rulers with whom it needed to deal.

Much of the discussion of early seventeenth-century relationships between England and India has depended upon interpreting another document-bearing Englishman. Sir Thomas Roe was the ambassador sent to the court of the Mughal emperor Jahangir by King James I and the East India Company

---

1. "List of documents delivered to Alexander Sharpie and Richard Rowles, March, 1607/8," in George Birdwood and William Foster (eds.) (1893) *The First Letter Book of the East India Company, 1600–1619* (Bernard Quaritch, London) pp. 264–65. For clarity I have standardized the uses of "i" and "j" and "u" and "v", and expanded some manuscript abbreviations in quotations. In addition, throughout the book dates before the change to the Gregorian calendar are given in old style form, but with the new year taken to commence on 1 January rather than 25 March.

in 1615. Roe's extensive account of his three-year mission, and his critical descriptions of the imperial court, have provided the basis for debates over how India and Indians were encountered by the English at the very beginning of the Company's attempts to get a toehold on the subcontinent.[2] Bernard Cohn has read Roe's refusals to participate in the rituals of political incorporation required by Jahangir, and his condemnation of salutation (*kūrnish*) and gift giving (*nazr*) as forms of debasement and bribery, as indications that Roe both understood quite clearly the politics of what was happening, and that he inhabited a quite different cultural realm that definitively shaped his readings of the Mughal court. As Cohn put it, "Europeans of the seventeenth century lived in a world of signs and correspondences, whereas Indians lived in a world of substances."[3] For Roe, therefore, the material world was a simple one: there was a more or less fixed relationship between things in that world and the words used to describe them, and, just as importantly, everything and everyone in that world had its price. This was, Cohn argues, quite different from the cosmology of Mughal India.

Understanding Thomas Roe through the production of differences between India and England, between "East" and "West," has also produced interpretations of his writings that draw upon and elaborate Edward Said's genealogy of the discourse of Orientalism. For Kate Teltscher and Jyotsna Singh, Roe's representation of the Mughal court as a theatrical spectacle and of Indian religion as a proliferation of superstitions denigrates India and, albeit with some moments of uncertainty, establishes European social and political organization and cultural identity as the norm. In this way the ambassador "lays the grid of a European system of differences—a ground for later colonial rule—even though the collective dream of imperial power is yet to take shape."[4] However, against the versions of Thomas Roe's mission that stress misunderstandings, incommensurable systems of meaning, and the ability of representation to create power-laden notions of "the other," William Pinch has argued that the differences between Jahangir and Roe should not be essentialized and used as the sole basis of explanation. The two men could, he argues, understand each other and the world the other

---

2. William Foster (ed.) (1899) *The Embassy of Sir Thomas Roe to the Court of the Great Mogul, 1615–1619* (Oxford University Press, Oxford).

3. Bernard S. Cohn (1996) *Colonialism and Its Forms of Knowledge: The British in India* (Princeton University Press, Princeton, NJ) p. 18.

4. Jyotsna G. Singh (1996) *Colonial Narratives/Cultural Dialogues: "Discoveries" of India in the Language of Colonialism* (Routledge, London) p. 40; Edward W. Said (1978) *Orientalism* (Routledge and Kegan Paul, London); and Kate Teltscher (1997) *India Inscribed: European and British Writing on India, 1600–1800* (Oxford University Press, Delhi).

inhabited. They were both part of court societies based on the "cult" of the monarch, and while those societies were different in many ways "these were primarily differences of detail, not of substance. The differences were translatable."[5] Indeed, this process of translation was, as Pinch puts it, based on a "feature of their relationship [that] was emblematic of all human relationships. Their initial deep incomprehension about the other's cognitive world impelled each to seek a language in which to achieve a mutual understanding." Instead of only seeing difference, Pinch emphasizes what Roe's account tells us about interaction, exchange, and understanding as "the outgrowth of proximity, fuelled by basic human curiosity, and achieved by means of fortuitous cultural convergences." What is required, he argues, are "historical discussions of how sameness *and* difference produce global cultural interaction."[6]

A concern with the negotiation of difference has, indeed, impelled recent reinterpretations of Roe's embassy. However, they reject Pinch's notions of "sameness"—with its history of ready translation, basic human curiosity, and an easy reciprocity between Jahangir and Roe—for a more power-laden sense of negotiation, translation, and exchange. Colin Mitchell has sought to demonstrate how Roe's specific prior experience as a courtier and an ambassador—as a gentleman of James I's Privy Chamber, and as part of the diplomatic missions to Spain in 1604 and the Palatinate in 1613—shaped his understandings of the workings of the Mughal empire in quite particular ways.[7] This was not a matter of mutual incomprehension, but of the ways in which a translation of Mughal courtly ritual and social hierarchy through the patronage politics of the Jacobean court, or of Islamic diplomacy through the protocols of European ambassadorial etiquette, produced a specific series of oversimplifications and misunderstandings. Jacobean England and Mughal India may have both been courtly societies but they were different, and the differences were significant in structuring Roe's interpretations of the nature of Mughal politics. Indeed, the problems of translation can be seen to cut both ways. For example, an early seventeenth-century Mughal watercolor with the Persian title *Pādshāhzāde-i farang* ("European prince") shows a figure with prominent symbols of European power, learning, and civility: an open book, a jeweled sword, and the rich clothing of an Englishman of the

5. William R. Pinch (1997) "Same difference in India and Europe," *History and Theory*, 38:3 p. 404. Pinch draws on C. A. Bayly (1996) *Empire and Information: Intelligence Gathering and Social Communication in India, 1780–1870* (Cambridge University Press, Cambridge).

6. Pinch, "Same difference," pp. 406–7.

7. Colin P. Mitchell (2000) *Sir Thomas Roe and the Mughal Empire* (Area Study Centre for Europe, University of Karachi, Karachi).

1630s (Figure 2). However, in what was probably the process of copying a Western painting, the prince seems to have been transmuted into a princess, and an orientalized one at that, with all of the implications that had for Mughal understandings of the gender politics of English courtly power and polish.[8]

In Roe's case these uneasy translations were based on significant problems of language. Sanjay Subrahmanyam notes that at one point Roe was speaking broken Spanish to an Italian interpreter, who translated into Turkish for a Safavid prince, who then rendered it into Persian for Jahangir's courtiers. These cross-cultural difficulties, evident also in Roe's "periodic uncertainty with how to deal with payments as well as how to receive gifts," are illuminated in Subrahmanyam's account by comparison with the shared frame of reference that guided discussions between Jahangir and visitors and envoys from the Ottoman domains, Iran, or Samarqand.[9] It is, he notes, telling that Roe, who was in a position of so little power, produced a lengthy account that insisted on comparing the political systems of England and India through the terms of an Oriental despotism that condemned the Mughal state. In contrast, it is striking that the usually curious Jahangir "found so little of interest in Roe."[10] One sign of the fact that, as Richmond Barbour puts it, the English were "utterly incidental" to the Mughal court is that the ambassador, and the English mercantile venture more generally, barely appear at all in its archival records.[11] The affairs of merchants were inappropriate material for courtly poetry and epic, although Europeans did feature more prominently in the work of miniaturist painters.[12] Indeed, the English Company's only appearance in the *Tūzuk-i-Jahāngīrī*, the emperor's autobiographical chronicle, was when Nicholas Downton's fleet defeated the Portuguese viceroy off Surat in 1615.[13] In this light, for Subrahmanyam and Barbour, Roe's account becomes one that says most about the

8. Rosemary Crill (2004) "Visual responses: depicting Europeans in south Asia," in Anna Jackson and Amin Jaffer (eds.) *Encounters: The Meeting of Asia and Europe, 1500–1800* (V&A Publications, London) pp. 188–99.

9. Sanjay Subrahmanyam (2002) "Frank submissions: the Company and the Mughals between Sir Thomas Roe and Sir William Norris," in H. V. Bowen, Margarette Lincoln, and Nigel Rigby (eds.) *The Worlds of the East India Company* (Boydell Press, Woodbridge) p. 82.

10. Subrahmanyam, "Frank submissions," p. 83.

11. Richmond Barbour (2003) *Before Orientalism: London's Theatre of the East, 1576–1626* (Cambridge University Press, Cambridge) p. 146.

12. John F. Richards (1995) *The Mughal Empire* (Cambridge University Press, Cambridge) p. 287; and J. M. Rogers (1993) *Mughal Miniatures* (British Museum, London).

13. Even then it came third in a list of items of news after the capitulation of Amar Singh, Rana of Mewar, to imperial authority and the death of the troublesome Bahadur; see Henry Beveridge

Figure 2 *Pādshāhzāde-i farang* (European Prince). (Private collection, Porto.)

self-justificatory maneuvers of an English ambassador whose embassy might be accused of failure.

This discussion serves to situate Thomas Roe's embassy, and his account of it, as a specific moment in, and a particular record of, the ongoing encounter between Europe and Asia. It emphasizes the ways in which such encounters were both the site of attempts at exchange, shaped by the interests and understandings of each side, and moments when either party might seek to dominate the encounter, in words or deeds. All those engaged in such encounters were involved in something new and, with their different degrees of engagement and control, were seeking to interpret what was happening and to benefit from it. However, the discussion also emphasizes that understanding this process is only partially served by focusing on the rhetorical and narrative devices of accounts such as the one that Roe produced for his political masters, with all their demonstrations of ontological certainty and a will to power, as well as their ambivalences and signs of weakness. Instead, this chapter suggests that this encounter can be better understood through another set of writings interpreted in a different way. These are the letters sent from English monarchs to Asian rulers, understood as material, mobile objects. They were texts with which Roe was himself very familiar since he was charged with delivering a letter from James I to Jahangir. Such documents were the mechanism by which the mercantile interaction between England and Asia was to be initially effected. They were texts that spoke of English self-representation and power, but they also sought to initiate exchange and, in order to work, required the mediation of differences. Understanding them allows the negotiation over texts and their meanings to be put right at the heart of the active work of making connections between different parts of the globe.

These writings also require a different mode of interpretation. Drawing on the ideas set out in the previous chapter, and on studies of material culture more broadly, my argument is that most attention has been devoted to understanding travelers' cultural baggage and not enough to what was in their luggage. It is productive, therefore, to interpret not only "travel writing" but also "how writing travels": considering the materiality of texts and endeavoring to understand their changing geographies. This means asking how and why such written objects move around the world by attending to their different modes and patterns of movement in terms of the social relations that are constituted around and through them as they are made, made

---

(ed.) (1978 [1909–14]) *The Tūzuk-i-Jahāngīrī, or Memoirs of Jahāngīr*, translated by Alexander Rogers, 2 vols. (Munshiram Manoharlal, New Delhi) vol. 1 pp. 272–75.

mobile, transferred, and make the world.[14] If all this sounds like these inscriptions are being fetishized, then the chapter shows that that is part of their history too.

More specifically, in terms of the English East India Company in the early seventeenth century the argument is that following the travels of writings such as those given to Alexander Sharpeigh can show the active and problematic construction of a particular mode of global connection. This was a "revolutionary departure from the current commercial practices," an integrated trading network operating between London, the Levant, India, and the East Indies, and made up of differentiated forms of power, knowledge, and ritual, and a variety of agencies.[15] In detailing the production of this global historical geography, the relationships between writing, "East" and "West," are not limited to what East India Company representatives like Thomas Roe wrote about Asian "barbarism" or "exoticism"—however ambivalent—that might be read as "some traces of an incipient colonial ideology."[16] Instead, writings as material objects are understood as always involved in highly differentiated and negotiated cultural politics that gradually entangled the agency of Europeans and Asians in the active making of an intercontinental network that required constant material and symbolic work.

Global Writing

It is useful to begin by detailing what was in Alexander Sharpeigh's packet, since it demonstrates the key uses of writing by the Company in its early years. The documents are listed in the first letter book as follows:

i.   Sharpeigh's commission from the Company setting down rules and directions for the voyage.

ii. Twelve letters in English and Portuguese from James I to "princes" and city governors in the Indies asking for the establishment of regular trading relations and trading privileges. Some were addressed to particular individuals—the

14. For early modern Europe and Asia, see Anna Jackson and Amin Jaffer (eds.) (2004) *Encounters: The Meeting of Asia and Europe, 1500–1800* (V&A Publications, London). More broadly, see Nicholas Thomas (1991) *Entangled Objects: Exchange, Material Culture, and Colonialism in the Pacific* (Harvard University Press, Cambridge, Mass.). For objects as well as for people, "travel" may not always be the most appropriate term; see Janet Wolff (1993) "On the road again: metaphors of travel in cultural criticism," *Cultural Studies*, 7, pp. 224–39.

15. Kirti N. Chaudhuri (1965) *The English East India Company: The Study of an Early Joint-Stock Company, 1600–1640* (Frank Cass, London) p. 27.

16. Singh, *Colonial Narratives/Cultural Dialogues*, p. 24.

"King" of Bantam or the "Governor" of Aden—others had the addressee left
blank and were to be filled in on the voyage and used where necessary.

iii. A "saffe Conduct" from the Ottoman "Grand Signior" allowing two factors
(or merchants) to return to England overland from the Red Sea with important
commercial information.

iv. A commission from King James investing Sharpeigh, as general (or leader) of
the voyage, and his lieutenant Richard Rowles with certain powers.

v. A copy of the trading privileges that James Lancaster had brought back from
the sultanate of Aceh in northern Sumatra on the Company's first voyage
(1601–3), accompanied by some geographical descriptions and observations
drawn up as navigational aids.

vi. A letter from the Company's Court of Directors recommending the usefulness
of one Joseph Salbancke.

vii. An order of the Court of Directors emphasizing that the fourth voyage was to
be a trading voyage. Since England was now at peace in Europe, force might
only be used against Portuguese and Dutch ships in self-defense.

viii. Letters for the factors of the previous voyage, who were hopefully trading
profitably at Bantam in Java, instructing them to help the fourth voyage but
not to mix their accounts.

ix. An invoice of all the money (around £20,000 worth of bullion) and merchandise
put on board, ciphers for writing secretly overland, and a list of measures,
weights, and scales for use when buying and loading spices.[17]

These forms of writing are open to a purely functional reading. How else
would long-distance trade be organized at this time? Yet, as has been argued
in chapter 1, writing as a technology of European expansion cannot be so
unproblematically treated. As Stephen Greenblatt has it, sixteenth-century
Europeans in the New World "shared a complex, well-developed, and, above
all, mobile technology of power: writing, navigational instruments, ships,
warhorses, attack dogs, effective armor, and highly lethal weapons, including gunpowder." Within this list the political nature of writing as a "technology of preservation and reproduction" becomes clear.[18] Yet, as has already
been noted, the politics of writing technologies are also complicated by the
recognition that traveling east and traveling west from early modern Europe
involved very different intercultural relations of power and knowledge.

17. Birdwood and Foster, *First Letter Book*, pp. 227–66.
18. Stephen Greenblatt (1991) *Marvelous Possessions: The Wonder of the New World* (Clarendon Press, Oxford) p. 9. Samuel Purchas (1905–7 [1625]) *Hakluytus Posthumus, or Purchas His
Pilgrimes*, 20 vols. (James MacLehose & Sons, Glasgow) vol. 1 p. 486, called writing the Europeans'
"literall advantage" in the Americas.

The circumstances of European involvement in trading within and to early modern Asia are quite specific within the history of global trade. Unlike trade with Africa, the Americas, or, later, the Pacific islands, Europeans could profit only by becoming part of the already existing, high-volume, and very extensive intra-Asian trades in bulk commodities such as cloth and porcelain, trades that Europeans neither controlled nor fundamentally transformed.[19] European shipping did not make up a very significant proportion of the goods exchanged; what was carried back to Europe was also insignificant in relation to what was circulating between parts of Asia. Europeans were unable to control the terms of trade until after the middle of the eighteenth century. This lack of European control was partly a result of the fact that Asian rulers and merchants were more powerful than the European trading companies; for the English it was also a product of unsuccessful economic competition and military engagements with the Dutch East India Company (the VOC).[20] In Asia there was neither the decimation of indigenous populations by disease nor the establishment of extensive settlement and plantation agriculture that transformed the Americas. While trading on the African coast and among the Pacific islands before 1800 was facilitated by indigenous polities and local traders who adapted themselves to and profited from the changing circumstances, these were rarely as established, powerful, and able to set the terms of trade as they were in maritime Asia.[21]

Therefore, in the early seventeenth century, going east meant becoming just another part of a developing Indian Ocean trading economy within which "there was virtually nothing to be discovered or constructed."[22] Going east also meant traveling to domains controlled either by territorial

19. Karl R. Haellquist (ed.) (1991) *Asian Trade Routes* (Curzon Press, London); and Marie A. P. Meilink-Roelofsz (1962) *Asian Trade and European Influence in the Indonesian Archipelago between 1500 and about 1630* (Martinus Nijhoff, The Hague).

20. Holden Furber (1976) *Rival Empires of Trade in the Orient, 1600–1800* (University of Minnesota Press, Minneapolis).

21. See, for example, discussions of the role of Africans in the slave trade in Herbert S. Klein (1999) *The Atlantic Slave Trade* (Cambridge University Press, Cambridge); and Paul E. Lovejoy (2000) *Transformations in Slavery: A History of Slavery in Africa*, 2nd ed. (Cambridge University Press, Cambridge). For the Pacific, see Greg Dening (1980) *Islands and Beaches: Discourse on a Silent Land, Marquesas 1774–1880* (University Press of Hawaii, Honolulu); and Thomas, *Entangled Objects*. For Asia, see Kirti N. Chaudhuri (1985) *Trade and Civilization in the Indian Ocean: An Economic History from the Rise of Islam to 1750* (Cambridge University Press, Cambridge); Susil Chaudhuri (1975) *Trade and Commercial Organization in Bengal, 1650–1720: With Special Reference to the English East India Company* (Firma K. L. Mukhopadhyay, Calcutta); and Derek Massarella (1990) *A World Elsewhere: Europe's Encounters with Japan in the Sixteenth and Seventeenth Centuries* (Yale University Press, New Haven).

22. Niels Steensgaard (1981) "The companies as a specific institution in the history of European expansion," in Leonard Blussé and Femme Gaastra (eds.) *Companies and Trade: Essays*

empires—Ottoman, Mughal, Chinese, and Persian—more powerful than European states, or to the smaller, more recent and more cosmopolitan political entities of the Asian islands and coasts that were at least comparable to European powers. Finally, going east meant traveling to literate cultures with complex and developing traditions and procedures for conducting diplomacy, administration, and trade in writing. For example, under Akbar in the second half of the sixteenth century a Mughal imperial apparatus was constructed that depended upon both an ethnically heterogeneous "political-military administrative elite"—a nobility—and a corps of lower-status Hindu and Muslim officials skilled in the bureaucratic protocols of state administration. Both groups were bound to the emperor by an ethic of imperial service and honor.[23] As Muḥammad Bāqir Najm-i S̱ānī put it in his *Mauʿīẓah-i Jahāngīrī* of 1612–13—a treatise on the art of government for Akbar's successor Jahangir—the people of the pen (*arbāb-i qalam*) were one of the "four pillars of the state." Indeed, he argued that they were superior to the people of the sword since their skills were useful "for the benefit of friends as well as in warding off enemies." Without them, he informed the emperor, imperial rule could not be consolidated and preserved, and the treasury would be emptied rather than replenished.[24]

The pen was, therefore, a tool of Asian diplomacy. In conducting their relationships with other polities, the chancelleries of the Islamic "gunpowder empires" shared traditions of diplomatic writing that were already several centuries old by the 1600s, even if they were being turned to new uses.[25] This *inshā'* tradition, which also governed the relationships between the nobility and the emperor (see chapter 1), sought to prescribe—through essays on the principles of epistolography, the qualifications of scribes, and books of model letters, orders, and decrees—the forms that the extensive and minutely differentiated imperial documentation should take. The forms of

on *Overseas Trading Companies during the Ancien Régime* (Leiden University Press, Leiden) p. 253.

23. John F. Richards (1984) "Norms of comportment among imperial Mughal officers," in Barbara D. Metcalf (ed.) *Moral Conduct and Authority: The Place of Adab in South Asian Islam* (University of California Press, Berkeley) pp. 255–89; John F. Richards (1986) *Document Forms for Official Orders of Appointment in the Mughal Empire* (E. J. W. Gibb Memorial Trust, Cambridge); and John F. Richards (1998) "The formulation of imperial authority under Akbar and Jahangir," in Muzaffar Alam and Sanjay Subrahmanyam (eds.) *The Mughal State, 1526–1750* (Oxford University Press, New Delhi) pp. 126–67.

24. Sajida S. Alvi (ed.) (1989) *Mauʿīẓah-i Jahāngīrī of Muḥammad Bāqir Najm-i S̱ānī: Advice on the Art of Government—An Indo-Islamic Mirror for Princes* (State University of New York Press, Albany) p. 63.

25. Colin Mitchell (1997) "Safavid imperial *tarassul* and the Persian *inshā'* tradition," *Studia Iranica*, 26, pp. 173–209.

texts changed, therefore, depending upon both the purpose of the document and the relative rank of the sender and the recipient.[26] On the one hand, these *inshā'* works emphasized poetic skill in the use of Qur'anic phrases and rhetorical allusions that displayed a deep knowledge of Persian literature. On the other, the codification and standardization of documentary forms and formulas were being reinforced by the increasing bureaucratization of Ottoman, Mughal, and Safavid chancelleries in the seventeenth century, and the rise of a body of scribes employed as professional state specialists. This meant that diplomatic letters (*tarassul*) took certain predictable forms while also working through elaborate poetic phrases. They traditionally began with a *taḥmīdiya*, words of praise to God, sometimes written in red or gold. This was followed by the *iftitāḥ*, a series of carefully chosen honorifics, ending with the name of the addressee, which was written above and to the right of the main text. The main body of the letter would open with the *khiṭāb*, a lengthy set of titles, epithets, and blessings designed to exalt the royal recipient and shaped by the relative status of the addresser and addressee and the traditions of the chancellery from which it came. This was followed by greetings, expressions of eagerness, and guarantees of mutual concord and friendly relations. Here, as well as in the body of the letter, the munshi who composed the text could adorn it with elaborate examples of his fine rhetorical skills. The missive would end with a validating reference to the seal (*muhr*) and an invocation to God or the Prophet Muhammad. As well as the seal, letters might carry a ruler's signature or a *ṭughrā*, an intricate calligraphic design in vermillion or gold bearing the ruler's name and a few short titles.[27] Indeed, the type and quality of the paper and the calligraphy also carried important messages to the recipients.[28] Such letters were a crucial part of the exchanges of ambassadors and envoys that structured the politics of early modern Asia, and were also the mechanism for making agreements between empires and the smaller states of maritime southeast Asia, which were themselves undergoing processes of centralization and scribalization in the seventeenth century.[29]

26. Momin Mohiuddin (1971) *The Chancellery and Persian Epistolography under the Mughals, From Babur to Shah Jahan (1526–1628)* (Iran Society, Calcutta).

27. Mitchell, "Safavid imperial *tarassul*."

28. Mohiuddin, *The Chancellery and Persian Epistolography*; and Annemarie Schimmel (1990) *Calligraphy and Islamic Culture* (I. B. Tauris, London).

29. Riazul Islam (1970) *Indo-Persian Relations: A Study of the Political and Diplomatic Relations between the Mughal Empire and Iran* (Iranian Culture Foundation, Teheran); and Anthony Reid (1975) "Trade and the problem of royal power in Aceh: three stages—c. 1550–1700," *Monographs of the Malaysian Branch of the Royal Asiatic Society 6: Pre-colonial State Systems in Southeast Asia*, pp. 45–55.

Thus, King James's letters could be sent via men like Alexander
Sharpeigh to Asian "princes" and "Potentates" with the expectation of a
reply.[30] Moreover, the importance of Asian polities in safeguarding and guaranteeing trade, the weakness of the English, and the Company's reliance—at
home and abroad—on politically grounded trading agreements meant both
that such letters needed to be sent and that the replies could be expected
to have some meaning and efficacy since these princes could enforce their
will and their word.[31] This put these royal letters at the heart of the Company's early strategy and at the center of their relationships with Asian
polities. They were the technologies of power that constituted the sinews of
the Company's long-distance connections in its early years. By establishing
trade agreements, these texts became a vital part of making a world of voyages strung between London and the ports and factories (Company trading
stations) in the Red Sea, India, and the East Indies. The letters were a global
political technology; they were made to travel and make connections. However, the local mechanism by which these global connections were to be
made was potentially a very weak one: a person a long way from home with
a piece of paper. Therefore, I argue that making the person carrying the piece
of paper into a key element of this new global geography was not dependent
only on what was written on that paper. Instead, by following the letter as
it traveled, it is possible to understand what had to cohere around it and
within it to produce a strong and effective relationship at a distance.[32]

Setting this out involves dividing the letter's journey into three parts:
first, the production of the royal letters within England as part of the local
political and economic relationships between Crown and Company; second,
the letter's voyage across the oceans, considering the material, financial, and
political security of the text, the ship, and the venture that allowed the letters to be delivered; finally, the actual delivery of the letter from ship to
shore, and the local intercultural negotiations, translations, and rituals that

30. Henry Stevens (ed.) (1886) *The Dawn of British Trade to the East Indies as Recorded in
the Court Minutes of the East India Company, 1599–1603* (Henry Stevens, London) p. 91. This is
a printed version of the Company's first minute book.

31. James D. Tracy (ed.) (1991) *The Political Economy of Merchant Empires: State Power
and World Trade, 1350–1750* (Cambridge University Press, Cambridge); and Om Prakash (1981)
"European trade and south Asian economies: some regional contrasts, 1600–1800," in Blussé and
Gaastra, *Companies and Trade*, pp. 189–205.

32. The formulation of these questions owes a lot to Bruno Latour (1987) *Science in Action:
How to Follow Scientists and Engineers through Society* (Harvard University Press, Cambridge,
Mass.); and John Law (1986) "On the methods of long-distance control: vessels, navigation and
the Portuguese route to India," in John Law (ed.) *Power, Action and Belief: A New Sociology of
Knowledge* (Routledge and Kegan Paul, London) pp. 234–63.

involved. This account shows how these letters and the writings that traveled alongside them were a vital part of making the relationships necessary to establish a global network.

Within England

First there had to be a letter to be delivered. The production of this textual object depended upon the relationship between the monarch and the Company; this shaped the form, content, and uses of the letter in particular ways.

The English East India Company was established in the late 1590s by Levant Company merchants who saw their trade threatened by Dutch voyages via the Cape of Good Hope that traded directly for spices that otherwise came overland to Aleppo and Constantinople.[33] Their response required a new form of organization to combine the capital and ships concentrated into a few metropolitan mercantile hands by the privateering war against Spain and the commercial and diplomatic expertise gained by trading in the Levant.[34] The result was a joint-stock company with a royal charter. These two key elements meant that the Company's existence and effectiveness depended upon written contracts that provided a unique combination of delegated powers: "The government delegated certain sovereign rights to the company as a corporation, while at the same time a number of private citizens delegated the right to dispose of part of their property to the same body."[35]

Because it was a joint-stock company, the funds subscribed were managed by the Company's elected directors. This allowed those without mercantile experience to invest, although their more active involvement was discouraged. It also allowed the risks of a trade "so farre remote" to be shared as widely as possible.[36] Such a trade required large and lumpy capital investments in exports with a turnover time of two years or more during

---

33. Chaudhuri, *English East India Company*; and Robert Brenner (1993) *Merchants and Revolution: Commercial Change, Political Conflict and London's Overseas Traders, 1550–1653* (Cambridge University Press, Cambridge).

34. Kenneth R. Andrews (1964) *Elizabethan Privateering: English Privateering during the Spanish War, 1585–1603* (Cambridge University Press, Cambridge); and P. W. Klein (1981) "The origins of trading companies," in Blussé and Gaastra, *Companies and Trade*, pp. 17–28.

35. Steensgaard, "The companies as a specific institution," p. 247.

36. Stevens, *Dawn of British Trade*, p. 8. Brenner, *Merchants and Revolution*, p. 4, argues that the Company was run by London's "merchant establishment." In contrast, shareholders were "members of the aristocracy, high office-holders, the gentry, the shopkeepers, widows and orphans, and foreign merchants." Chaudhuri, *English East India Company*, p. 33.

which the trading capital would be tied up. It also required fixed capital investment in ships, factories, and factors, which would be tied up for even longer.[37] The history of English ventures eastwards during the 1580s and 1590s was a "saga of trial and error" replete with financial ruin, shipwreck, and diversion into plundering and privateering.[38] So, even operating on the basis of a joint stock, the Company's first voyage "was little more than a hesitant, semi-speculative financial venture."[39]

Between 1600 and 1613 the East India Company sent out twelve separately capitalized, and generally profitable, voyages. The Company then moved to financing groups of voyages on terminable joint stocks, and did not institute a permanent joint stock until 1657. As a joint-stock company, this corporate body existed more than anywhere else in the written contracts that bound adventurers into the Company, the adventure books into which they subscribed, and the written minutes—or court books—that detailed the decisions and orders of the Court of Committees and the management of the funds.[40] These were all necessary as administrative and legal devices that allowed the Company's directors to act, as symbolic representations of the delegation of private property rights to the Court of Committees, and as part of the legitimation of the work of the managers to the investors.

The joint-stock organization was also adopted because it was a necessary condition for the grant of a royal charter. This was given to the Company in December 1600 and, as another piece of parchment, it is best understood as a contract struck between a group of private property owners and the Crown.[41] The merchants received the monopoly privileges seen as necessary

37. Peter J. Marshall (1998) "The English in Asia to 1700," in Nicholas Canny (ed.) *The Oxford History of the British Empire*, vol. 1: *The Origins of Empire* (Oxford University Press, Oxford) pp. 264–85.

38. Kenneth R. Andrews (1984) *Trade, Plunder and Settlement: Maritime Enterprise and the Genesis of the British Empire, 1480–1630* (Cambridge University Press, Cambridge) p. 2; John C. Appleby (1998) "War, politics, and colonization, 1558–1625," in Canny (ed.), *Origins of Empire*, pp. 55–78; Philip Lawson (1987) *The East India Company: A History* (Longman, London); and Glyndwr Williams (1997) *The Great South Sea: English Voyages and Encounters 1570–1750* (Yale University Press, New Haven).

39. Chaudhuri, *English East India Company*, p. 38.

40. Stevens, *Dawn of British Trade*, pp. 54, 144, 186.

41. "A Priviledge for fifteene yeeres granted by her Majestie to certaine Adventurers, for the discoverie of the Trade for the East-Indies, the one and thirtieth of December, 1600." Reproduced in Purchas, *Hakluytus Posthumus*, vol. 2 pp. 366–91. This was a piece of paper that cost much money and effort: 200 "angelles" to pay persons to attend the issuing of the charter, and "100^{li} in gold and...50^{li} more in silver to pay other Charges for the passing of the Pattent and the Warraunt for Coyninge the Signet the privy seale and great seale." Stevens, *Dawn of British Trade*, p. 107. In 1610 the Company planned to have it copied onto vellum and bound "in some decent manner." W. Noel Sainsbury (ed.) (1862) *Calendar of State Papers: Colonial Series, East*

for overseas trade in recognition of the fixed costs they would incur in establishing it. This gave them control over who entered the trade and over the supply of merchandise, reducing risks in both Asian and European markets. They also gained the right to export bullion to Asian markets where there was no demand for English woollen cloth—a privilege that was seen as damaging to the domestic economy. Finally, the charter gave the Company the expectation that the monarch's government would support it against its European rivals. The charter was an example of what Robert Brenner calls "politically constituted forms of property," created by the Crown and given to private property owners. He likens it to church offices and courtiers' privileges.[42]

The monarch received several political and economic benefits in return. The Company became a source of loans and bribes for the Crown throughout the seventeenth century, extorted by threatening the removal of monopoly privileges.[43] Successive monarchs also needed the increased customs revenue the Company generated. Customs made up the largest part of the royal income and had certain other advantages. They did not squeeze powerful landowners, they were more easily gathered from corporate bodies than from individual merchants, and they were not subject to the same troublesome forms of parliamentary approval as other revenues. This was especially important in the early seventeenth century, when royal finances were badly stretched by the costs of war, and by James I's unprecedented need to spend in order to consolidate political support; according to Mark Kishlansky, James had "the financial acumen of a child in a sweetshop."[44] The monarch could also expect political support from the Company's influential merchants.

The relationship between Crown and Company was not without tension. The two sides' interests did not entirely coincide, either at home or overseas. Taken together, James I and Charles I sold permissions to trade to other

---

Indies, China and Japan, 1513–1616 (Longman, Green, Longman & Roberts, London) [hereafter *Cal. S.P. 1513–1616*] entry 479.

42. Brenner, *Merchants and Revolution*, p. 666.

43. In 1614 Thomas Smythe suggested that the Company's need for protection against rivals and the state of the royal finances made it a good time to give the king a financial gift. *Cal. S.P. 1513–1616*, 739. There was also a £20,000 loan following the withdrawal of the royal patent for a Scottish East India Company (1618), and bribes of £10,000 to the king and the lord high admiral after the capture of Hormuz in 1622. Later in the century, the Protectorate got £15,000 in 1659; Charles II extorted £10,000 in 1662, £50,000 in 1666, and £70,000 in 1667. See Chaudhuri, *English East India Company*; and M. N. Pearson (1991) "Merchants and states," in Tracy, *Political Economy of Merchant Empires*, pp. 41–116.

44. Mark Kishlansky (1996) *A Monarchy Transformed: Britain 1603–1714* (Penguin, Harmondsworth) p. 83.

parties, supported amalgamation with the rival Dutch company (see chapter 4), and failed to provide military and diplomatic support. For its own part, the Company resisted royal interference, took Parliament's side over taxation increases, and evaded the Crown's strictures against hostility with European rivals in the East.[45]

Seventeenth-century trading companies always involved a relationship between merchants and the state, between "public and private rights."[46] However, unlike the Portuguese Estado da Índia, and the French and Dutch companies, which were in their different ways arms of the state, the English East India Company had the Crown as a powerful but "unreliable ally." They were locked together in a relationship of partial dependence and "mutual mistrust," albeit with the advantage generally on the side of the monarch.[47] Since "it was generally understood that individual Europeans, acting solely on their own authority, could not negotiate effectively with Asian princes, great or small," it was within this relationship, at once both supportive and undermining, that the royal letters were produced.[48]

These letters were provided through the Privy Council. Each batch was "solicited" by the Company, having been drafted by it following intelligence from knowledgeable and well-traveled merchants on addressees, proper titles, and content.[49] When agreed by the Crown's representatives, the Company had the letters translated and given to the chief herald to be "lymmed and faire engrossed . . . fitt for his Maties Signiture."[50] This was a matter of creating a particular sort of valued and valuable object. It was achieved through

---

45. See Lawson, *East India Company*, Andrews, *Trade, Plunder and Settlement*; Chaudhuri, *English East India Company*; Brenner, *Merchants and Revolution*; and Niels Steensgaard (1974) *The Asian Trade Revolution of the Seventeenth Century: The East India Companies and the Decline of the Caravan Trade* (University of Chicago Press, Chicago).

46. Steensgaard, "The companies as a specific institution," p. 247.

47. Marshall, "The English in Asia," p. 276; and Andrews, *Trade, Plunder and Settlement*, p. 278. On other companies see Leonard Blussé and Femme Gaastra (eds.) (1981) *Companies and Trade: Essays on Overseas Trading Companies during the Ancien Regime* (Leiden University Press, Leiden); and Jonathan I. Israel (1989) *Dutch Primacy in World Trade, 1585–1740* (Clarendon Press, Oxford) p. 70.

48. Furber, *Rival Empires of Trade*, p. 185.

49. Stevens, *Dawn of British Trade*, p. 91; *Cal. S.P. 1513–1616*, 356, 682, 702. Birdwood and Foster, *First Letter Book*, contains "Petition for letters from the King to certain Eastern Princes and Governors for the Third Voyage [1606/7]," pp. 103–5; "Copies of the Royal Letters thus obtained, dated 23rd February 1606/7," pp. 105–10; "Petition to the Earl of Salisbury for royal letters to carry out in the Fourth Voyage [1607 or 1608]," p. 216; and "Note of royal letters carried out in the Fourth Voyage, 1607/8," p. 231.

50. IOR B/3 *East India Company Court Minutes: 31st December 1606 to 26th January 1611*, ff. 18^r, 14^v.

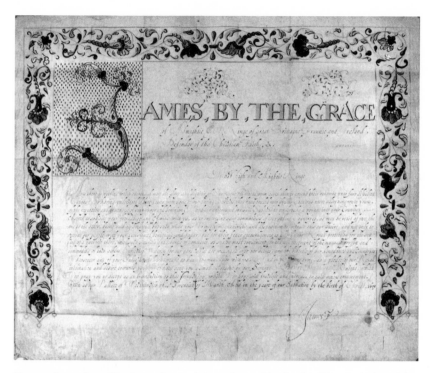

Figure 3 Letter from King James I to an Asian ruler, 20 March 1610. (By permission of the British Library, Add. Ch. 56456.)

the combination of an heraldic writing style addressing a "Most High and Mightie King" and recommending merchants to him in elegant phrases and carefully laid-out calligraphy; the appropriate parchment; elaborately decorated capitals and borders in blue, red, and gold; and the authenticating and authorizing armory of the royal signature and seals (Figure 3).[51] These letters were objects whose materiality was meant to display their political valence and their potential economic value.[52] To fend off any mis-

51. Three examples of these letters have survived. One is BL Mss. Add. Ch. 56456 *Letter from James I to an Eastern Prince concerning a factory, 1610*; and two are in the James Ford Bell Library, University of Minnesota, 1611 mJa *James I, King of England, 1566–1625, Manuscript Letter signed*, addressed "To the high and mightie prince, the Emperor of Japan": Westminster, *10 January [1611]*, and 1614 mJa *James I, King of England, 1566–1625, Manuscript Letter signed*, addressed "To the high and mightie monarch, the great Emperor of China": Westminster, 7 *February 1613 [i.e., 1614]*.

52. Peter Beal (1998) *In Praise of Scribes: Manuscripts and Their Makers in Seventeenth-Century England* (Clarendon Press, Oxford) p. v, argues that manuscripts should be considered "first and foremost as physical artifacts which have their own peculiar nature and mode of being."

reading, the Company purchased special boxes to keep the letters in and gathered presents, without which they feared the letter "woulde but little bee esteemed."[53] On the first voyage these were gifts of appropriately rich materials, or that were finely wrought: a 250-ounce "silver fountaine" and basin along with silver cups of various sizes; embroidered belts; a shield and helmet; a case of daggers; fans and plumes of feathers; and "faire Costlie Looking glasses."[54] These objects were not fixed in a circuit of ritual gift giving but were potential offerings whose value might be realized in other more commercial exchanges if the need or chance arose. As the Company put it, such objects "shalbe prepared for presentes or otherwyse disposed."[55]

However, the production of these letters was not unproblematic. There were tensions between Crown and Company here too. The Company had to take care not to request too much political or military intervention, and to leave any "matter of State" to the appropriate authorities.[56] Moreover, as William Segar, the herald who was paid £13 6s 8d for writing the letters taken on the first voyage, would have realized there was a contradiction in promoting trade and lauding merchants through the heraldic codes of honor in which he was a specialist, and that the very form of his writing was meant to convey.[57] As he put it in his *Honor Military and Civill* (1602), "The scope and marke of each mans endeavour, is either profit or honour. The one proper to men in bare or meane fortune, the other to persons of vertue and generous minde."[58] The letters, which drew their meaning from the herald's sense that "[a] man is also honoured, when his Prince or other superior is pleased to salute him by word or writing," confused these categories of honor and profit by using the codes and hierarchies of honor that flowed from the monarch and were based on military achievement to promote the pursuit of profit "proper to vulgar people, and men of inferior Fortune."[59] Following Richard Helgerson's reading of Richard Hakluyt's *Principal Navigations*, these letters can be understood as another textual form that dressed merchants "in figurative robes of honor" and only partially stabilized the

---

53. W. H. Moreland (ed.) (1934) *Peter Floris, His Voyage to the East Indies in the Globe, 1611–1615* (Hakluyt Society, London) p. 34; and Stevens, *Dawn of British Trade*, p. 139.

54. Stevens, *Dawn of British Trade*, pp. 116, 151–52.

55. Stevens, *Dawn of British Trade*, p. 118.

56. "Petition for letters," in Birdwood and Foster, *First Letter Book*, p. 104.

57. Stevens, *Dawn of British Trade*, p. 139.

58. William Segar (1602) *Honor Military and Civill* (London) p. 209; see also William Segar (1590) *The Booke of Honor and Armes* (London).

59. Segar, *Honor*, p. 212.

disruptions of aristocratic and monarchic power that produced by presenting their corporate profit seeking as a work of the nation as a whole.[60]

The royal letters—especially those left unaddressed—projected, therefore, an imaginary set of relations between the parties involved. They attempted to use personal relationships between powerful individuals who were unlikely to meet to establish trading relations between "nations." They offered the monarch's personal recommendations of the "Justice & Civillitie" of those who carried the letters, and asked that "amitie & friendshipp" or "mutuall amitie & entercourse" be accorded through written guarantees—a reply.[61] They also promised the same advantages to any subjects of these "foreign princes" who might visit England. This monarchical reciprocity constituted sovereignties as autonomous and in parity, what James Boon terms "reciprocal monarchs serving no pope."[62] It was a representation of Asian polities in terms of an English form of sovereignty and legitimate kingship, a complementarity designed to exclude the Iberians and the Dutch while it legitimated Stuart rule.[63] Boon argues that in Samuel Purchas's compilation of English voyages, *Hakluytus Posthumus, or Purchas His Pilgrimes* (1625), this "tied exotic courts to the monarch, over and above any companies, parliaments, or other forces of interest."[64] Within the letters it served to constitute trade between nations as "mutuall profitt," and presented the English East India Company's representatives as the monarch's direct agents.[65] Thus the form and content of each letter as it was produced in England through the relationship between merchants and the state both contained and tried to establish these imaginary relations, simplifying the real relationships of trade and agency, obscuring the tensions between

60. Richard Helgerson (1992) *Forms of Nationhood: The Elizabethan Writing of England* (University of Chicago Press, Chicago) p. 178.

61. "Circular Letter from Queen Elizabeth to the Kings of the East Indies [February, 1600/1?]," and "Letter from James I. to the King of Bantam, carried out in the *Second Voyage* [1604]," in Birdwood and Foster, *First Letter Book*, pp. 20, 47.

62. James A. Boon (1982) *Other Tribes, Other Scribes: Symbolic Anthropology in the Comparative Study of Cultures, Histories, Religions and Texts* (Cambridge University Press, Cambridge) p. 167.

63. The letters encouraged Asian leaders to deal with the English by suggesting that they would otherwise be thought to be dominated by the Portuguese; see "Circular letter from Queen Elizabeth," and "Royal Letters to Eastern Princes and Governors carried out in the *Sixth Voyage* [1609/10]," in Birdwood and Foster, *First Letter Book*, pp. 20–21, 349.

64. Boon, *Other Tribes, Other Scribes*, p. 157.

65. "Royal letters carried out in the Seventh Voyage [1611]," in Birdwood and Foster, *First Letter Book*, p. 422; and Richmond Barbour (2000) "'There is our commission': writing and authority in *Measure for Measure* and the London East India Company," *Journal of English and German Philology*, 99:2, pp. 193–214.

Crown and Company, and demonstrating once again the English merchants' substantial investment in written guarantees and binding contracts.

Across the Oceans

Having been written, the royal letter had to reach its destination. Once the letter was on board ship its holder needed both the power to carry it and the capacity to deliver it. Previous royal letters to the Chinese emperor had ended up frozen in Arctic ice along with their carriers in 1553, and disappeared with all ships in 1596 after Benjamin Wood's voyage was wrecked.[66] This capacity was a matter of the security, in broad terms, of the ship that carried the letter. This question of security can be examined in a series of interconnected ways, some obvious and some less so, in order to establish how various elements could be combined to strengthen the position of the man with the piece of paper as he traversed the oceans. This involves understanding the ship as a material space, as an accounting space, and as a political space. In each case, varied forms of writing attempted to construct and secure the boundaries of these spaces from leakage and collapse. Navigational aids sought to prevent shipwreck and capture; accounts and rules tried to secure products, provisions, and profits; and royal and Company commissions endeavored to provide a governable political order. Together they worked to enable the ship to act as a viable vessel for carrying the royal letters eastwards.

The Ship as a Material Space

The ship's security was a matter of the defense of its wooden walls against both natural and human forces. First, and most obviously, the ship was endangered by nature. It was threatened by storms, high seas, currents, shoals, and sand banks; by getting lost in vast oceans, or by losing rudders, masts, yardarms, or sails; through being eaten into by worms or rot, or by the ageing and stretching of planks and seams if its sides were left uncaulked for too long. The ship was also endangered by other people. The rival Ottoman, Portuguese, and Dutch "empires" at various times sought to sink or capture English East India Company ships.[67] The people and political authorities of

66. Richard Hakluyt (1907 [1598–1600]) *The Principal Navigations, Voyages, Traffiques & Discoveries of the English Nation*, 8 vols. (J. M. Dent and Sons, London) vol. 1 pp. 241–44, vol. 8 pp. 312–14; and *Cal. S.P. 1513–1616*, 252.

67. The Portuguese took a Company ship at Surat. Clements R. Markham (ed.) (1878) *The Hawkins' Voyages during the Reigns of Henry VIII, Queen Elizabeth and James I* (Hakluyt

the coasts and islands of Africa, Europe, and Asia also had an eye to getting their hands on valuable cargoes.[68]

These natural and human threats can be treated in parallel because of the spatial organization of power on these voyages. Niels Steensgaard has argued that the initial decisive advantage of the Dutch and English trading companies over their Asian and Iberian rivals lay not in the efficiency of their economic organization—their lowering of transaction costs and transport costs—but in their internalization of protection costs. These were the payments, as customs, bribes, or passes, that other long-distance traders, such as Armenian merchants leading caravans of camels to Aleppo or seagoing Gujarati merchants, paid to those, including the Portuguese Estado da Índia, who had control of the means of violence in the territories and seas across which they passed. Internalizing these costs was a matter of making protection predictable and economical by transforming uncertainty into risk and providing security at cost price.[69] The weakness and disinterest of the English state meant that the Crown "could not always provide the military and diplomatic shell within which trading activity could be carried on by private, civilian, commercial interests."[70] Instead, the Company had to provide for themselves the means of violence—cannon, shot, powder, small arms, swords, armor, and the fighting men who could use them—to protect their ships and to attack others, or to take hostages where dictated by the interlocking logics of trade and politics.[71] This transformed the geography of power. On the international scale, armed trade both was "a radical departure from the normal commercial practices followed in contemporary Europe" and meant that "it was usually the Europeans who broke violently into Asian trading systems that had been relatively peaceful before their

Society, London) pp. 392–93. The Turks tried to capture the *Darling* on the Company's sixth voyage. "Thomas Love's Journal" in Clements R. Markham (ed.) (1877) *The Voyages of Sir James Lancaster to the East Indies* (Hakluyt Society, London) p. 148. Dutch threats against the Company's eighth voyage were part of the prelude to war. Sir Ernest M. Satow (ed.) (1900) *The Voyage of Captain John Saris to Japan, 1613* (Hakluyt Society, London) pp. 40–41.

68. For example, attempts on a Company ship at Madagascar and its subsequent detention at Waterford and Duncannon to force the sale of cargo are recorded in "Journal of the Sixth Voyage, kept by Nicholas Downton," in Markham, *Voyages of Sir James Lancaster*, pp. 151–227. Breton villagers also took the cargo from the wreck of the *Union* in 1611. See Purchas, *Hakluytus Posthumus*, vol. 3 pp. 79–81.

69. Steensgaard, *Asian Trade Revolution*.

70. Michael J. Braddick (1998) "The English government, war, trade, and settlement, 1625–1688," in Canny, *Origins of Empire*, p. 293.

71. See, for example, the lists of armaments included in the inventories of the ships procured for the first voyage, in Stevens, *Dawn of British Trade*, pp. 15–20, 42–44.

arrival."[72] At a smaller scale, the envelope of political or military power within which traders who had successfully paid others for protection moved was more extensive than their defenses against the natural forces of winds, currents, and sandstorms. However, it was also more unpredictable whether, when, and where this envelope would be breached or collapse entirely. In contrast, for the trading companies that provided their own envelope of protection, the boundary that had to be defended against hostile human and natural forces was the same: the boundaries of the ship were everything.

A range of practices that countered these conjoined human and nonhuman dangers, and to which writing was central, was navigation. The course a ship took and where it anchored was a matter of "reading" economic possibilities, natural aids and hindrances, and political forces. In finding suitable anchorages, keeping a ship out of range of the guns of a castle or fort was as much a consideration as the potential trade available or the depth of water and solidity of the seabed.[73] Indeed, early English East India Company navigation was a promiscuous mixture of forms of knowledge and practice. It included astronomical and solar observation.[74] It involved consulting written accounts and maps, either from Dutch and Portuguese sources or produced by previous East India Company pilots, captains, and generals.[75] Luminaries such as Richard Hakluyt were paid to provide maps and "required to sett downe in wryting a note of the principall places in the East Indies wher Trade *is* to be had," while Englishmen and Indian sailors who knew the coasts "where none of our people formerly have been" were consulted on soundings

72. Kirti N. Chaudhuri (1991) "Reflections on the organizing principle of premodern trade," in Tracy, *Political Economy of Merchant Empires*, p. 437; James D. Tracy (1991) "Introduction," in Tracy, *Political Economy of Merchant Empires*, p. 9; and Kirti N. Chaudhuri (1981) "The English East India Company in the seventeenth and eighteenth centuries: a pre-modern multinational corporation," in Blussé and Gaastra, *Companies and Trade*, p. 30.

73. See, for example, "Journal of . . . Nicholas Downton," in Markham, *Voyages of Sir James Lancaster*, p. 167; Satow, *Voyage of Captain John Saris*, pp. 51–54; and "Commission from the Company for the [Third] voyage," in Birdwood and Foster, *First Letter Book*, pp. 114–36.

74. Nathaniel Marten's journal in Purchas, *Hakluytus Posthumus*, vol. 3 pp. 304–18; John Davis (1595) *The Seamen's Secrets* (London); Moreland, *Peter Floris, His Voyage*, p. 39; and William Foster (ed.) (1934) *The Voyage of Thomas Best to the East Indies, 1612–14* (Hakluyt Society, London) p. 49.

75. One Dutch source, Jan Huygen von Linschoten (1596) *Itinerario* (Amsterdam), was available in a less lavish English translation within two years, as *John Huigen Van Linschoten, His Discours of Voyages into Ye East & West Indies* (John Wolfe, London, 1598). Its use is shown in Moreland, *Peter Floris, His Voyage*, pp. 31, 104; Satow, *Voyage of Captain John Saris*, p. 188; and Foster, *Voyage of Thomas Best*, pp. 48, 49, 88. For the collection and use of other Dutch, Portuguese, and English sources, see *Cal. S.P. 1513–1616*, 357, 362, 479; Stevens, *Dawn of British Trade*, p. 125; Markham, *The Hawkins' Voyages*, p. 379; and Satow, *Voyage of Captain John Saris*, p. 190.

and fortifications.[76] When at sea, advice, information, and maps were sought from captains of European and Asian ships met along the way, and "local" pilots—Indians, Africans, Arabs, Javanese, and Japanese—were either paid, pledged, or kidnapped, depending on the circumstances, and relied upon to keep ships away from dangerous coasts and people.[77]

That such an assemblage of navigational techniques made little distinction between the natural and political environments is evidenced by the journals that the Company required be kept by captains, masters, merchants, and pursers. These were to be compared to one another during the voyage "soe as a perfect discourse may be sett downe."[78] As the written record of the voyage, these journals were to be a guarantee to the adventurers in London of the performance of what they had ordered to be done, or at least that decisions made on the voyage did not contradict those orders. Where possible, they were to provide the foundation of succeeding voyages by providing knowledge of winds and shoals, useful ports and places of refreshment, good routes to take, supposedly friendly or treacherous peoples, good commodities and markets, and the extent of Portuguese and Dutch power as well as the orientations of Asian polities. These journals were collated, archived, and used by the Company in increasingly systematic ways in order to provide "navigational" knowledge for subsequent voyages.[79] Such

---

76. Stevens, *Dawn of British Trade*, pp. 124, 143; *Cal. S.P. 1513–1616*, 479, 474. See Heidi Brayman Hackel and Peter C. Mancall (2004) "Richard Hakluyt the younger's notes for the East India Company in 1601: a transcription of Huntington Library manuscript EL 2360," *Huntington Library Quarterly*, 67:3, pp. 423–36.

77. On sources from other ships, see "The voyage of Captain Sharpeigh, 1608–1609," in Markham, *Voyages of Sir James Lancaster*, p. 128; Markham, *The Hawkins' Voyages*, p. 386; William Foster (ed.) (1905) *The Journal of John Jourdain, 1608–1617* (Hakluyt Society, Cambridge) pp. 41, 226; and Satow, *Voyage of Captain John Saris*, p. 186. On pilots, see Markham, *The Hawkins' Voyages*, p. 387; Foster, *Journal of John Jourdain*, pp. 24, 56–57, 191–92, 311; Moreland, *Peter Floris, His Voyage*, p. 102; Satow, *Voyage of Captain John Saris*, pp. 27, 78; and Foster, *Voyage of Thomas Best*, pp. 42, 105, 107, 184.

78. "Commission from the Company to Alexander Sharpie and Richard Rowles for the Fourth Voyage, 29th February, 1607/8," in Birdwood and Foster, *First Letter Book*, p. 242.

79. After the return of the third voyage in 1609 the Company resolved "[t]o consider the commission given to David Middleton and others, how it has been observed, and to take the journals of the voyage before the mariners are paid." *Cal. S.P. 1513–1616*, 430. The journals' importance is also signaled by the prohibition on factors writing to anyone other than the Company about trading opportunities; see "Commission from the Company for the [Third] voyage," in Birdwood and Foster, *First Letter Book*, p. 131. On the collation of written material by the Company, see *Cal. S.P. 1513–1616*, 744, and the "motion made to have all journals first written in the Company's books before they be lent to any man, and none to use them without consent of the committees." *Cal. S.P. 1513–1616*, 831, 1016. For subsequent use, see Foster, *Voyage of Thomas Best*, p. 177.

writings had a key role in attempts, even if they were only attempts, to secure the ship as a material space.[80]

The Ship as an Accounting Space

The ship's boundaries were also used in accounting for the coming and going of merchandise, and therefore in safeguarding the security of the voyage by guaranteeing its profitability. The ship was to be defended against the defrauding of the Company and the embezzlement of the cargo, provisions, and equipment by both Asian traders and its own employees. This was done via a carefully constructed set of relationships between writing, authority, property, and space within which a true and verifiable knowledge of the contents of the ship and its ownership could, in theory, be ascertained at any point.

To prevent Asian merchants from overstating the quantity of spices sold, "the sufficientest & best advised" factors were to ensure that the weights, and therefore prices, assigned to each bag were correct. This meant understanding how different systems of weights and measures corresponded to each other. Company factors were instructed that "in ev^{r}y place where you came and trade" they were to use the scales provided in London to "Compare their weightes & measures wth o^{rs} before you buie, to thend that you may better know what to doe and howe to make yor reckoninges."[81] They were also to check the quality of the goods before they were loaded. This meant ensuring that the spices were adequately "garbled," with the valuable spices separated from worthless husks and dirt. As early as the second voyage ships sailed from London carrying "garblers wth sives and ffans."[82]

The most elaborate regulations, however, were designed to protect the Company from its own crews. Provisions were increasingly tightly controlled with a system of invoices, signed warrants, and accounts that made captains and pursers accountable to the Company on their return.[83] Private trade in certain goods was banned from the first voyage. To secure this the

---

80. Barbour, "'There is our commission,'" p. 212, sees both the Company's orders and the journals as attempts "to micro-manage the unmasterable."

81. "Commission from the Company for the [Third] voyage," in Birdwood and Foster, *First Letter Book*, p. 128.

82. "Commission from the Company for the [Second] voyage," in Birdwood and Foster, *First Letter Book*, p. 57.

83. Birdwood and Foster, *First Letter Book*, allows comparison of the lack of discussion of this in the "The Company's Commission for the [First] voyage [February 1600/1]" (pp. 4–9), with the increasingly sophisticated regulations in "Commission from the Company for the [Third]

Company gave itself the right to "search of all such Chestes, Boxes, Pack-
ettes, Bookes, [and] Writinges" for evidence of concealed cargoes, and stipulated that anything found would become its property.[84] By the sixth voyage,
crew members were allowed to take out or bring back only as much permitted goods as "will be conteyned wth in His proportioned chest of allowance,"
and their holdings had to be registered in the purser's book or forfeited. Indeed, the problem was such that the Company ordered "that there be noe
dealinge by Exchadge betwixt party & party neither money for mony, Co-
moditee for money, or Comodity for Comodity neither outwardes nor home-
wardes."[85] These trading voyages were intended to be trade-free zones so that
the Company's monopoly could be protected from its own employees.

When loading spices onto ships in the Indies the Company sought to ensure that "all that hath beene bought hath beene received" via instructions
stating that every bag was to be weighed on shore by a trustworthy factor
who should note the weights in a book.[86] The bags should then be numbered and sent on board under the care of a "trustie person" carrying a letter
detailing bag numbers and weights. Each bag was to be reweighed by the
purser—who would be standing with the master by a single open hatch—
checked against the letter, which was to be kept, as well as the details being
entered into the purser's book. Only then was the bag to cross the boundary
into the hold. Finally, the general was to check the purser's book every night,
and it was to be compared against the factor's account from the shore. The
purser's book was also to contain an account of everything that was carried
on and off the ship by everyone, including the general and the purser himself.[87] This would also allow the contents of each ship arriving in London
to be known in detail, permitting an accurate account of the voyage to be
given to customs officers and the proportioned disbursal of the cargo to the
adventurers. This, however, was only possible if adequate measures were
immediately taken by the Company's committees to prevent goods from
being run off the ship for private profit as soon as it arrived at the western

---

voyage" (p. 117) and "Commission from the Company to John Saris and Gabriel Towerson, for
the Eighth Voyage, 4th April, 1611" (pp. 417–18).

84. "The Company's Commission for the [First] voyage" and "Commission from the Company for the [Second] voyage," in Birdwood and Foster, *First Letter Book*, pp. 5–6, 58.

85. "Commission from the Company . . . for the Sixth Voyage," in Birdwood and Foster, *First
Letter Book*, pp. 344, 332; and *Cal. S.P. 1513–1616*, 472.

86. "Commission from the Company . . . for the Sixth Voyage," in Birdwood and Foster, *First
Letter Book*, p. 341.

87. "Commission from the Company for the [Third] Voyage," in Birdwood and Foster, *First
Letter Book*, p. 127.

ports, and, once it got to the Thames, the supervised unloading of cargo by porters wearing "sewtes of canvas dublett and hose wth. out pockettes."[88] Without these safeguards the ship's boundaries and the Company's accounts were dangerously permeable.

It is uncertain how far all these instructions could be carried through. The Company's inability to act against private trade beyond the home ports, and its rather lax attitude to its generals' personal profits as long as overall returns remained high, suggest that they were not fully enforced.[89] What is clear, however, is that it was writing as both rule-making and verifiable accounting that was an essential tool in attempts to establish the boundaries of the ship as the crucial ones in making the vessel a "calculable space" and in ensuring the underlying financial security of the voyage.[90]

The Ship as a Political Space

As the discussions of the materiality of the vessel and its role in the accounting of profit and loss have already begun to show, the ship's internal social and political organization was part of its effectivity and capacity at sea and while trading, and was therefore part of ensuring the security of the voyage and the letter. Again, in writing, the commission from the sovereign guaranteed the powers of the general. It enabled him to carry the monarch's letter and to show that he acted on his sovereign's behalf. David Middleton's demonstration to the Dutch when confronted in the Moluccas in 1609 that he "had the Kings Majesties broad Seale to my Commission" declared that it made its holder someone who could not be dismissed as a pirate or other mere private interest.[91] It also gave the general a monopoly of legitimate violence over the ship's crew. Those who led East India Company fleets held, by virtue of their commissions, the power of martial law to punish those who did not "beare themselves one towardes another in all good order and quietnes for avoeiding any occasion that might breed mutiny quarrells

88. Stevens, *Dawn of British Trade*, pp. 245–48, quotation from 247; and *Cal. S.P. 1513–1616*, 474.

89. See Moreland, *Peter Floris, His Voyage*, pp. lv, 16; and *Cal. S.P. 1513–1616*, 884. For the treatment of generals see *Cal. S.P. 1513–1616*, 755; and Satow, *Voyage of Captain John Saris*, pp. lxvi–lxx.

90. Peter Miller (1992) "Accounting and objectivity: the invention of calculating selves and calculable spaces," *Annals of Scholarship*, 9, pp. 61–86.

91. Purchas, *Hakluytus Posthumus*, vol. 3 p. 98. Middleton refused to show them any more than his name and the seal. It was, however, no guarantee. William Hawkins was told by a Portuguese captain that he did not give "a fart for his Comission." Markham, *The Hawkins' Voyages*, p. 393.

or dissencion amongest them." They were empowered to deploy the punishments used by "armies at sea" and, for capital offenses of murder and
mutiny, had the power to order execution after an offence was "trulie and
justlie proved."[92] What Edmund Burke later characterized as "in reality a
delegation of the whole power and sovereignty of this kingdom sent into
the East" was, therefore, directed not only outwards as armed trade, but
also directed inwards to govern the crew.[93] The ship of state became the
ship as state.

Members of the crews on the early voyages were punished for murder, sodomy, attacking officers, dueling, absconding, going or staying ashore
without permission, drinking on the Sabbath, theft, and insubordination.
They were punished by hanging from the yardarm, duckings, beatings, whip-
pings, being tied to the capstan or nailed to the mast through their hands, or,
for lesser offences, spells in irons below decks.[94] Minor offenses were dealt
with by the general's summary justice. Major offenses occasioned trials
where witnesses were examined, depositions written, and cases conducted
before juries of twelve mariners "according to our English lawes."[95] This
shipborne polity was to be self-sufficient in justice and state-sanctioned violence exported from England. It was intended to be bound back together
by the symbolism of major punishments. The hangings on the early voyages, one for sodomy and two for mutinous murder, were performed aboard

92. "Royal Commission to Alexander Sharpie and Richard Rowles for the Fourth Voyage, 8th
February, 1607/8," in Birdwood and Foster, *First Letter Book*, p. 229.

93. Burke is quoted in Janice E. Thompson (1994) *Mercenaries, Pirates, and Sovereigns:
State-Building and Extraterritorial Violence in Early Modern Europe* (Princeton University Press,
Princeton, NJ) p. 32.

94. See William Foster (ed.) (1940) *The Voyages of Sir James Lancaster to Brazil and the East
Indies, 1591–1603* (Hakluyt Society, London) p. 125; Moreland, *Peter Floris, His Voyage*, pp. 51,
125, 133; Satow, *Voyage of Captain John Saris*, pp. 90, 91, 94, 150; and Foster, *Voyage of Thomas
Best*, pp. 57, 37–38, 103, 109, 116, 117, 141, 142, 152.

95. Robert Coverte (1612) *A True and Almost Incredible Report of an Englishman that
(Being Cast Away in the Good Ship Called the Assention in Cambaya the Furthest Part of the
East Indies) Travelled by Land Through Many Unknowne Kingdomes, and Great Cities* (London)
p. 21, reports the trial and execution of Thomas Clarke and Francis Driver for murdering John
Lufkin, the master of the pinnace *Good Hope*, on the fourth voyage. Two others were acquitted.
See also Foster, *Journal of John Jourdain*, pp. 79–81. The trial of Nicholas White for sodomy before
"a Jury of the cheifest of the Marryners" is in "A Journall kept by John Jourdain [1607–1617],"
BL Sloane Mss. 858 ff. 20ᵛ–21ʳ, and "Journal of William Revett" IOR L/MAR/A/VII ff. 23ʳ–23ᵛ.
It does not appear in the "transcription" of Jourdain's journal published by the Hakluyt Society.
White was sentenced to death. The cabin boy William Acton, also accused of sodomy, pleaded
"nonage," and since his age could not be proven the sentence was deferred until they returned to
England. Nicholas Cowan, who had also had some sexual contact with Acton but "not...in the
same nature as the other" (f. 20ᵛ), was whipped at the capstan. Other trials appear in Moreland,
Peter Floris, His Voyage, p. 65; and Foster, *Voyage of Thomas Best*, p. 162.

ship in front of the assembled crew and accompanied with sermons, moral lessons, and, where possible, repentant speeches from the condemned.[96]

The power of life and death that the general had over the crew resided in the paper—the royal commission—as "sufficient warrant and discharge."[97] Two copies were prepared to enable the general's lieutenant to rule his ship when they parted company, but not before. There could be no division of the "absolute authoritie of justice" by which the ruler of this polity was constituted.[98] Likewise, the names of the general's successors in case of death were "to be kept close" in "a boxe sealed wth hard waxe," and to be opened with due ceremony only when the time had come.[99] However, if this piece of paper made the man who held it powerful, then that power had to be enacted and it could be challenged. There is a tantalizing glimpse of this from the eighth voyage. When the *Hector* returned to England from Japan, the general, Captain John Saris, was accused of mistreating his crew. This included an incident during which he had been struck by Thomas Fuller, the ship's master, in a dispute over the provisions that Saris was allowing. As it was reported in the Company's minutes, Fuller, "forgettinge himselfe, tooke him [Saris] by the throate, kickt him and strooke his Comission out of his hand against the Missenmaste."[100] The paper, then, in Saris's hand as the guarantee and symbol of his power over the crew, was swept aside by Fuller. The Company decided, although not without dissent, that they had to support Saris's punishment of "some whoe contempned there Commission, and the authoritie given him by his ma[tie]," even though he had "exceeded his authoritie a little."[101] The alternative was that their commanders would be prevented from punishing for fear of the actions that might be taken against

---

96. The final repentance was not always forthcoming: see "Journal of William Revett," f23[v]. Daily religious services that all men were to attend and prohibitions against blasphemy, theft, and gambling also explicitly aimed to prevent disunity among the crew; see "Commission from the Company for the [Second] Voyage," in Birdwood and Foster, *First Letter Book*, pp. 53–54.

97. "Royal Commission . . . for the Fourth Voyage," in Birdwood and Foster, *First Letter Book*, p. 229.

98. Foster, *Voyages of Sir James Lancaster*, p. 76.

99. "Commission from the Company . . . for the Sixth Voyage," in Birdwood and Foster, *First Letter Book*, p. 345. For transfers of power, see Moreland, *Peter Floris, His Voyage*, pp. 38–39; and Markham, *The Hawkins' Voyages*, p. 392. There were potential problems when no successor was named; see "Company's Commission for the [First] Voyage," in Birdwood and Foster, *First Letter Book*, pp. 8–9.

100. IOR B/5 *East India Company Court Minutes: December 1613–10th November 1615*, f. 303.

101. IOR B/5 f. 354. There is some ambiguity over which commission Saris was holding. The dispute, however, was over the forms of authority and punishment, which were a matter of the commission from the king.

them. Overall, however, incidents such as this heightened the Company's sense that the royal commission's powers of punishment were a necessary guarantee of the security of the ship from mutiny and mishap, but not a sufficient one. Their own commission for each general outlined a more collective set of powers and relationships.

The Company directors argued that their ideal general was a composite creature: "partlie a Navigato^{r}, partelie a Merchaunt to have knowledge to lade a shipp, and partlie a man of fashion and good respect."[102] The power he was granted to run the voyage depended not only upon the royal commission, but upon his accumulated maritime and mercantile knowledge, and on specific social and cultural qualities.[103] He was, like a wise monarch, to "soe behave himselfe as he may be both feared and loved," while relations between all on the voyage were to be governed by mutual respect, obedience to superiors, and "love and kindnes" on all sides.[104] Indeed, the general was to exercise "a moderacion of gov^{r}nm^{te}," and was not to rule by dictat. Major decisions were to be taken in consultation with chief officers and factors, "whome we wish you to imbrace & use wth a loveinge respect, & freindlie & diligentlie to hearken to their Counsells."[105] Consequently, the debate over David Middleton's "misgovernment" concluded that "to avoid the like hereafter he is aquainted with a resolution that a council of the officers of the ship be called together upon any great cause to give their verdict," and later generals were reminded that "he is accompted of more fortitude that over cometh himself then he that prevaileth against a multitude."[106]

These councils were one part of a series of negotiations between the general, his officers, and the crew that also recognized that the power of seamen to abscond or mutiny shaped the governance of the ship. They worked alongside inquiries by the generals into problems aboard ships over

102. They were preventing a Captain Harris from being forced on them by the lord privy seal in 1614; IOR B/5 f. 52. "Fashion" here means "high quality or breeding . . . eminent social standing or repute" (*Oxford English Dictionary*).

103. See, for example, the disagreements over whether David Middleton was fit for the job. *Cal. S.P. 1513–1616*, 702 (entries for 14, 15, 18, and 19 March 1614).

104. "Commission from the Company . . . for the Fourth Voyage," in Birdwood and Foster, *First Letter Book*, p. 241. This political formulation is a familiar part of early seventeenth-century political culture; see Kevin Sharpe (2000) *Remapping Early Modern England: The Culture of Seventeenth-Century Politics* (Cambridge University Press, Cambridge) pp. 38–123.

105. "Commission from the Company for the [Second] Voyage" and "Commission from the Company . . . for the Fourth Voyage," in Birdwood and Foster, *First Letter Book*, pp. 53, 264.

106. *Cal. S.P. 1513–1616*, 702; and "Commission from the Company . . . for the Sixth Voyage," in Birdwood and Foster, *First Letter Book*, p. 344. See also Peter Floris's attempts to get the dictatorial and distrustful Thomas Essington to consult his council on the *Globe*. Moreland, *Peter Floris, His Voyage*, pp. 92–93.

victualling or into damaging conflicts between merchants and seafarers, and
petitions, or round robins, sent to the general by his crews complaining of
abuses.[107] These forms of governance relied more upon the respect that a
general might command than upon his power to punish; they reveal that
the security of the ship as a political space was also a matter of the careful
management of status, respect, justice, and mercy. For example, Thomas
Best's capitulation to being "earnestlie entreatt[ed]" by his crew to commute
punishments for mutiny as a reward for their valor in battle made mercy
as powerful a tool of the voyage's security as the hangman's rope.[108] In
contrast, after the mutiny and murder of the *Good Hope*'s master, John
Lufkin, on the fourth voyage, Alexander Sharpeigh was warned "that if
he did winke att such a fowle matter the next boute would be his." That
the mutineers were hanged in this case was perhaps because their actions
and words raised the specter of an illegitimate exercise of collective power
by the crew that challenged rather than reinforced the general's authority.
When asked who had killed Lufkin, "they awnswered: One and all of them;
that it was better for one to dye then all."[109] The general's leadership came
in combining and judging the different forms of maritime, mercantile, and
political expertise needed to fulfil the orders outlined in the Company's commission.

This commission was, therefore, carried in its own particular way. On
the first voyage, for example, it was "agreed to be written in fowre several
partes and unto everie part the Comon seale of the Companie to be added to
thende that *in* everie ship one of the said Comissions may be kepte in the
handes of the principall merchauntes."[110] On other voyages the general was
to "give often notice to all persons … what they are to observe by this o^r
Comission & direccions."[111] However, Nicholas Downton's comment that
the "sharing of the authority of a commander among divers doth much
increase pride, cause divisions, and greatly hinder the common business"

---

107. William Foster (ed.) (1943) *The Voyage of Sir Henry Middleton to the Moluccas,
1604–1606* (Hakluyt Society, London), p. 9; Foster, *Journal of John Jourdain*, p. 208; "Journal
of … Nicholas Downton," in Markham, *Voyages of Sir James Lancaster*, p. 162; and Foster, *Voyage of Thomas Best*, pp. 11, 107, 137–38, 140.

108. Foster, *Voyage of Thomas Best*, pp. 38, 126–27. For other examples of mercy see Foster, *Voyages of Sir James Lancaster*, p. 132; Moreland, *Peter Floris, His Voyage*, pp. 51, 126;
Satow, *Voyage of Captain John Saris*, pp. 89, 92–93; and Foster, *Voyage of Thomas Best*, pp. 115,
162.

109. Foster, *Journal of John Jourdain*, pp. 79–80.

110. Stevens, *Dawn of British Trade*, p. 132.

111. "Commission from the Company for the [Fifth] Voyage," in Birdwood and Foster, *First
Letter Book*, p. 301.

points towards the continuing tensions within the royal and Company commissions' combination of absolute power and collective consultation.[112]

Overall, therefore, the man with the letter had the power to carry it
due to a potentially potent combination of power delegated by the Crown
in writing and a cultural authority based on knowledge, status, and respect
that combined and adjudicated between different forms of interest and expertise. On every voyage these political considerations and the materiality
and mathematics of the ship's boundaries were made to coincide by the
geography of power implicit in the Company's strategy. Through the combination of writing and authority in these three areas of practice the ship
could become the crucial vessel to be carefully and conjointly managed as
a material space (through the practicalities of navigation and the use of violence), as an accounting space (through the protocols of bookkeeping and
the bounding and banning of exchange), and as a political space (through
the commissions' absolutism of punishment and the heeding of wise counsel). Work in all these registers at once, and combining writing with other
forms of action, could give the Company's agents the capacity to convey the
sovereign's letter to its destination.

From Ship to Shore

The delivery of the monarch's letter by the general on behalf of the Company was shaped by these relationships between writing, knowledge, and
power. In large part this was because negotiating over the circumstances of
trade and exchange necessitated working from the strongest position possible to try to guarantee "mutual respect" a long way from home.[113] First,
knowledge and experience gathered in the Company's journals and codified
in their commissions gave the general a framework for what to expect and
what to do. Second, the royal commission and the monarch's letters gave
the general a delegated power that stated that what he promised would be
held to by the Crown.[114] Finally, the ship served as a space to act from and
withdraw to. It was the general's own territory, and its integrity was actively
produced and maintained through the many forms of work outlined above
rather than simply being given. However, handing over the letter also meant

112. William Foster (ed.) (1897) *Letters Received by the East India Company from its Servants
in the East*, vol. 2 (Sampson Low, Marston & Company, London) p. 184. Thomas Essington also
saw councils as likely to "diminishe his autoritie." Moreland, *Peter Floris, His Voyage*, p. 93.

113. Furber, *Rival Empires of Trade*, p. 314.

114. See "Circular Letter from Queen Elizabeth," in Birdwood and Foster, *First Letter Book*,
p. 21.

leaving the ship and entering into more uncertain forms of encounter, ritual, and exchange that were mediated by the vital work of interpretation and translation. It was apparent that the Company's men would have to improvise, and that they could only imperfectly enact the delegated authority of the king and Company, who might then deny their actions.[115] These questions of representation were certainly a matter of concern. King James I's instructions to Thomas Roe on his appointment as ambassador recognized the problem, and ordered him "in your Carriadge, to be Carefull of the preservaccion of our honour and dignity, both as wee are a soveraine Prince and a professed Christian, aswell in your speeches and presentacion of our letters as in all other Circumstances as farre as it standeth with the Custome of those Countries."[116] Treading the fine line between other peoples' customs and maintaining status and respect meant that the delivery of letters could be managed well or badly. The Company's early history contains examples of both.

The Company's first voyage took James Lancaster to the trading port and sultanate of Aceh on the northern tip of Sumatra carrying letters from Queen Elizabeth. By the early seventeenth century this was becoming one of the key entrepôts of western Indonesia, where the products of China, Japan, and southeast Asia were traded for those of India and west Asia. Traders came to Aceh from Constantinople, Venice, Aleppo, the Red Sea, and Arabia; from Gujarat, Malabar, Coromandel, Bengal, and Arakan; and from the Malay peninsula, Siam, China, Borneo, and Java.[117] Along with this expanding trade, and against the incursions of the Portuguese after 1509, had grown an increasingly centralized Islamic sultanate that had formed an alliance with the Ottoman Turks and was able to channel all commerce through the port capital Banda Aceh, and largely through the hands of the sultans themselves. Indeed, after a period of control by a local elite of nobles and merchants in the 1580s, there had been a strong reassertion of royal power. Sultan Ala'ad-din Ri'ayat Syah Sayyid al-Mukammil (who ruled between 1589 and 1604) had both violently destroyed the existing elites and increased the state's control over trade via the routines of a new port bureaucracy

---

115. Barbour, "'There is our commission.'"

116. Foster, *The Embassy of Sir Thomas Roe*, vol. 2 p. 552. The "Commission from the Company . . . for the Sixth Voyage," in Birdwood and Foster, *First Letter Book*, p. 336, noted "yf any thinge be presented to the kinge, Prince or greate Potentate, yt is to be done (as you knowe) with state and Comlines."

117. Ashin K. Das Gupta (1962) "Acheh in the seventeenth-century Asian trade," *Bengal Past and Present*, 81, pp. 37–49.

that "primarily took care of the ruler's rights and interests in trade."[118] As well as creating new state forms, the sultan was keen to use alliances with other European powers against his rivals the Portuguese. By 1601 he had already exchanged diplomatic letters with the Dutch and knew of the English queen's victories over Spain.[119] It was into this dynamic political world that the Company's ships sailed.

After sending a messenger to negotiate safe passage, Lancaster landed with thirty men and was met by one of the sultan's "noblemen." Lancaster refused to relinquish her majesty's letter—although he did allow the seal, superscription, and signature to be studied—"saying he would deliver it to the King himselfe; for it was the order of Ambassadors, in those parts of the World from whence he came, to deliver their letters to the princes owne hands, and not to any that did represent the kings person." In time the sultan "sent six great elephants, with many trumpets, drums and streamers, with much people, to accompany the general to the court . . . The biggest of these elephants was about thirteene or fourteene foote high; which had a small castle (like a coach) upon his back, covered with crimson velvet. In the middle thereof was a great bason of gold, and a peece of silke exceeding richly wrought to cover it, under which Her Majesties letter was put. The generall was mounted upon another of the elephants."[120] In this fashion the letter completed its journey in a way that was both customary for merchants arriving at Aceh and presenting themselves to the sultan, and sufficiently reminiscent of Elizabethan and Acehnese royal ceremonial to impress both the Company's men and the sultan's subjects.[121]

118. Takeshi Ito (1984) *The World of the Adat Aceh: A Historical Study of the Sultanate of Aceh*, unpublished Ph.D. thesis, Australian National University, p. 278; Reid, "Trade and the problem of royal power"; and Takeshi Ito and Anthony Reid (1985) "From harbour autocracies to 'feudal' diffusion in seventeenth-century Indonesia: the case of Aceh," in Edmund Leach, S. N. Mukherjee, and John Ward (eds.) *Feudalism: Comparative Studies* (Sydney Association for Studies in Society and Culture, Sydney) pp. 197–213.

119. The oldest surviving document from Aceh is a 1601 letter from the sultan to Prince Maurice of the Netherlands. L. F. Brakel (1975) "State and statecraft in seventeenth-century Aceh," *Monographs of the Malaysian Branch of the Royal Asiatic Society 6: Pre-colonial State Systems in Southeast Asia*, pp. 56–66; and Purchas, *Hakluytus Posthumus*, vol. 2 p. 314.

120. Foster, *Voyages of Sir James Lancaster*, pp. 91–92.

121. See Augustine Spalding (1614) *Dialogues in the English and Malay Languages* (London) pp. 1–10; for a similar ceremony at Patani, see Moreland, *Peter Floris, His Voyage*, pp. 33–34. On Elizabethan and Acehnese iconography and pagentry, see Louis A. Montrose (1999) "Idols of the Queen: policy, gender, and the picturing of Elizabeth I," *Representations*, 68, pp. 108–61; John Nicols (1823) *The Progresses and Public Processions of Queen Elizabeth*, 3 vols. (John Nicols and Son, London); and G. W. J. Drewes and P. Voorhoeve (eds.) (1958) *Adat Atjèh: Reproduced in Facsimile from a Manuscript in the India Office Library* (Martinus Nijhoff, The Hague).

At the palace, "the pivotal seat of power of the Sultanate," the delivery of the letter was accompanied with exchanges of gifts and compliments, and with banqueting and entertainment by dancing and music-making women who "are not usually seene of any but such as the king will greatly honour."[122] Lancaster played his part well and was granted the right to confer with the sultan's nobles over trade. This was done in Arabic, but Lancaster had brought with him from England a Jewish interpreter from the Barbary Coast "who spake that language perfectly" and had, it seems, been christened at the Cape of Good Hope with Lancaster as godfather.[123] While other accounts of this boundary figure show distrust of his mobility across religious categories, the services he rendered at Aceh "stood him [Lancaster] in good steed at that time."[124] Therefore, Lancaster was able to keep hold of the letter, convey it to the sultan in a manner appropriate to both parties, use a trustworthy and competent interpreter, and engage in courtly rituals. He got what he went to get.

In contrast, the fourth voyage in 1608 took Alexander Sharpeigh and his parcel of papers to Aden on the southern coast of Arabia. The city was still in ruins after the Turks had taken it in 1538, and much of its trade had been lost to the Red Sea port of Mocha. Sharpeigh kept his ship out of the range of the castle's guns and made arrangements to deliver a letter he had from James I to the city's governor. The governor, Rajab Agha, received Sharpeigh warmly and promised trading privileges. Then he suggested that Sharpeigh should carry the king's letter to his ruler, the pasha of San'a, whose palace was three hundred miles to the north, or at least send one of his men. Sharpeigh refused to do this. Perhaps this was because he had no knowledge of who the pasha was. He had not been mentioned by the Company, and there was certainly no letter for him. Moreover, none of the blanks could be adapted for the task. They were either written for city governors or autonomous princes, whereas the pasha was a part of the Ottoman hierarchy and therefore somewhere in between.[125] Sharpeigh decided to send the letter—probably the one to the governor of Aden—with one of the governor's own footmen. Once at San'a it was translated by a French renegade who "understood some words, and made the Bashaa believe that he understood all." He "fained an interpretacion,"

---

122. Ito, *World of the Adat Aceh*, p. 21; and Foster, *Voyages of Sir James Lancaster*, p. 93.

123. Foster, *Voyages of Sir James Lancaster*, p. 124.

124. Foster, *Voyages of Sir James Lancaster*, p. 97. A later encounter produced the opinion that "[w]ith the English he was of their religion; with the Mahometans, of theirs; whereas he was all the while a Jew." Albert Gray (1887) *The Voyage of François Pyrard of Laval to the East Indies, the Maldives, the Moluccas and Brazil* (Hakluyt Society, London) vol. 1 p. 284.

125. "Note of Royal Letters," in Birdwood and Foster, *First Letter Book*, p. 231.

saying that the ship was full of "cloth of gold and silver and much silks" rather than the woollens, lead, and tin that the Company wanted to sell there.[126]

Therefore, Sharpeigh lost control of the letter and its translation, refused to engage in accepted rituals of power, and picked a mode of travel for the letter's final journey "which was," according to John Jourdain, one of the voyage's factors, "much misliked both of the Governor and of our owne people, in soe much that the Governor, perceivinge in the Generall therein some weaknes, made not so much reckoninge of him as in former time."[127] Sharpeigh's poor delivery had rendered him a weak man. The governor took advantage of this and detained him ashore to try to extort taxes and tribute, the protection costs that Sharpeigh had inadequately internalized. He got free by having the ship's company take hostages themselves.[128] However, Sharpeigh was unable to maintain leadership of his crew, despite having executed some of them for mutiny and murder. A bad and unpopular decision, "done without councell" and under the influence of the ship's "headstrong" master—whom other officers strongly suspected of sodomy—led to the sinking of his ship off Surat and the destruction of its hierarchical polity.[129] Since "nowe there was no respect of persons, that it was everyone for himself," the crew ended up roaming northern India spending the Company's gold.[130] As a weak man with a piece of paper, Sharpeigh managed to lose the rest of the letters with which he had been entrusted. They were never delivered.[131]

---

126. Foster, *Journal of John Jourdain*, p. 73.

127. Foster, *Journal of John Jourdain*, p. 73, who also notes Sharpeigh's refusal to accept the present of a horse from the governor (pp. 73–74). Coverte, *True and Almost Incredible Report*, p. 20, records that the governor found Sharpeigh "to be a plaine and simple man."

128. See "Coppies of L[rs] of . . . Alexand[r] Sharpeigh wrytten to the Marchauntes and M[r] aboard at the tyme of his detaynment in Aden by the Governor," IOR L/MAR/A/VII ff. 34[r]–42[r].

129. Foster, *Journal of John Jourdain*, p. 115. On the master, Philip de Grove, "headstrong" is from Frederick C. Danvers (ed.) (1896) *Letters Received by the East India Company from its Servants in the East*, vol. 1: *1602–1613* (Sampson Low, Marston & Company, London) item 12. Thomas Jones notes that Sharpeigh was "ruled in most things by our Master." Purchas, *Hakluytus Posthumus*, vol. 3 p. 68. Jourdain refers to him as "a crooked apostle." Foster, *Journal of John Jourdain*, p. 138; the manuscript version of this journal notes that Nicholas White, hung for sodomy (see n. 95), "had alwayes hope to be reprived . . . because he should not confesse some fowle matters w[ch] he knewe by the M[r]: Phillipp de Grove: as afterwards was proved." BL Sloane Mss. 858 f21[r]. Coverte, *True and Almost Incredible Report*, p. 68, is less coy, calling Grove "a detestable buggerer."

130. This is what John Jourdain reported Robert Coverte saying to him during the sinking of the *Ascension* as Coverte sat on Jourdain's back laden with Company gold. Foster, *Journal of John Jourdain*, p. 118; see also pp. 133–34.

131. Sharpeigh's account is in IOR L/MAR/A/VII ff. 43[r]–45[v].

Between them these quite different stories suggest that the royal letters that the Company's generals sought to deliver could certainly make sense within the worlds of Asian trade and politics, just as the replies that they solicited could make sense in London. However, making them make sense was a matter of interpretation and translation on both sides. This was not simply the reliable interpretation and translation of the words they contained.[132] Both the sender and the recipient of such letters had to engage in interpreting and translating the form of the letter (or reply), its carriage, and the mode of its delivery in order to construct an understanding of the encounter and the nature of the exchange. This process was shaped in particular ways by different cultures of writing. For example, the letters the Company brought to Jahangir (akin to Figure 3) would have looked, to Mughal imperial eyes, like poor versions of the *tarassul* that they exchanged with other powers. They contained an opening phrase invoking God (*taḥmīdiya*), in red and gold, but one that gave undue emphasis to the English king. There was an honorific (*iftitāḥ*), albeit one that was pitifully short and did not place the addressee's name above that of the addresser and the main text. There were words of praise for the recipient (*khiṭāb*), and expressions of the desire for friendly relations, but they used none of the carefully crafted poetic language to exalt the sovereign being addressed. Seals, signatures, and calligraphy would all have been recognizable, but also recognizably different.[133] Just as the English Company delivered poor presents, and often delivered them badly, so they brought poor letters.[134] It is, however, important to understand that this paucity was also part of Mughal expectations and practice. The letters exchanged between Islamic empires were routinely more fulsome in their use of honorifics, titles, and poetry than those sent from their imperial chancelleries to European powers. There was less advantage in evoking the full weight of the shared epistolographic tradition in these cross-cultural communications.[135] It is also notable that attempts were made to translate the form of these letters as well as the content. As Sir Thomas Roe testily noted to the Company, sensing another slight to himself and King James, "Another terrible inconvenience that I suffer: want of an interpreter. For the

---

132. Although reliable translation between languages was always a problem on the early voyages, and attempts to resolve it were made using prior experience, local interpreters, specified "linguists," and books. See, for example, Foster, *Journal of John Jourdain*, pp. 108, 267; "Journal of . . . Nicholas Downton," in Markham, *Voyages of Sir James Lancaster*, p. 168; Moreland, *Peter Floris, His Voyage*, p. 45; and Spalding, *Dialogues in the English and Malay Languages*.

133. Mitchell, *Sir Thomas Roe.*

134. Subrahmanyam, "Frank submissions"; and Barbour, *Before Orientalism.*

135. Mitchell, "Safavid imperial *tarassul*."

Broker's here will not speak but what shall please; yea they would alter the Kings letter because his name was before the Mughals, which I would not allow."[136] He did not consider the possibility that adjusting an English letter to Mughal expectations might have done him a favor.

This meant that these exchanges of documents were always active, and to a certain extent collaborative, processes of the construction of meaning in which both sides participated, albeit unequally. That this way of making texts meaningful across differences of culture and interest always brought with it problems of interpretation is vividly demonstrated by a final example. The tenth voyage, led by Thomas Best, arrived on the west coast of India in 1612, intending to establish a factory. With their knowledge of the willingness of the English to use force against Indian and Portuguese shipping—both Best and Henry Middleton on the voyage before him had turned their guns to good use—the merchants and port authorities of Surat were willing to come to an agreement.[137] Yet the long-promised *farmān*, or imperial order, confirming the trading privileges they had agreed on was so delayed that Best, suspicious of Indian motives, threatened to leave the coast without delivering the letter and presents that he carried from King James. When the *farmān* finally arrived Best suspected that it was "a counterfaite," since it had been delivered only as "a private letter."[138] As the voyage's surgeon, Ralph Standish, put it, Best "stood in doubt wether yt was the firma [*farmān*] or no, for that yt was brought in no statt nor fashion, nott beseaming the letter of so greatt a kinge as the Magolle was."[139] The general demanded "the cheife men of Suratt to come downe and deliver it to me, with those rights to which it belonged." Five days later this was done "with great solemnity." The dignitaries arrived from Surat with a company of horsemen, and "the cavellero that brought the firma from the courtt was, in verie rich apparrell, reedy with the firma in his hand, covered with read silke and maid up in cloth of gould."[140] Without the presence of a Persian translator, Best received the *farmān* on being assured that it contained the articles he had previously agreed to. His reception of the document was marked by a blast of English trumpets and a volley or two of small shot. The day was then given over to celebrations, Best "being verie mirrie and

---

136. William Foster (ed.) (1926) *The Embassy of Sir Thomas Roe to India, 1615–1619*, rev. ed. (Oxford University Press, London), p. 100.

137. Phanindranath Chakrabarty (1983) *Anglo-Mughal Commercial Relations, 1583–1717* (O.P.S. Publishers, Calcutta).

138. Foster, *Voyage of Thomas Best*, p. 40.

139. Foster, *Voyage of Thomas Best*, pp. 143–44.

140. Foster, *Voyage of Thomas Best*, pp. 40, 144, 238.

joyfull, nott onlie he but all the marchantts and company generallie, that this bussines was in such desent maner accomplished."[141] Their sense that a vital and profitable part of the Company's long-distance global trading network had been secured was based on their interpretation of the appropriate rituals for a binding written agreement between kings: a matter of solemn rites with cloth, costumes, trumpets, and cannon, as well as with pieces of paper.[142]

Unfortunately, it later turned out that the document was not judged sufficient to secure the Company's trade in India. The specific articles had never been signed by the emperor, who had issued only a general *farmān*, and would not agree to be bound to the demands of mere merchants. Once the other officials who had put their names to the particulars were dead, the document (the original of which had also been mislaid) was "of small valliditye."[143] Indeed, it was in an attempt to rectify this that Sir Thomas Roe was eventually dispatched to the Mughal court with his own letter from the king.

Conclusion

The production, transportation, and exchange of royal letters were part of a particular moment in the making of a global trading network. They were part of a world where mercantile capacity needed to enlist the monarch's political power to preserve markets, enforce authority aboard ship, and make agreements with distant polities. Some of these elements lasted longer than others. By 1615 the permanent establishment of overseas merchants served by regular shipping meant that neither royal letters nor royal commissions for each voyage were appropriate for governing trade. While still relying upon its monopoly privileges, the Company sought to move towards a less immediate dependence on the Crown and towards other written instruments.[144]

In the early seventeenth century, however, the royal letters had, through their making, carriage, and delivery, crystallized, consolidated, and changed

---

141. Foster, *Voyage of Thomas Best*, p. 145.

142. For a similar incident, see Foster, *Voyage of Sir Henry Middleton*, pp. 57–58.

143. Chakrabarty, *Anglo-Mughal Commercial Relations;* and Foster, *Voyage of Thomas Best*, p. 259 n. 2.

144. The Company wanted "power to authorize their servants sent abroad to govern their men with sufficient and absolute authority under some great seal, to be procured without troubling the King for every fleet as they hitherto have done." *Cal. S.P. 1513–1616*, 765, 704. See also "Form of commission from the Company to the commander of a fleet" and "Form of commission from the Company to the captain of a ship," in Birdwood and Foster, *First Letter Book*, pp. 493–99.

the forms of global connection between England and the Indies. Understanding these texts as material objects allows a focus on their production within the specific relationship between merchants and the monarchy; their transportation alongside other documents on vessels formed of planks, mathematics, and political power; and their delivery within intercultural contexts of ritual, translation, and exchange whose outcomes were shaped by participants on all sides. Following the letter as it traveled enables the understanding of writing, East and West, to be extended beyond the consideration of seventeenth-century Europeans' representations of the East to examine the forms of the global networks that got them there and back as intricate cultural constructions of power, knowledge, ritual, and other materials.

Instead of emphasizing only the differences between East and West, this way of interpreting the texts of global encounter, like the voyages, brings them into the same orbit. Bernard Cohn's discussion of Sir Thomas Roe's cultural difficulties in India, with which we began, was also a discussion of writing. In order to pursue his argument that "Europeans of the seventeenth century lived in a world of signs and correspondences, whereas Indians lived in a world of substances," Cohn argued that writing for Europeans was a simple and practical matter of messages, orders, contracts, and entitlements. For Indians, by contrast, rulers' letters "were a sharing in the authority and substance of the originator, through the act of creating the document. Hence, in drawing up a document, a letter, or a treaty, everything about it was charged with a significance that transcended what might be thought of as its practical purpose. The paper, the form of address, the preliminary phrases of invocation, the type of script, the elaboration of the terminology, the grammar, the seals used, the particular status of the composer and writer of the document, its mode of transportation, and the form of delivery—all were meaningful."[145] Tracing the royal letters' journeys shows that this was also true of the Europeans who made, carried, and delivered them. It was true because they were messages about monarchic power; this governed their material production, defined the appropriate modes of transportation, and made for forms of delivery that could be "much misliked" by all sides. This is not, however, to collapse the differences that these letters were intended to mediate. English royal letters and the *tarassul* of Islamic empires were formed within distinct social and political traditions of diplomacy and epistolography. They could be translated one to the other, but that translation was always an active and grounded process of interpretation. Translation involved not only the words the documents contained, but the ways in which

145. Cohn, *Colonialism and Its Forms of Knowledge*, p. 19.

they were carried and their mode of delivery. This process was shaped by exertions of power, from both sides, as well as feelings of suspicion and weakness. It was also animated by the desire for exchange. Significantly, these letters were always intended as instruments of exchange. They were there to be handed over, translated, and acted upon. They were there to make connections and open negotiations, however unequal, between different parts of the world. The process that they began, the networks that they helped to establish, were then consolidated and animated by other forms of correspondence. These are the subject of the next chapter.

CHAPTER THREE

Streynsham Master's Office: Accounting for Collectivity, Order, and Authority at Fort St. George

By the second half of the seventeenth century the circumstances of the English East India Company had changed. Driven out of the Indonesian archipelago and direct participation in the spice trade by the Dutch, the Company had begun to concentrate its efforts on trading cloth and other products from the Indian subcontinent. This produced a fourfold increase in the value of Asian imports to Britain between the 1660s and the 1680s.[1] The seats of this unprecedented process of capital accumulation were first the Coromandel Coast and, later, the Bay of Bengal. In order to tap into this trade the Company founded a series of factories, or trading stations, in India (Figure 4). Those on the west coast were controlled from Surat, and later from Bombay. On the east coast the key location was Fort St. George, the land for which had been given to the Company by the raja of Chandragiri in 1639. At various times those running this factory controlled not only its own trade, but also that of subordinate factories along the coast—notably at Masulipatnam and Madapollam (and, after 1682, at Vizagapatam, Cuddalore, Petapoli, and Conimere)—and in the Bay of Bengal and its hinterland, at Hugli, Balasore, Kassimbazar, Dacca, Maulda, and Patna. Establishing and operating these factories led to the elaboration and regulation of a range of procedures designed to enable the trade to happen and to manage the relationships between the Court of Directors in London and its employees in India. In 1675, Streynsham Master (Figure 5), whose opinions on the state of religion in India we have already encountered in Robert Boyle's

1. Kirti N. Chaudhuri (1978) *The Trading World of Asia and the English East India Company, 1660–1760* (Cambridge University Press, Cambridge).

Figure 4 English East India Company factories, 1670–90.

Figure 5 Sir Streynsham Master. (National Portrait Gallery, London.)

correspondence, was sent out from London to take up the position of agent at Fort St. George and, as he put it, to undertake "y^e Regulating and new Methodiseing their Factorys & Acco^{tts} upon ye Coast & Bay."[2] This process of reorganizing the ways in which the Company's servants made decisions and kept their consultation books, pursued trade and tallied their accounts, and explained their actions in the letters they wrote back to London is the subject of this chapter.

How then are these processes that connect power, knowledge, and exchange through the routine forms of writing undertaken in the Company's Indian factories to be understood as part of the making of global trade? In many accounts of what we might call early modern globalization, these processes are too small-scale to figure in the analysis. They are also judged to be incidental. Within world systems theory or the history of the unfolding of the capitalist mode of production, the development of the trading companies' operations are presented as the working-out of the profit-seeking "logic" of mercantile capital.[3] Products, markets, and exchange mechanisms linking different parts of the world—the core and the periphery—are part of the process of mercantile capital investment and return that operates in ways that are seemingly impersonal and inexorable, a driving force for action that affords little space for consideration of the accomplishment of the specific institutional forms and social relationships through which it worked.[4]

Those historians and economists who have studied the institutional organization of the trading companies have done so primarily in order to highlight the effectiveness, or otherwise, of their forms of operation for producing action at a distance. This has been undertaken in pursuit of arguments about the modernity of these organizations based on their profit-seeking activity and capacity, and their resolution (or not) of the "principal-agent problem": whether the directors in London could effectively control those

2. OIOC Eur Mss. E210 *Sir Streynsham Master* 1/10: "A Memorandum of ye good services done ye East India Company by Streynsham Master," f. 51^v.

3. Immanuel Wallerstein (1980) *The Modern World System II: Mercantilism and the Consolidation of the European World-Economy, 1600–1750* (Academic Press, London); Eric R. Wolf (1982) *Europe and the People Without History* (University of California Press, Berkeley); and Jan Nijman (1994) "The VOC and the expansion of the world-system 1602–1799," *Political Geography*, 13:3, pp. 211–27.

4. For a similar critique, see David Hancock (1995) *Citizens of the World: London Merchants and the Integration of the British Atlantic Community, 1735–1785* (Cambridge University Press, Cambridge) p. 15.

they had working for them overseas.[5] These debates over economic principles share some problems with K. N. Chaudhuri's influential modeling in
the 1970s of the English East India Company using the protocols of systems
theory to process vast amounts of quantitative archival data.[6] First, they
both have a set of prior assumptions of how the Company should have operated against which its actual workings are judged, rather than attempting
to understand the messy and contingent history of institutional change, and
the forms of power and knowledge that it brought together. For Chaudhuri,
this does mean acknowledging that a systems approach cannot adequately
deal with processes of institutional reorganization and particularly with
how the "system" was constructed in the first place.[7] Second, the ways in
which the Company worked, particularly the forms of writing and accounting that it used, are understood as "mechanisms" for achieving systemic
ends that are understood as separate from them: making a profit, achieving
compliance with orders, communicating decisions, and disciplining subordinates. In contrast, this chapter argues that a detailed reconstruction of
the process of institutional change can show how these forms of writing
and accounting, as forms of power and knowledge and as material practices,
played a vital part in the process of making what were taken to be decisions,
well-run factories, and appropriate orders.

   This is based upon an understanding of these forms of writing (indeed,
all forms of writing) and the ways in which they were made, deployed, and
read, as complex cultural artifacts, or "literary technologies," that helped
make the world they inhabited.[8] As Bruno Latour argues in his discussion
of the role of inscriptions, or "paper forms," in the making of scientific
knowledge, any attempt to provide a "hard" answer, one that involves a fac-
tual or accurate statement, will mean "bringing to the surface one of these

---

5. For example, Kirti N. Chaudhuri (1981) "The East India Company in the seventeenth
and eighteenth centuries: a pre-modern multinational organization," in L. Blussé and F. Gaastra
(eds.) *Companies and Trade: Essays on Overseas Trading Companies during the Ancien Régime*
(Leiden University Press, Leiden) pp. 29–46, and the debate in Ann M. Carlos (1994) "Bonding
and the agency problem: evidence from the Royal African Company, 1672–1691," *Explorations
in Economic History*, 31, pp. 313–35; S. R. H. Jones and Simon P. Ville (1996) "Efficient transactors or rent-seeking monopolists? the rationale for early chartered trading companies," *Journal
of Economic History*, 56:4, pp. 898–915; Ann M. Carlos and Stephen Nicholas (1996) "Theory
and history: seventeenth-century joint-stock chartered trading companies," *Journal of Economic
History*, 56:4, pp. 916–24; and S. R. H. Jones and Simon P. Ville (1996) "Theory and evidence:
understanding chartered trading companies," *Journal of Economic History*, 56:4, pp. 925–26.

   6. Chaudhuri, *Trading World*.

   7. Chaudhuri, *Trading World*, p. 41.

   8. The term is taken from Steven Shapin and Simon Schaffer (1985) *Leviathan and the Air-
Pump: Hobbes, Boyle, and the Experimental Life* (Princeton University Press, Princeton, NJ).

forms. Without them we simply *don't know.*" He then traces a geography of the spaces in which these inscriptions are produced, the networks through which they travel, and the centers of calculation (archives and collections) in which they are combined and superposed to provide answers. Significantly for the discussion here, Latour insists that all of this applies to economics (and politics) as much as to the natural sciences. Indeed, for him accounting and basic economics are "the most classical of calculations."[9] The calculative practices that make up economic knowledge are therefore put to work in a paper world, thereby constituting and transforming economic events and processes, and organizational forms and spaces.[10]

The modes of writing that concerned Streynsham Master are, therefore, understood as active in the construction of economic and political relationships—of collectivity, order, and authority—rather than simply being the more or less practical and mechanical, albeit logistically problematic, means of representing them to others.[11] This chapter argues that Master's "Regulating and new Methodiseing" sought to institute writing practices that, in their repeated performance and reinscription, were intended to constitute a distinction between the "public" world of the Company's business and the "private" actions of its servants. This was important because although East India Company employees like Streynsham Master were paid a salary for their work for the Company, they went east in order to make their fortunes through various forms of private trade. This trade was both within India and the Indian Ocean and back to Europe, and might involve buying and selling commodities—cloth, spices, and precious stones—for themselves or others, or providing insurance for the ships that carried these goods. Private trade bound these employees into trading networks that involved other Company servants, the servants of other European companies, Indian merchants, and mercantile interests in the City of London.[12] On

9. Bruno Latour (1987) *Science in Action: How to Follow Scientists and Engineers through Society* (Harvard University Press, Cambridge, Mass.) pp. 226, 239, 252; and Bruno Latour (1999) *Pandora's Hope: Essays on the Reality of Science Studies* (Harvard University Press, Cambridge, Mass.).

10. Peter Miller (2001) "Governing by numbers: why calculative practices matter," *Social Research,* 68, pp. 379–96.

11. For similar ideas, see Michel Callon (ed.) (1998) *The Laws of the Markets* (Blackwell, Oxford and the Sociological Review); and Timothy Mitchell (2002) *Rule of Experts: Egypt, Technopolitics, Modernity* (University of California Press, Berkeley). For a good historical example, see David Hancock (2000) "'A world of business to do': William Freeman and the foundations of England's commercial empire, 1645–1707," *William and Mary Quarterly,* 3rd ser., 57:1, pp. 3–34.

12. Elizabeth Lee Saxe (1979) *Fortune's Tangled Web: Trading Networks of English Entrepreneurs in Eastern India, 1657–1717,* unpublished Ph.D. thesis, Yale University; Ian Bruce Watson (1980) *Foundation for Empire: English Private Trade in India, 1659–1760* (Vikas, New

the one hand, this private trade has been understood as the main incentive that kept Company servants loyal. It is seen as the reward that came with Company employment, keeping the Company's orders filled and preventing worse abuses that might lead to dismissal.[13] On the other hand, the Company has been interpreted as surviving only because it provided an "umbrella" within which these dynamic private trading networks could operate. The Company is seen as a shell that existed to make possible the private profit seeking of the City of London's wealthy merchants.[14]

Streynsham Master was an architect of the Company's incentive structures, and was involved in extensive and profitable private trade (particularly in diamonds) with City of London merchants as well. Yet neither explanation of the "function" of private trade or of the East India Company captures his part in the business of institutional change. During the 1660s and 1670s the Company was still working out how its work was to be done, including what forms of private trade should be allowed to its servants and others. Master's "Regulating" of writing practices was part of that process of attempting to construct distinctions between what was permissible and what was not; what was "private" and what was not. Interpreting what he was doing means tracing out the relationships being made between inscriptions, organizational forms, and institutional spaces. The reorganization through documentation reshaped the relationships between people and took material shape in a transformation of the space of the factory to carve out and furnish writing offices. These were sites that grounded the processes of inscription and located their production of social collectivity, order, and authority. These inscriptions, organizational changes, and new spaces could not in themselves guarantee the Company's profitability. However, their new classification and parceling out of actions and intentions, as well as money and goods, made real a distinction between the Company's (public) interest and private interests that structured their operation and, in the end, caught Streynsham Master himself out. More broadly, this historical geography of inscription can suggest ways in which capital's "logic" might be reconceived in the working of global trade.

Delhi); and Søren Mentz (2005) *The English Gentleman Merchant at Work: Madras and the City of London, 1660–1740* (Museum Tusculanum Press, Copenhagen).

13. Santhi Hejeebu (2005) "Contract enforcement in the English East India Company," *Journal of Economic History*, 65:2, pp. 496–523; and Ann M. Carlos and Santhi Hejeebu (2006) "Specific information and the English chartered companies," paper presented at the 14th International Economic History Congress, Helsinki, Finland.

14. Mentz, *The English Gentleman Merchant at Work*, p. 262.

Inscribing Global Trade

How did the Company's trade with the east coast of India work, and what roles was writing expected to play within it? Every year, in about December or January, a fleet of ships would be loaded with cargo. This was predominantly American bullion gathered in European markets, along with some English broadcloth. The voyage around the Cape of Good Hope took about eight months. On arrival, the cargo would be distributed among the various factories, the ships moving up and down the coast and sometimes further afield to trade Indian goods in the Indonesian archipelago. The bullion was coined into local currencies and used to trade for cloth—cottons, silks, and mixtures of the two—with substantial Hindu and Muslim merchants from thriving commercial ports such as Masulipatnam who organized production in the weaving villages of the hinterland. These rich and well-connected merchant communities were part of the vibrant and long-standing trading world of the Indian Ocean that the Company was attempting to profit from. The most successful—men such as Kasi Viranna, Beri Timanna, and his younger brothers Pedda and Chinna Venkatadri—were "the Indian equivalent of the Medici family or Fuggers, and the Tripps."[15] Their extensive trading networks, often organized through family connections or those of caste and religion, already took shipments of the many different varieties of Coromandel cloth out to markets in the Burmese kingdoms of Arakan and Ava, where they were exchanged for rice and slaves; the Thai kingdoms, where they brought in elephants, Chinese goods, and Japanese copper; and in the sultanates of the Malay peninsula, Sumatra, Java, and the Moluccas, where they were traded for spices, pepper, opium, indigo, steel, tin, gold, and yet more elephants.[16] The Company also bought saltpeter, sugar, turmeric, and cinnamon from these merchants with which to lade their ships. In order to find good winds and a safe passage home, the Company's fleet had to be dispatched the following December or January. However, this gave only a short time in which to gather the return cargo, and any imminent deadline forced prices up. This made it necessary to use contracts with Indian traders and local systems of credit and finance to help gather cargoes prior

15. Chaudhuri, *Trading World*, p. 138.

16. Sinnappah Arasaratnam (1986) *Merchants, Companies and Commerce on the Coromandel Coast, 1650–1740* (Oxford University Press, Delhi); and Sinnappah Arasaratnam (1987) "India and the Indian Ocean in the seventeenth century," in Ashin Das Gupta and Michael N. Pearson (eds.) *India and the Indian Ocean, 1500–1800* (Oxford University Press, Calcutta) pp. 94–130.

to the ships' arrival. On their return, the brightly patterned cloth and strong smelling spices were sold at quarterly sales at East India House.[17]

In purely practical terms, the problems of coordination and organization were substantial ones. The Company sought to match, as profitably as possible, supply from India and demand in European markets and needed to orchestrate the movements of its ships and cargoes accordingly. As part of attempts to achieve this, and as chapter 2 already began to show, various forms of writing were required almost every step of the way. In the simplest cases, the Company required the labeling and listing of cargoes sent and their prices. Where this information—on tickets tucked into bales of cloth, marks on bags of spices, and musters detailing whole cargoes—was missing, profits were rendered insecure.[18] They were also threatened when it was overdone, with buyers wanting to open all the bales to see the differences in quality that the written words, numbers, or symbols suggested.[19] There were also lengthy letters sent from London to India that gave orders for goods to be supplied and instructions for the deployment of shipping. In return came not only vessels full of merchandise, but boxes of documents including replies to these letters, requested information, and copies of the accounts, letterbooks, and diaries that detailed the factories' operations. These often mundane forms of writing were a key technology in conducting long-distance trade because they provided a means to shape its nature and functioning.

Yet both this trade and the inscriptions that tracked and shaped it did not simply develop and operate smoothly, particularly during the massive expansion of the Company's business in the late seventeenth century. The Company's directors faced difficult questions over how they should organize their array of factories and often had little effective control over their operation. These difficulties were in part a matter of the sheer distance and length of time involved, and the problems of coordinating and controlling the work of their many trading posts. But they were also caused by the disputes among Company employees, and between the directors and their servants in India, resulting from private trade. Consequently, the Court of Directors tried to reorganize how the Company's business was run during this period. They made new rules on private trade, and effectively withdrew from trying to

17. Chaudhuri, *Trading World*.

18. IOR E/3/88 *Despatch Book: 5th July 1672 to 9th September 1678*, f. 118[v] (24 December 1675), and E/3/89 *Despatch Book: 6th November 1678 to 28th June 1682*, f. 43[r] (3 December 1679).

19. IOR E/3/86 *Despatch Book: 6th February 1661 to 12th January 1666*, f. 33[r] (10 November 1661).

keep trade within Asia (known as the "country trade") a Company monopoly. They issued "indulgences" in 1667, 1670, 1674, 1675, and 1679 that opened trade with almost every part of Asia in almost every commodity to private English ships.[20] After a failed experiment with an enforced monopoly in the early 1680s, they also opened up the lucrative but risky diamond trade between India and England to private enterprise, along with trades in other luxuries.[21] However, the counterpart to these opportunities was that the shipment of their prime commodities to Europe, particularly Indian cloth, was to remain closed to anyone but the Company itself. As some forms of private trade were declared legitimate, others remained, or became, illegitimate.

During the same period the Company also reorganized its own operation in India. It ordered the subordination of the factories in the Bay of Bengal, and therefore all factories on the east coast as well, to Fort St. George in 1661, and their separation out again (although only for three years) when William Hedges was sent out in 1681 to run the agency in the bay.[22] They also attempted to regulate how these factories worked. In 1674 William Puckle was sent to investigate the Company's affairs.[23] More significantly, Streynsham Master arrived in 1676 as the new agent at the fort (and with extensive private trading privileges granted by the Company), but was charged with traveling to all the subordinate factories to ensure their proper operation before assuming control from William Langhorn.[24] Central to Master's work was attention to how writing and accounting were to be done. His changes were ones that, his nineteenth-century biographer noted, "had a commanding influence on the public accounts of the English in India for a long time afterwards."[25]

This chapter considers three forms of inscription that Streynsham Master sought to regulate. These were the factories' consultation books (or diaries); the account books that each of the factories also kept; and the letters that passed between the Court of Directors and the Company's servants in India. In each case it is demonstrated that the forms of writing

---

20. Peter J. Marshall (1987) "Private British trade in the Indian Ocean before 1800," in Das Gupta and Pearson (eds.) India and the Indian Ocean, p. 281.

21. Mentz, The English Gentleman Merchant at Work.

22. IOR E/3/85 Despatch Book: 9th November 1657 to 28th January 1661, f. 183ʳ (28 January 1661); and Colonel Sir Henry Yule (ed.) (1887) The Diary of William Hedges, 3 vols. (Hakluyt Society, London).

23. IOR G/26/12 Diary of William Puckle, while at Masulipatam and Fort St. George, June 1675 to Jan 1676.

24. Saxe, Fortune's Tangled Web.

25. Sir Richard Carnac Temple (1911) The Diaries of Streynsham Master, 1675–1680, 2 vols. (Indian Record Series, John Murray, London) vol. 1 p. 2.

adopted were actively shaped by the social, cultural, and political relations that they were written and read within. In turn, it can also be shown that these modes of inscription, and the material geographies of their production, storage, and consumption, were an integral part of making those social relations, and with them the economic and political practices of trade. Thus, consultation books worked to constitute what counted as legitimate decisions; factory accounts and the management of bookkeeping were part of the social and moral order of the factory; and letter writing established relations of authority and subordination via the rhetorics of respect. Understanding what Streynsham Master was doing, and how he was doing it, involves considering these modes of writing as cultural and material practices that make up relationships that structure power and exchange, or what are considered to be the political and the economic. Inscriptions were constructed out of a cultural politics of collectivity, order, and authority, and, in turn, structured those relationships into the working of global trade. In doing so these ways of putting pen to paper both depended upon and shaped the organizational forms and spaces of the Company and the factory. In each case they tried to draw a boundary, or rather a series of boundaries, between the public world of the Company's trade and the private world of its servants' interests. This was a new conceptual geography that found material form in the factories' writing offices.

Consultation Books and Collective Decisions

The first part of the Company's business to be considered is how decisions were made in the factories, and the role that their consultation books were to play in that. Questions of access to these books—their presence within a restricted public sphere of collective decision making—were crucial to how effective factories were meant to function, and to what it meant to make a decision.

In what was effectively an extension of the shipboard practices of the early voyages (see chapter 2), decisions on all Company business were to be made by the head of the factory—the chief, agent, or governor—acting in council. In the 1660s the council at the fort might simply be selected by the agent "as in your Judgements are best accomplished with abillities to doe us service."[26] However, Streynsham Master brought changes that defined council membership on the basis of appointment to specific positions—as chief, bookkeeper, secretary, or warehousekeeper—and established rules of

---

26. IOR E/3/85 f. 183ᵛ (28 January 1661).

succession between the factories that bound their councils together. For example, on the death or removal of the second in council at Hugli, the most senior of the chiefs at Kassimbazar, Patna, or Dacca was to take his place, being replaced by the third in council at Hugli, the fourth moving up to third, and being himself replaced by the senior merchant at the fort.[27] While agents in council were now able to suspend members for "unfaithfulness or other great misdemeanour," all dismissals or appointments had to follow Master's rules and be ratified in London.[28] Councils were not, even temporarily, to include those not entitled to sit, although they might gauge the opinion of the wider body of factors and merchants in specific cases.[29]

Company policy was that councils were to operate through free and open debate. When Thomas Chamber took over as agent at Fort St. George on the death of William Greenhill, the Company advised him "that all matters bee debated and concluded of by Consulta[ti]on and not as formerly singly by our Agent and such as he hath deputed, wch wee utterly dislike and will by noe meanes allow of in the future."[30] Conclusions were to be reached by majority vote. Dissenters were not obliged to give their consent, although they were required to follow the majority decision, and they were encouraged to enter their reasons for disagreement.[31] In the event of a stalemate, the chief cast the deciding vote.[32] Agents felt, therefore, compelled to apologize and attempted to show general consent when external conditions such as war meant "it was impossible to reduce all into formall consultations."[33] Master sought to routinize this even further by stipulating that there should be regular meetings at every factory—on Mondays and Thursdays—with all members summoned at eight in the morning by the secretary, and absences noted.[34]

27. IOR E/3/88 f. 146ᵛ (16 December 1675).

28. IOR E/3/88 f. 113ᵛ (24 December 1675).

29. Records of Fort St. George (1911) *Diary and Consultation Book, 1678–1679* (Government Press, Madras) [hereafter RFSG, *Diary, 1678–1679*] p. 82 (1 July 1678); Records of Fort St. George (1914) *Despatches from England, 1680–1682* (Government Press, Madras) [hereafter RFSG, *Despatches, 1680–1682*] p. 22 (5 January 1681); and Records of Fort St. George (1913) *Diary and Consultation Book, 1681* (Government Press, Madras) [hereafter RFSG, *Diary, 1681*] p. 71 (1 December 1681).

30. IOR E/3/85 f. 183ᵛ (28 January 1661).

31. IOR E/3/88 f. 74ᵛ (23 December 1674).

32. IOR H/803 *Home Miscellaneous Series: Early Papers Relating to Bengal, 1670–1708*, p. 302 (29 June 1681).

33. IOR E/3/35 *Original Correspondence: 28th March 1674 to 20th March 1675*, f. 153ʳ (20 November 1674).

34. RFSG, *Diary, 1678–1679*, p. 8 (31 January 1678); Temple, *Diaries of Streynsham Master*, vol. 1 p. 290, vol. 2 p. 205; and IOR E/3/39 *Original Correspondence: 25th March 1678 to 24th March 1679*, f. 153ʳ (8 August 1678).

The Company's aim was simple. It sought to ensure by debate, in imitation of its own practice, that corporate interests were paramount. The directors saw the greatest threat to their profits to lie in the power of the agent or chief, and his own private interests. Indeed, it was the Company's powerful servants in India—factory chiefs such as William Jearsey and Richard Mohun at Masulipatnam, Matthias Vincent at Hugli, and William Langhorn, Streynsham Master, and Thomas Pitt at Fort St. George—who were at the heart of private trading networks. They were certainly able to use the power that their Company positions had given them to profit from private trade. This could involve keeping other, less powerful Company employees out of profitable markets by fair means or foul.[35] It could involve exploiting their status to gather lucrative work as commissioners for City of London merchants wanting to trade diamonds or ships' captains seeking to finance and insure private cargoes carried to eastern ports or back to Europe.[36] It could also mean using their influence over the deployment of Company resources to broker private deals with the Indian merchants with whom they co-owned merchant ships and cofinanced voyages as far afield as Malacca and Manila.[37] In each case, whether or not those private trading activities were ones permitted by the Company, they raised the question of whether the chief was putting himself or the Company first. Indeed, there was always the suspicion that private interests were paramount. As Streynsham Master put it, "our Company are very ungrateful and cruel to a man that has got an estate in their service, although he has done them a hundred times more service."[38] Therefore, it was not so much a question of whether the chief at any factory would come under suspicion for shaping the work of the factory to his own ends, but when that would happen. The councils' debates were the means to prevent those interests from becoming dominant, and to detect them when they were.

The language that was used to discuss consultations, both in England and in the Company's factories, was familiar from seventeenth-century political

35. Saxe, *Fortune's Tangled Web*.

36. Søren Mentz (1996) "English private trade on the Coromandel Coast, 1660–1690: diamonds and country trade," *Indian Economic and Social History Review*, 33:2, pp. 155–73; and Emily Erikson and Peter Bearman (2004) "Routes into networks: the structure of English trade in the East Indies, 1601–1833," *Institute for Social and Economic Research and Policy Working Paper 04–07*, Columbia University, New York.

37. Sinnappah Arasaratnam (1986) "Society, power, factionalism and corruption in early Madras, 1640–1746," *Indica* 23, pp. 113–34, which is reprinted in Sinnappah Arasaratnam (1995) *Maritime Trade, Society and European Influence in Southern Asia, 1600–1800* (Variorum, Aldershot).

38. Quoted in Watson, *Foundation for Empire*, p. 179.

controversies about the relationship between the English king and Parliament.[39] Thus, despite the advice given him on taking office, there were complaints from the fort that Chamber's rule meant "liveing under the Arbitrary governemente of One Man, which will not advise with his Councill in matters which concerne the gouvernmente of yo[r] honour's towne," and fears that the council might be overawed by fear or respect into relinquishing their "free vote."[40] In the same vein, Richard Mohun was presented by his accusers as a chief whose "imperious carriage soe overawes the Hono[ble] Comp[as] people there that they are ready to Signe any thing he shall tender to them," and one, it was suspected, that "has threatened to Ruin some of them: when displeased, thereby to make them Submitt to his arbitrary Power."[41]

Consultation was the defense against this abuse of authority; its recording in writing was therefore of great importance. In the mid 1670s the Company had Streynsham Master require of their factories that "All Transactions of buying and selling and all other Our Affaires are to be resolved and concluded in Councell, to w[ch] purpose you are to keepe dayly or frequent consultations and to take care that the Secretary do dayly and truely register all things in the Booke of Consultations."[42] Entries were to be made whether decisions had been arrived at or not, and the books were to be given marginal notes and an index for future reference. Moreover, the consultation entries, as well as any amendments and the copies of the books made for dispatch to London, had to be signed by all council members to show their consent.[43]

There was, therefore, a required format for these records. In 1679, "that none may be to Seeke how to perform what is now apoint'd," Streynsham Master sent the factory at Hugli "Extracts of the forme practised by our selves here in Entering these things in our Consultation booke ... w[ch] we recommend to your observation."[44] Later that year Hugli's council also received from London copies of the Madapollam consultation books, judged "agreeable to y[t] methode ... to w[ch] wee require your conformity."[45] In 1681,

---

39. David Wootton (ed.) (2003) *Divine Right and Democracy: An Anthology of Political Writing in Stuart England* (Hackett, Cambridge).

40. IOR E/3/26 *Original Correspondence: 26th March 1659 to 23rd March 1661*, ff. 168[r] (24 January 1660), 206[r] (no date).

41. Records of Fort St. George (1910) *Diary and Consultation Book, 1672–1678* (Government Press, Madras) [hereafter RFSG, *Diary, 1672–1678*] pp. 61 (7 May 1675), 148 (15 May 1676).

42. IOR E/3/88 f. 147[r] (16 December 1675).

43. RFSG, *Diary, 1672–1678*, p. 110 (16 February 1677); and RFSG, *Diary, 1678–1679*, p. 85 (4 July 1678).

44. IOR E/3/39 f. 154[v] (24 February 1679).

45. IOR E/3/89 f. 89[r] (3 December 1679).

they passed them on to all the subordinate factories for their "imitation."[46] Albeit slowly, uniformity was spread.

The consultation books also worked within a hierarchy of examination and accountability. In the late 1650s consultation books were sent annually from the subordinate factories to the fort, where they were examined and packed with the fort's own documents for shipment to London.[47] From 1676, on Master's orders, they were to be copied and sent monthly from the subordinate factories to Hugli or Masulipatnam, and twice annually, in May and November, from there to Fort St. George for annual dispatch to London.[48] This allowed the decisions made to be more closely scrutinized. Consultation books were used within the factories to inform future practice on the basis of past decisions. Entries from them might also be copied and sent as orders to subordinate factories.[49] They were certainly read carefully by Puckle and Master as part of their factory investigations and reorganizations.[50] Finally, the Company's requirement that they be used as the place for justifying decisions that contravened Company orders meant that those returned to London became the basis for the directors' demands for action over events occurring in India two or three years previously.[51] All these issues of keeping, signing, and sending consultation books in the proper way and at the proper time became points of contention between those who kept the books and those further up the hierarchy to whom they were sent for scrutiny.[52]

This was not simply a practical matter. It was an issue of Company politics. The consultation books did not just record decisions for consideration in another place or time, they also constituted certain otherwise indeterminate actions, events, forms of words, and modes of social practice as decisions, assigning as they did so responsibility and authority. Therefore, these books acted as the guarantee, within each factory and elsewhere, of the

46. IOR H/803 p. 287 (25 March 1681).

47. IOR E/3/85 f. 25^r (27 February 1658).

48. Temple, *Diaries of Streynsham Master*, vol. 2 pp. 78–79.

49. RFSG, *Diary, 1672–1678* p. 8 (27 June 1672); and Temple, *Diaries of Streynsham Master*, vol. 2 p. 198.

50. IOR G/26/12 p. 4 (1 and 2 July 1675); and Temple, *Diaries of Streynsham Master*, vol. 1 p. 245.

51. For factory justifications see, for example, Records of Fort St. George (1911) *Diary and Consultation Book, 1679–1680* (Government Press, Madras) [hereafter RFSG, *Diary, 1679–1680*] pp. 1–2 (3 January 1679); and Records of Fort St. George (1913) *Diary and Consultation Book, 1681* (Government Press, Madras) [hereafter RFSG, *Diary, 1681*] p. 2 (20 January 1681). On Company actions on the basis of the consultation books sent home, see IOR E/3/89 f. 100^v (7 and 9 January 1680).

52. For example, see IOR E/3/41 *Original Correspondence: 25th March 1681 to 20th March 1682*, ff. 82^r (17 September 1681), 220^r (20 February 1682).

decision-making process. They guaranteed decisions as emanating from the legitimate social practice of the chief's consultation in council. This, however, depended upon their publicity, since consultation books could be such a guarantee only if they and other writings were made available to the appropriate Company servants. These books had to belong to the factory and not be the property of any individual. They were, the Company stipulated through Streynsham Master, "not to be carried away upon the remove of the Cheife," but handed over to his successor.[53] Moreover, the Company's "Marchants, Factors, Writers & Apprentizes" were "not to be debarred or kept from ye sight of Our Bookes and affaires."[54] This served several purposes. First, the books were there "to trayne them up."[55] Second, it meant that other employees could more easily take over the running of the factory in case of sudden death. Finally, and most importantly, it was necessary for free debate and voting in council that all "writeings bee at all times Communicated, & left to ye publique view of ye Counsell."[56] Consequently, Thomas Chamber was chastised in 1662 for keeping such information from his council. As the Company noted, "to which practice wee can give no other construction then that it hath very much tended to our prejudice, for faire and honest dealings need not Shunn the light."[57] The books were to act as the guarantee and check against those chiefs who might act "arbitrarily and absolutely" against the Company's interests in order to feather their own nests.[58]

The consultation books, as a vital part of the process of making legitimate decisions and of guaranteeing the Company's interests, were to be kept public. This was, however, in what might be called a "restricted public space."[59] The intention was never that the books should simply be open to all. Having them accessible within the limits of the council—who both witnessed the books with their signatures, and had their authority and decision-making capacity established by them as a record—was a necessary insurance against the private interests of each one of them, particularly the chief who held a power that might be misused. The consultation books,

53. Temple, *Diaries of Streynsham Master*, vol. 2 p. 6. John Brewer (1989) *The Sinews of Power: War, Money and the English State, 1688–1783* (Unwin Hyman, London) argues that the fiscal-military state was instituting similar rules at the same time.

54. IOR E/3/88 f. 147^r (16 December 1675).

55. IOR E/3/88 f. 147^r (16 December 1675).

56. IOR E/3/85 f. 183^v (28 January 1661).

57. IOR E/3/86 f. 50 (20 February 1662).

58. IOR H/803 p. 354 (1 December 1680). This concerned the case of Matthias Vincent, whose books showed him to be "doing all things by himself and according to his own will without consultacon."

59. Shapin and Schaffer, *Leviathan and the Air-Pump*, p. 39.

therefore, sketched a line between the "public" and the "private" in Company decision making.

There were, however, differences over how consultation should be done. These differences raised important political questions about the authority accorded to those who ran the factories and the discretion they had to act as they saw fit. William Puckle, Streynsham Master, and those who scrutinized the books in London called into question the way William Langhorn ran his consultations at Fort St. George. There were complaints that there were no "frequent Stated times," and they did not "then sitt & orderly & fairely debate."[60] Master was impatient to replace Langhorn as agent and keen to suggest that he often acted without consultation. He reported to London that "when yᵉ Councill did meet there was nothing done regularly, but one walked one way & another walked another way in yᵉ Roome, when yᵉ busyness was moved, wᶜʰ was spoken to but indifferently, afterwards [Langhorn] himselfe drawes up in writeing what he thinkes fit, & all the Councell signe it."[61] However, Langhorn and his council defended themselves by setting out a model of consultation that offered the positive version of Master and Puckle's accounts of disorder and arbitrariness. Here Langhorn's practice was presented as "rather encouraging and entreating us of Councell clearely to deliver our minds . . . then any wayes over awing of the rest." It was a collective rather than authoritarian epistemology that involved a socialized hermeneutics, a division of intellectual and administrative labor, and a sting in the tail for those who prioritized Company hierarchy over effective factories:

> [M]atter of moment and difficulty requiring Consultation, appearing to him as some solid substance beyond the circuit or penetration of the Eye at once, and wᶜʰ a single man attempting though by degrees shall loose as fast as he changes aspects; But wᶜʰ by multitudes of Councelloʳˢ taken in parts; Collation and discourse shall search & sift out every scruple. It has ever been commendable to use our own judgemᵗˢ yet best when examined and tried (according to the Owners discretion) by that of others. But to subject every triviall matter to these formalitys, were to turne either business or Servants into ridicule.[62]

60. Records of Fort St. George (1911) *Despatches from England, 1670–1677* (Government Press, Madras) [hereafter RFSG, *Despatches, 1670–1677*] p. 129 (12 December 1677).

61. OIOC Eur Mss. E210/1/10: "The Character of Government at Fort St. George from 1672 to 1677," f. 29ʳ.

62. IOR E/3/37 *Original Correspondence: 5th April 1676 to March 1677*, f. 56ᵛ (15 January 1677).

As Langhorn concluded, "Neither is all business of a like nature. Playn cases requiring little more than their hearing, matters of difficulty their debate, and publick transactions their formallitys...At his [Puckle's] trifling Rate the Hono^ble Comp^a had need have two Agents and Councells, the one to consult the other to act."[63]

For Master, Puckle, Langhorn, and the Company, the shared acceptance of the principle of consultation in council, and its necessary recording in the consultation books, still raised questions of the proper authority of the agent or chief within the hierarchy of the factory—what was his relation to his subordinates to be?—and of the discretion allowable to chief and council in relation to the Company's orders. These two issues also shaped and were shaped by the inscriptions through which orderly factories and orders for factors were made and guaranteed: accounting and letters.

Accounts and the Social Order of the Factory

Streynsham Master went to settle the factories of the Coromandel coast and the Bay of Bengal with two sets of printed orders that were made in 1667 and sent to India, but which seemed to have fallen into disuse. These covered, first, "ye Manadgement of Our Affaires and keeping Our Bookes" and, second, "Certain orders and rules for promoting of sobriety and piety."[64] Through them the Company sought to intervene in how the business of each factory was conducted and in the personal lives of its servants. Considering how accounting was done can demonstrate that the practice of bookkeeping and the social and moral order of the factory were connected through ideas of hierarchy and the role of the chief or agent.

Master's prime role, and his first enthusiasm, was putting the accounts in order. His two journeys around the factories involved extensive practical work balancing books that were in a confused and disorderly state and setting out ways of keeping them in the future. This reform of accounting procedures was based on the "plaine & cleare Method" he had devised at Surat in the 1660s, and which he had already used in 1668 to reorganize the factories at Karwar and Calicut.[65] Extending those methods to the chaotic

63. IOR E/3/37 f. 66^r (15 January 1677). He also commented on Puckle's own practice: "his irregular whisperings in time of Consultations both here and at Metchlep^m: and pulling those of Councell aside, and into other roomes, never practised that we know of by any but himself" (f. 66^r).

64. IOR E/3/88 f. 144^r (16 December 1675); and Ethel Bruce Sainsbury (ed.) (1925) *A Calendar of the Court Minutes of the East India Company, 1664–1667* (Clarendon Press, Oxford) p. 418 (18 December 1667).

65. IOR E/3/88 f. 145^r (16 December 1675); and *Oxford Dictionary of National Biography*.

accounts of the coast and bay was based on a few simple principles.[66] The first was uniformity across the different factories. As those in charge at Fort St. George put it to the Hugli factory some years later, the Surat method was "the forme wch the Hono^{ble} Company have aproved & ordered to be observed . . . and not any one to follow his own fancy for should that be allowed of there would be every year as many different wayes of keeping acco^{ts} as there are Men and books, wch ought not to be in the Hono^{ble} Companys business and therefore they have prescribed one set forme."[67] The second principle was what might be called legibility. This was not simply a matter of whether the books could be read, but of shaping bookkeeping practices so that the accounts made visible what the Company wanted to be able to see.[68] This was particularly important in London, where the accounts were, with all their faults, the most complete and reliable picture of Company trade in India. As recent histories of accounting have argued, accountancy does not simply mirror preexisting economic practices, representing them more or less well. Accounts should be understood as constituting economic practice through modes of inscription and analysis, enabling new forms of intervention that would not otherwise be possible.[69]

Streynsham Master's reforms of accounting aimed to change how Company trade could be understood by introducing new ways in which places and commodities, and their relative profitability, could be made visible and legible, and therefore the basis for action. Each factory was to have its own accounts.[70] For each of them Master's bookkeeping practice required the itemization of all trading and packing costs and their proportional allocation to each commodity. All factory charges were to be detailed and presented in annual tables to "be more readily compared wth ye factory bookes," and, when balanced, the overall amounts on the debit and credit sides were to be subdivided to show what was fixed capital, what lay as bad debts, goods as yet unsold, or money advanced or borrowed at interest by the factors— so that "business [could be] directed and Governed accordingly."[71] More

66. It was also based on leaving copies of "the Surratt Bookes letter M" at the factories; see Temple, *Diaries of Streynsham Master*, vol. 2 pp. 1–2 (quote on p. 2), 102–3; and RFSG, *Diary, 1678–1679*, p. 6 (21 January 1678).

67. IOR E/3/39 f. 155^{r-v} (24 February 1679).

68. Although for a bookkeeper whose figure 8 was "hardly to be known," see RFSG, *Despatches, 1680–1682*, p. 11 (4 January 1681).

69. Anthony G. Hopwood and Peter Miller (eds.) (1994) *Accounting as Social and Institutional Practice* (Cambridge University Press, Cambridge). This perspective also forms the basis of the journal *Accounting, Organizations and Society*.

70. Temple, *Diaries of Streynsham Master*, vol. 1 pp. 498–99, vol. 2 pp. 165–66

71. IOR E/3/40 *Original Correspondence: 28th March 1679 to 5th April 1681*, ff. 192^{v}–193^{r} (12 December 1679); and Temple, *Diaries of Streynsham Master*, vol. 1 p. 282. The Company

significantly, Master proposed changing the date on which all the accounts were to be balanced from the end of November to the end of April: "For [30 November]...is almost in the midst of the season for Europe shiping, and so divides one yeares business into two paire of bookes [each pair a Journal and Ledger]:...And soe, againe, part of two yeares business comes in one paire of bookes...Thus confusedly comes one yeares returnes to be opposed to the following yeares receipts." As he put it, the new system "will certainly comprehend one shippings business." It would also reveal the relationship between charges and investments and allow the books to be balanced at a quieter time for the bookkeepers.[72]

The final principle was that the accounts should be able to be combined within each factory and across the hierarchical system of factories. This was aided by the changes already discussed, such as the uniform method and the single date for balancing the books. In addition, the Company's rating of the pagoda at 9s and the rupee at 2s 3d for the purposes of bookkeeping allowed accounts to be combined and compared even in the face of regionally and temporally differentiated exchange rates.[73] Master's rules also specified how books should be continued from one year to the next, and how, in each factory, the accounts of the warehousekeeper, the steward, and the purser should be combined into the general books.[74] Most importantly, he aimed to establish a nested hierarchy of accounts that would include all the general accounts of all the factories of the coast and bay into one system of accounts whose balances and differentiations would reveal both the geography of Company trade and account for all investments and returns within a single bookkeeping system. So, just as they sent them the goods and received from them the bullion, the accounts of all the subordinate factories in the bay would be cleared into—and therefore "inferior" to—the accounts at Hugli, which was to occupy the same position in relation to Fort St. George's accounts. In theory, any parcel of goods or money could be traced through the books as it moved through the factory system, changing hands and changing form as it did so. This arrangement meant that adjustments to the inferior factories' accounts were done on the basis of those of the superior, but not vice versa. However, it also meant that "superior" factories had to wait for the accounts of "inferior" ones before their books could be closed. The aim

---

even pursued legal action in Chancery against a former chief in Bengal, William Blake, to clarify the principle that factory charges should be determined by the Company, clearly stated "& not be cast in or concealed under ye prizes of goods"; IOR E/3/88 f. 82^r (8 January 1675).

72. Temple, *Diaries of Streynsham Master*, vol. 1 pp. 280–81.

73. Temple, *Diaries of Streynsham Master*, vol. 1 p. 277.

74. RFSG, *Diary, 1678–1679*, pp. 6–8 (31 January 1678).

and the problem were succinctly stated in a letter from the fort to the bay in 1681:

> Your Bookes are so long coming to us y^t wee are forced to shutt up ours wthout them for it is very inconvenient to keep them open till ye ships come when wee are so full of other businesse and wee must tell you it is a great neglect in ye Inland factoryes not to send you theirs for y^r Bookes coming wthout theirs included will signify no more than ye Acc^t Currant now Rec[eive]d from you for as your Bookes are to Include all ye Subordinate factoryes acc^{ts} so ours must take in y^{rs} and theirs y^t so ye Comp^a may see their whole charges and Proffitt both upon ye Coast and Bay and therefore a defect herein tends to nothing but confusion.[75]

The whole system depended, of course, upon the books being kept properly at each factory. It was intended that any defects would be sorted out by making bookkeepers responsible for their books, along with the chiefs who were named in their titles and also had to sign them. Penalties were instituted, to be enforced by the agent at the fort, for failing to balance and send the accounts promptly.[76] Overall, the aim was to ensure that London annually received a uniform, legible, and combined set of accounts revealing the nature of the Indian trade.[77]

In London, the accounts were part of decision making over orders and investments. They were also scrutinized by the auditor and Company accountant for unwarranted charges, mistakes, frauds, and shoddy bookkeeping. These problems then had to be rectified in India, often several years after they had first been committed to paper.[78] Keeping the books was inseparable from doing business in India, and reading the books in London was an important part of adjusting the business that was done, even if the rhetoric of control was often more significant than the reality.[79] As a result, "Counting-house worke" judged "confus'd irregular and disorderly" was generally attributed to badly run factories, "to loose and ill methods

---

75. IOR E/3/41 f. 107^r (17 November 1681).

76. Temple, *Diaries of Streynsham Master*, vol. 2 pp. 2, 331; and IOR E/3/89 ff. 33^v–34^r (3 January 1679).

77. This was also pursued by sending "able" and "honest" accountants to India: IOR E/3/89 f. 85^r (3 December 1679); and RFSG, *Despatches, 1680–1682*, p. 21 (5 January 1681).

78. RFSG, *Despatches, 1670–1677*, pp. 3–17 (29 November 1670), 55 (15 December 1676), 144–49 (12 December 1677); and IOR E/3/89 f. 43^r (3 January 1679).

79. This is argued for the later colonial period in Arjun Appadurai (1996) "Number in the colonial imagination," in *Modernity at Large: Cultural Dimensions of Globalization* (University of Minnesota Press, Minneapolis) pp. 114–35.

and habits wch o^r Servants there had contracted by long Sloth, carelessness and neglect."[80] This drew on a hard-won set of metaphorical and material connections between balanced accounts, merchants' fidelity, and social order.[81] It meant that William Puckle and Streynsham Master had to concern themselves both with how the books were kept and with problems of moral and social order in the Company's factories.

Puckle's instructions sent him to examine whether the Company's orders on religious duties were being followed in Masulipatnam and to discover the "refractory," "idle, and debauched," as well as to regulate the business there.[82] One of Master's tasks was to ensure that the so-called "Company Commandements," there being ten of them, for "Christian and sober Comportment" were displayed publicly and followed in each factory.[83] As a result of his investigations Puckle suggested that the young men at Masulipatnam, particularly the writers who spent all day copying the books, might be put under a new regime that would discipline them for absences from prayers, work, the factory, or their chambers after ten in the evening, or for frequenting punch houses and for beating servants. Master, following Langhorn's rules for the fort, sought to regulate the keeping and use of punch and arrack houses, and to enforce fines for profanity, blasphemy, lying, gambling, drunkenness, absence from prayers, adultery, fornication, and uncleanness in all factories. He also established the first Anglican church in India.[84]

This idealized Christian moral order was combined with a strongly hierarchical sense of social order. At Masulipatnam, Puckle found that the Company's replacement of Richard Mohun by four equal commissioners was causing more problems than it was solving. There were "great contests" among the commissioners and, as a result, the young writers were drinking in their rooms, blaspheming, and "declaring themselves for parties, & y^t they are not obliged to obey ye order of any one Com^r when another Com^r differs from him."[85] His remedy was to reestablish the hierarchy of chief and

---

80. IOR E/3/89 f. 88^v (3 December 1679) condemning all the factories in the bay except Kassimbazar.

81. Mary Poovey (1998) *A History of the Modern Fact: Problems of Knowledge in the Sciences of Wealth and Society* (University of Chicago Press, Chicago); and Margaret R. Hunt (1996) *The Middling Sort: Commerce, Gender, and the Family in England, 1680–1780* (University of California Press, Berkeley).

82. IOR E/3/88 f. 80^r (23 December 1674).

83. Yule, *Diary of William Hedges*, vol. 2 p. cccvi; and Temple, *Diaries of Streynsham Master*, vol. 1 p. 274.

84. RFSG, *Diary, 1672–1678*, pp. 82–83 (16 February 1676), 131–33 (21 February 1672); RFSG, *Diary, 1678–1679*, pp. 108–10 (19 August 1678), 126–27 (24 October 1678).

85. IOR G/26/12 pp. 26, 44.

council, the placing of each writer under a member of the council, and the removal of status symbols—native footmen and umbrella carriers—that the lesser employees had assumed in contravention of the practice at the fort.[86] Yet this hierarchy was to be one that also encouraged progression. Puckle recommended that "the order for succession be stricktly observed, chief-ship being the hopes, aymes, & encouragem^t of all subordinate."[87] Master was later sent out with the new rules for succession, which would move able employees around the factory councils and up the hierarchy, and with a strict grading of positions—apprentice, writer, factor, merchant, senior merchant—via which a Company servant would progress, learning the job from the bottom up, from copying the books to running a factory.[88] Those already in India were to be reassured that they would be rewarded on merit and that none would be sent out from England and put above them.[89]

Of course things were never so harmonious. The vicissitudes of rise and fall in the hierarchy, and the machinations generated by personal rivalries and the competition for private trade, caused Langhorn to comment on the "em-ulation envy & backbiting so much in use . . . that we tax poore Machiavelly."[90] Yet it was this idea of moral order, hierarchy, progression, and tutelage that was meant to structure the world of the factory and its business. These small groups of Europeans—Fort St. George, the largest, counting only about twenty-five Company servants from writers upwards—were, ideally, to live together in the factory. This was, as contemporary representations show, understood as an enclosed world, even if it was, in reality, a highly permeable one (Figure 6). Most significantly, all employees were meant to eat together at a "publicque table" that was strictly organized by position in the factory.[91] At the head was the agent, or chief, who was expected to be given due respect and to act in a way that required it.[92] His position was crucial. As William Langhorn put it, "when the chief in place and power, who is at the head &

86. RFSG, *Diary, 1672–1678*, p. 82 (16 February 1676); and RFSG, *Diary, 1678–1679*, pp. 8–9 (31 January 1678).

87. RFSG, *Diary, 1672–1678*, p. 82 (16 February 1676).

88. IOR E/3/88 ff. 145^v–146^r (16 December 1675); and RFSG, *Diary, 1678–1679*, p. 55 (5 March 1678).

89. RFSG, *Despatches, 1670–1677*, p. 53 (15 December 1676); and IOR E/3/88 f. 145^v (16 December 1675).

90. IOR E/3/37 f. 58^v (15 January 1677).

91. IOR E/3/88 ff. 35^v (29 September 1673), 113^v–114^r (24 December 1674). For disputes over position, see the chaplain at Masulipatnam's complaint to William Puckle "that he hath not that respect & place of preference at Table & elsewhere that is due unto him He being a Minister & M^r of Art of the University of Oxford" (IOR G/26/12 p. 14).

92. The most damning accusations related, it seemed, to sexual impropriety, Catholicism, and defrauding the Company. See, for example, the accusations against Matthew Mainwaring (RFSG,

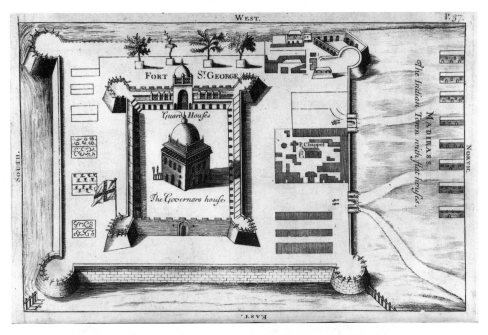

Figure 6 Fort St. George in the late seventeenth century. From J. Fryer (1698) *A New Account of East-India and Persia* (London). (By permission of the British Library, 567.i.20.)

heart of all, is as he ought to be, the whole body can hardly be out of tune, and so the contrary."[93] At the other end, the writers were to be sat at the same table for the "more orderly Government of the Youthes."[94] Doing so would give them "So good an example and preventing them from keeping ill Company," since they might otherwise "stragle into Punch howses & other inconvenient places."[95] There were, however, tensions here between the claims of an ordered hierarchy and the maintenance of social position and authority. Despite what his views on being chief may have suggested, Langhorn was also concerned that his status might be damaged by being "tyed to sitt like a pedant amongst his boyes."[96] In censuring him for this, Master was clear that his rival's exclusion of the writers from the factory's

Despatches, 1670–1677, pp. 62–67) and Master's accounting of Langhorn's Catholic sympathies (OIOC Eur Mss. E210/1/10 f. 29ᵛ).

93. IOR E/3/37 f. 58ᵛ (15 January 1677).

94. Temple, *Diaries of Streynsham Master*, vol. 2 p. 8.

95. Temple, *Diaries of Streynsham Master*, vol. 1 p. 247; and IOR E/3/88 f. 35ᵛ (29 September 1673).

96. IOR E/3/35 f. 63ʳ (20 August 1674).

public table "was ye Direct way to ruine all ye young men & consequently the Companys busynes, by keeping their Factors in Ignorance."[97]

These connections between the Company's business and its servants' morals turned upon this vision of the factory as a hierarchical and artificially enclosed social and moral order. This made the personal lives of the Company's employees an important part of the working of a set of "public" arrangements in the factory, built upon notions of hierarchy, authority, and respect. As with the consultation books, the threat to the orderly and profitable conduct of the Company's business, both disclosed by the accounts and made possible through them, came from illegitimate private action. The integrity of the accounts was threatened by Company servants' private trade, and by many forms of inexperience, carelessness, and trickery.[98] Because of this the directors in London felt threatened by the Indian merchants (known as "banians") upon whom the fortunes of the English depended. These were the men that the Company relied on to supply goods to ship to Europe. Successive agents at the fort designated Beri Timanna, Kasi Viranna, and the Venkatadri brothers as "chief merchant" and allowed them to monopolize the Company's business. They were powerful figures with fingers in many pies, including significant political influence within the Golconda kingdom. They were attracted by the profits that might be made by dealing with the English but, certainly before the end of the seventeenth century, were not dependent upon them.[99] As one Company servant said of Kasi Viranna, emphasizing his independent power in Madras, "Sir William [Langhorn] governs within the Fort and Verrona without."[100] They were also joined to Englishmen in India through private trading ventures in "hybrid Indo-European" ships with both Asian and European financing, cargoes, and crews.[101] For example, Streynsham Master and Kasi Viranna were both investors, along with groups of Company factors from Masulipatnam and Fort St. George and Pedda Venkatadri, in the *Trivitore*, which sailed under Viranna's name to Manila in 1679. Indeed, this private joint-stock venture seems to have continued into the 1680s as a group of "English

97. OIOC Eur Mss. E210/1/10 f. 29^r.

98. For private trade and its disruption of accounting at Masulipatnam, see RFSG, *Despatches, 1670–1677*, p. 63 (20 December 1674); RFSG, *Diary, 1672–1678*, p. 56 (7 May 1675); RFSG, *Diary, 1678–1679*, p. 19 (19 February 1678); for negligence and trickery, see RFSG, *Despatches, 1680–1682*, p. 11 (4 January 1681); and RFSG, *Diary, 1672–1678*, p. 85 (16 February 1676).

99. Saxe, *Fortune's Tangled Web.*

100. Quoted in Charles G. H. Fawcett (1952) *The English Factories in India*, vol. 2 (n.s.): *The Eastern Coast and Bengal, 1670–1677* (Clarendon Press, Oxford) p. 124.

101. Marshall, "Private British trade," p. 280.

and Coromandel merchants [who] traded together to China, Achin, Aracan, Manila, Bengal, and Sumatra."[102]

These Indian merchants were well versed in the arts of economic inscription. Viranna was the banker for the 1679 *Trivatore* voyage, collecting the shares and paying for the outfitting costs. Such men were used to forms of double-entry bookkeeping, contracts, and credit notes that were familiar to their European counterparts, although they understood their trading practices within a political and religious context that was not shared by the Company.[103] Where they could do so, the Company's servants, including Streynsham Master, sought to tie these merchants into joint-stock agreements with written contracts in order to limit their autonomy and bring down the prices of the goods they supplied to the Company. Significantly, this remained impossible in Bengal throughout the seventeenth century, and was achieved at Fort St. George only after Viranna's death in 1680.[104] The power of these merchants and their entanglement with Englishmen's interests were seen as a threat in London in several ways. Dependence upon them as chief merchants was seen to push up the prices for the goods the Company bought. More perniciously, the close personal ties between the Indian merchants and the Company's servants was thought to lead to corruption, as the higher prices the Company paid could be used as a sweetener for offering better terms in private deals. As the directors complained of Timanna and Viranna, "We know they gain vast estates by us and live accordingly and might Live Better, if they had not many hard Bargains forced upon them, and on others accounts, not ours."[105] Whatever was going on beneath the surface, conflict between these merchants' own interests and

102. Saxe, *Fortune's Tangled Web*, p. 105. Mentz, *The English Gentleman Merchant at Work*, doubts the importance of Indian finance to such voyages, but since the Manila trade was closed to the English but not to Asians, they were bringing more than just capital to the venture.

103. Sanjay Subrahmanyam and C. A. Bayly (1988) "Portfolio capitalists and the political economy of early modern India," *Indian Economic and Social History Review* 25:4, pp. 401–24; Sudipta Sen (1998) *Empire of Free Trade: The East India Company and the Making of the Colonial Marketplace* (University of Pennsylvania Press, Philadelphia); and Ashin Das Gupta (1998) "Trade and politics in eighteenth-century India," in Muzaffar Alam and Sanjay Subrahmanyam (eds.) *The Mughal State, 1526–1750* (Oxford University Press, New Delhi) pp. 361–97. For a specific example of accounting practice in Asian trade, see Levon Khatchikian (1996) "The ledger of the merchant Hovhannes Joughayetsi," in Sanjay Subrahmanyam (ed.) *Merchant Networks in an Early Modern World* (Variorum, Aldershot) pp. 125–58.

104. Temple, *Diaries of Streynsham Master*, vol. 2 pp. 146–52, 374–78; and Susil Chaudhuri (1975) *Trade and Commercial Organization in Bengal, 1650–1720: With Special Reference to the English East India Company* (Firma K. L. Mukhopadhyay, Calcutta).

105. Quoted in Saxe, *Fortune's Tangled Web*, p. 89.

those of the Company meant that their influence over the accounting process, and the conduct of trade, was considered a danger.[106] As the merchants in Hugli wrote to their subordinate factory in Balasore in 1682, complaining about the errors in their accounts with Indian merchants, "'tis not at all reasonable, (what ever opinions some may have) that a Banians mere affirmation should controul our books of Acc^{ts} & it may bee a caution hence forward for any not to rest too strongly on Country Servants Acc^{ts}."[107]

These threats meant that the accounts were read for signs of these disturbances of factory practice in ways that made auditing the books and a moral accountancy of employees' lives and interests inseparable. The construction of a uniform, legible, and interlocking system of accounting both served the ends of business and better revealed certain differences and discrepancies that then had to be accounted for in economic, political, or moral terms.[108] Notably, Streynsham Master also went to India to deliver new and, as the Company saw it, more indulgent rules on private trade. What came with these allowances for private profit making was a desire for a clearer line between Company business and the private dealings that often involved the same merchandise and the same Indian merchants (and, the directors suspected, the same capital). Consequently, the Company also demanded the "punctuall observance" of its oft-neglected order to keep a full register of private trade.[109] Ideally, when properly kept by appropriately trained and disciplined Company servants, the accounts would be a product of the public social order of the factory and bear no trace of illegitimate private concerns. These notions also shaped the ways in which the Company in London and its servants in India wrote to each other and read the letters they received.

Letters, Orders, and the Rhetoric of Respect

Communication between London and Fort St. George was based on an annual letter that went out with the fleet and a reply that came back with the cargo. However, these general letters were supplemented by others, on

106. Ian Bruce Watson (1987) "Indian merchants and English private interests: 1659–1760," in Das Gupta and Pearson (eds.) *India and the Indian Ocean*, pp. 301–16.

107. IOR E/3/42 *Original Correspondence: 25th March 1682 to 24th March 1683*, f. 31^r (27 May 1682).

108. It is, of course, the case that a more closely defined system of regulation and accountability also made it clearer what modes of evasion and deception were necessary for Company servants to hide what they did not want others to see. I owe this point to Santhi Hejeebu.

109. IOR E/3/89 ff. 38^r, 82^v (3 December 1679); and IOR E/3/88 f. 148^v (16 December 1675). Unsurprisingly, the reporting of private trade was always partial at best.

average about five a year, that were sent on ships to other Indian ports or overland.[110] This was less secure, but served as some insurance if the fleet was delayed or the letter did not get through. Once on the coast, under Master's instructions, a "good Correspondence" between the factories was to be ensured by having those at the fort peruse the Company's letters to the bay before sending them on with any supplementary advice and orders.[111] Masulipatnam simply received instructions from the fort. Replies from both places to the Company came back to the fort, for comments to be added, prior to sending them back to London. They would arrive in the capital sometime after the next letter had been sent to begin the process again.

The letters from London contained a range of material. They detailed, or amended, the orders for cloth and other goods, stating the types, colors, quality, and lengths required, often with patterns, and the prices that should be paid based on information from sales. They were full of information about the state of the market.[112] This meant detailed commentary on the previous year's cargo—what was too short, too rough, or the wrong color—angrily pointing out what had been delivered that was not what had been ordered. They identified what was damaged en route and, along with information about European wars, their orders for how business should be conducted, allocating or reallocating positions, ruling on suspensions and appointments, making rules for the paperwork, and passing comment on the performance of their employees. The responses from India sought to answer their queries, question or confirm instructions and orders, justify controversial decisions, and pose their own questions about the business. They were also required to provide information on supply, detailing new goods that the Company might find a market for; on demand, particularly for the woollen cloth the Company was compelled to export; and on the machinations of local rulers and European competitors.

In both cases the letters were written in a particular way. They were composites put together by a range of people. Those from London were compiled by the Committee of Writing Letters from materials provided by other committees—on shipping, treasury, or the coast and bay.[113] At Fort St. George, they were read and debated section by section by the agent and council over a matter of months, and replies were similarly composed, read,

110. IOR E/3/89 f. 88ᵛ (3 December 1679).

111. IOR E/3/88 f. 114ʳ (24 December 1674).

112. John Styles (2000) "Product innovation in early modern London," *Past and Present*, 168, pp. 124–69.

113. See, for example, Ethel Bruce Sainsbury (ed.) (1935) *A Calendar of the Court Minutes of the East India Company, 1674–1676* (Clarendon Press, Oxford).

amended, and signed collectively. When these long replies reached London, they were taken apart and the elements allocated to the appropriate committees for consideration. This structure of communication produced a particular format. For clarity, the Company insisted that the letters state the date and location they had been sent from, when letters were last sent (often enclosing copies), which ships had recently arrived, and which letters were the ones being answered. More specifically, as with the consultation books, they required that the letters be signed by all the members of the council, and by no others.[114] In the 1670s they began to insist that to "facilitate your Correspondence" the fort and bay must "answer ye Severall Paragraphs as they lye in Order," numbering them for ease of reference.[115] This revealed some of the tensions built into the relations of authority that the letters made possible across the oceanic space between London and Fort St. George. William Langhorn's response to a system that was clearly for the convenience of the committees in London rather than the councils in India was heavy with sarcasm: "We have confined ourselves to your own rules of answering the paragraphs in your letters in order as they ly except N° 102; our practice formerly being to extract and reduce all business of a like nature into method, for the more compendious way of replying, exacter coherence, and curranter revisall; but we shall allwaies be glad to learne."[116]

Tellingly, this innovation was enthusiastically adopted by Streynsham Master as part of his welding of the factories into a single system through their paperwork.[117] This was also pursued through the archiving of the letters in registers. Of course, mercantile practice had meant that these registers had always been kept in the factories. However, under Master, these letterbooks were given titles, marginal notes, and indexes, and were to "be all bound up handsomely, and covered with leather...."[118] They were to be closed at the same time as the accounts to be sent home on the fleet, as well as archived at each level, by having subordinate factories send copies of their letters to each other to Hugli, to have Hugli send two copies of all volumes of letters sent and received to the fort, and the fort to send one of these and copies of all its letters to London.[119]

114. IOR E/3/88 f. 49^v (13 March 1674); and Temple, *Diaries of Streynsham Master*, vol. 1 p. 251.

115. IOR E/3/88 f. 116^v (24 December 1674).

116. IOR E/3/37 f. 45^v (6 August 1676).

117. IOR E/3/39 ff. 395^v–396^r (24 February 1679).

118. Temple, *Diaries of Streynsham Master*, vol. 2 p. 7.

119. Temple, *Diaries of Streynsham Master*, vol. 2 pp. 5–6; and RFSG, *Diary, 1678–1679*, p. 8 (31 January 1678).

There were, of course, substantial problems with this mode of communication. Those in Hugli, Fort St. George, and London complained of letters delayed, mislaid, or not sent, and the gap between delivery and receipt could make orders irrelevant or troublesome.[120] The official correspondence between the directors and the factories also had as its constant shadow an extensive network of private communication, both within India and between India and London. This often used the same modes of conveyance, but sidestepped and disrupted the official channels to organize private trade and patronage systems, and was used to spread malicious gossip and to denounce other Company servants.[121] Finally, particularly for those in London, there never seemed to be enough information, or on the right subjects. In part, the Company's directors saw this as a problem of how those in India read. In 1672 they commented on both the councils of the fort and the bay, that "neither they nor you ... seeme to lay our Letters before you, when you goe to give answer unto them, wch wee require in the future to be amended."[122] Yet it was also a problem with how they wrote. As it was stated in 1661, "Our Letters from ye Coast & Bay ... are full of obscure passages ... [T]hey tell us all is subject to ye greate Kinge, & all at peace in ye Bay, but say not whoe ye greate King is, nor on what termes ye peace is concluded, as if because they know those things, it necessarily followes y^t wee must alsoe, though at soe greate a distance. Wee would have you advise them, & to take notice of it your selves, y^t wee shall expect punctuall & full advices of all passages, & in such language as may be under stood."[123]

These matters of reading and writing were a particular problem when it came to giving orders. In theory it was simple. Orders made in London had to be followed at the fort and down the hierarchy to the subordinate factories. Each level should expect compliance from those below on the basis of the orders from London.[124] By the mid 1670s, however, this hierarchy had become dislocated. William Langhorn noted that "the now contemned authority of the Agency ... is little better than laughed at, and despised" both at Masulipatnam and Hugli, "thereby unravelling the whole Oeconomy of the

120. RFSG, *Despatches, 1670–1677*, pp. 45–54 (15 December 1676); IOR E/3/86 ff. 77^r (11 July 1662), 87^r (27 October 1662); and E/3/39 f. 217^v (14 October 1678).

121. Mentz, *The English Gentleman Merchant at Work*; Yule, *Diary of William Hedges*, vol. 2 p. xii; IOR E/3/88 f. 74^r (23 December 1674); and E/3/37 ff. 53^v–58^r (15 January 1677).

122. IOR E/3/88 f. 5^v (9 August 1672).

123. IOR E/3/86 f. 25^r (31 August 1661).

124. Records of Fort St. George (1912) *Diary and Consultation Book, 1680–1681*, p. 60 (9 August 1680); and RFSG, *Despatches, 1680–1682*, pp. 32–33 (5 January 1681). In IOR E/3/88 f. 197^r (18 December 1676) the directors warned that "they that slight them that Act by our authority, doe the same to us."

Hono^ble Comp^as affaires in India."[125] Moreover, the distances involved, and the necessity of putting orders into practice in uncertain and changing conditions, introduced problems of interpretation. Orders could not, somehow, be separated from the way that they were written and from the changing contexts in which those words were read. It was clear, for example, that there were different readings of the same situation in Langhorn's complaint that the Company was always "taking all things in the worst sense."[126] This also meant that matters could be made better or worse by the way orders were composed. Master argued that, under Langhorn, "There hath noe regard been to keep them [the subordinate factories] in any order or quiett, . . . soe that all his time they have been in uncessant broyles and undetermined differences by many very obscure orders and directions."[127] Puckle, who had been sent to determine if orders were being followed at Masulipatnam, recommended "That all advices and orders . . . be in full and plaine termes, not dubious and uncertaine to be understood, wch hath sometime occasioned divisions amongest the Com^rs as to their opinions."[128] Plainness was, indeed, the order of the day. The Company's insistence on "positive orders in plain words" accorded with John Hill's instruction in his *Young Secretary's Guide* (1689) that letters to factors "ought to be no more than the plain sense of the Fact."[129] Yet, along with continuing differences over the interpretation of the facts, the Company also had to deal with the question of discretion.

The trade with India could not operate effectively only on the basis of orders from London. It was, for example, London's claim of absolute authority that caused problems in India when supposedly subordinate factories took it at its word and refused to take orders from the fort, wanting direct and unmediated communication with the Court of Directors.[130] More routinely, however, Company servants in all the factories had to be empowered to act in response to specific local economic, political, and environmental conditions where Company orders were nonexistent, unclear, or erroneous.[131] The directors needed the factories to assert this discretion where following orders would be detrimental to their interests, but they could not just let the

125. RFSG, *Diary, 1672–1678*, p. 48 (16 June 1675).

126. RFSG, *Diary, 1672–1678*, p. 99 (14 September 1676).

127. OIOC Eur Mss. E210/1/10 f. 29^r.

128. RFSG, *Diary, 1672–1678*, p. 82 (16 February 1676).

129. IOR E/3/89 f. 229^r (18 November 1681); RFSG, *Despatches, 1670–1677*, p. 54 (15 December 1676); and John Hill (1689) *The Young Secretary's Guide* (H. Rhodes, London) p. 5.

130. RFSG, *Diary, 1672–1678*, p. 48 (16 June 1675); and IOR H/803 p. 364 (17 April 1682).

131. IOR E/3/26 f. 54^r (11 15th June 1659), E/3/89 f. 34^r (3 January 1679), E/3/36 *Original Correspondence: 25th March 1675 to March 1676*, f. 188^{r-v} (19 June 1677), H/803 pp. 363 (17 April 1682), 399; and RFSG, *Diary, 1678–1679*, p. 87 (15 July 1678).

factories act as they wanted.[132] This was a recipe for conflict. As Langhorn put it, "In one breath you bid us use our judgem^{ts}, & disapprove the use."[133]

These conflicts were played out in the Company's letters as contests over and in the language of respect. Langhorn diagnosed the problem in 1676: "The shortnes of your orders, in the most important points of your business," and the lack of discretion and trust given to servants, "makes all sides wary" and "where the Treack is not thouroughly beaten, do either keepe to orders or tarry for them."[134] Indeed, as Master presented his enemy, he was notorious for not taking any initiative and simply waiting to be told what to do.[135] Langhorn, for his part, portrayed the agent's position as all responsibility with no reward or respect from on high, and argued with the directors that "one would think it should no less concern you that they be well satisfied of your good meaning than you of theirs."[136]

The Company was certainly keen on the language of respect. It was to run, however, in only one direction: towards London. Master, on his travels around the subordinate factories, instructed them "That the Letters and advices to the Honourable Company may be adressed with a become-ing respect, as becomes servants to their Masters, They are always to be wrote in a Submissive stile and Directed, 'To the Honourable Governour and Company of Merchants of London tradeing to the East Indies.'"[137] The Company's definition of misdemeanor included not only refusing to obey orders but also "bad language" to superiors.[138] As a result, the Company's responses to Langhorn and, later, to Master's letters were positively vitri-olic. It is clear that it was as much bad language as bad actions that were at stake. Langhorn was censured for his "higher Stile," "unhandsome & dis-respectful passages," "indecent language," and "haughty vaine unmannerly expressions," and was told to "manage your pen with more respect."[139] Mas-ter's sin was pride, wallowing in the "vaine ostentatious pomp of India" and

132. See, for example, RFSG, *Despatches, 1670–1677*, p. 137 (12 December 1677), whose para-graph 105 sanctioning discretion was later used in the consultation books where such deviations had to be justified (see n. 51 to this chapter).

133. IOR E/3/37 f. 41^r (23 July 1676).

134. IOR E/3/37 f. 47^v (23 July 1676).

135. Temple, *Diaries of Streynsham Master*, vol. 1 p. 494.

136. IOR E/3/37 f. 47 (23 July 1676).

137. Temple, *Diaries of Streynsham Master*, vol. 2 p. 7. Letters with "many expressions of disrespect" were judged "inconsistent wth our subordinate relation." RFSG, *Diary, 1672–1678*, p. 67 (30 January 1679).

138. IOR E/3/88 f. 251^r (12 December 1677).

139. IOR E/3/88 ff. 181^r (15 December 1676), 249^v (12 December 1677); and E/3/89 f. 31^v (3 January 1679).

thinking himself "too good or too bigg."[140] He was reminded that his "stile" of writing the relationship between London and the fort was in error, "wee having power without your leaving it to us." And if the Company punned against him on his dismissal, arguing that "we have at least the same power in our own affairs as every Master hath in his own family," it was because he had been guilty of misusing one of their own terms: "you say you crave our pardon for yo[r] plainnesse . . . [But] you betray your own weaknesse as well as your pride . . . in supposeing that wee cannot judge betweene plaineness and Insolence."[141]

The Company in London had, therefore, a way of reading that combined its concerns with abuses of office (illegitimate private trade, corruption, or excessive factory costs), following orders, and giving respect. Serious misdemeanors might be detected from a close scrutiny of the language of a letter. As they put it in 1681, "it is our Constant Maxim grounded upon a Long experience, that a wastefull haughty or prodigall person, can never be a good Servant for us, Let him be otherwise never so Crafty plausible or methodicall."[142] They concluded that "[w]e have seldome observed such peremptorines in Servants, but at length we find it accompanied with infidelity."[143] Yet, as with all such readings, these conclusions were produced by the interpretative community of readers, rather than simply being inherent in the texts themselves. The directors' interests cohered around the profitability of their collective concerns in India. Yet they diverged on how or by whom that would be best achieved, and on their own private interests, including the networks of private trade and patronage that also tied them to those in India.[144] This shaped how those in London read the letters they received.

While it has been noted that any agent's eventual fall from grace was perhaps inevitable, when and how that came about was a matter of managing the politics of patronage in the Court of Directors and other places, and of shaping the process of reading the letters that came from India. In Streynsham Master's case this was a family affair. His mother, born Anne Oxenden, came from a Kent family with strong connections within the East India Company. Indeed, Master had first gone to India in the 1650s under

140. IOR E/3/89 f. 148[r] (5 January 1681).

141. IOR E/3/89 ff. 82[r–v], 85[r] (3 December 1679), 148[r] (5 January 1681).

142. RFSG, *Despatches, 1680–1682*, p. 41 (15 March 1681).

143. IOR E/3/89 f. 225[v] (18 November 1681).

144. Mentz, *The English Gentleman Merchant at Work*.

the protection of an uncle, one of his mother's brothers. Sir George Oxenden subsequently became the president of the factory at Surat and the first governor of Bombay. He is buried in an elaborate mausoleum in Surat cemetery with his brother Christopher, who was also a Company servant. In London in the 1670s it was a cousin on his mother's side, Sir James Oxenden, who kept Master informed of the factions among the Company's directors, and of "the constant applications wee made to maintaine you in your designed and appointed Station." This involved not only attendance and argument every time Master's case had to be put, and Langhorn's denied, but ensuring that the opposition's attempts to enlist King Charles II on their side were countered. Oxenden had to have eyes everywhere. As he told Master, "It was my fortune to be att Court when our Gouvenor and Deputy were sent for by the King, and seeing Cooke in the long Gallery I suspected some Contrivance against you." Writing and reading were crucial here. Master's long memorial on the new factory regulations was "accounted very laborious and judicious." It helped Oxenden "to infuse an Esteem into ye Dissenters," while representing written charges against Master made by Joseph Arnold, one of the commissioners at Masulipatnam, as "Impertinent nonsensical and malicious," convincing others that they were "very idle and the effect of an overheated braine" caused by Arnold's "drinking and debauchery." Other writings were dealt with more privately. Oxenden was invited to dinner by "mr Brittaine and Natt Scotton," who he felt had offered Master an affront over Arnold's accusations. The purpose was "to reason the Case & show mee y^r Letters." Some missives, however, were for sympathetic eyes only. Oxenden informed Master that he had shown the postscript of Master's letter, relating a dispute with Langhorn, "to some of our Friends . . . wch they liked very well and advised me as you did to communicate it to few Persons." In all of this there was a salutary lesson from the decline of Gerald Aungier, former governor at Bombay. His fall from favor was hastened by writing to at least four men that he wished them alone to be his successor. As Oxenden put it, "If these persons to whom he had wrott had lived in Several Nations his Artifice might probably have been concealed, but they have all mett together and show their Letters to one another and some of the Company and despise him as the greatest Dissembler in nature."[145]

It was clear, therefore, that reading and writing needed careful management. Letters sent between London and Fort St. George contained much

145. OIOC Eur Mss. E210/1/7: a letter from Sir James Oxenden of Deane to Mr Streynsham Master at Madras, 12 January 1678, ff. 18^r–19^v.

more than orders for cloth and details of new products and markets. Their language, and the deployment of the rhetorics of respect, was also the making and management of relations of authority, submission, and discretion. Orders were not simple matters of fact, but questions of interpretation. It mattered who put pen to paper and how. Once again, the ways these letters were written, and particularly how they were read by the Company's directors, inscribed a line, albeit a flexible one, between appropriate Company service—a public realm of duty, deference, and fidelity—and a private sphere that the directors suspected was shaped by illegitimate private interests, corruption, dissembling politicking, and the sins of sloth, pride, and vanity. They scrutinized the letters they received for signs that the line had been crossed, and they acted on their interpretations.

Conclusion: Streynsham Master's Office

This chapter argues that understanding writing and accounting as crucial technologies for making global trade entails seeing them as the practical means for running factories, communicating at a distance, integrating supply and demand, and controlling subordinates. But, like all technologies, they are more than just mechanisms that get things done. This is because these inscriptions are made from an array of elements that bring with them particular relationships between power and knowledge, and among their writers and readers. Consultation books carried with them notions of collective decision making within a restricted public sphere; accounting was built within ideas of moral and social order; and letters to and from India deployed the languages of respect and disrespect. From these "cultural" elements, political and economic relationships were made. Indeed, that is the second reason why they are more than mere mechanisms. They did not just represent or record the decisions, flows of goods and money, and orders that made up this branch of global trade; they actively made them. Consultation books were required to establish what could be taken to be decisions; accounting practices made the recognition of profitable exchange a possibility; and the Company's letter books constituted what forms of authority were permissible. This made writing important. Company officials at all levels, particularly Streynsham Master, attempted to control the responsibility for documents, the construction and use of archives, the formats of writing, the ways in which written material was produced, how documents should be read and by whom; their content, but also their style. Doing so meant that relationships constructed around and through these inscriptions and actions shaped the world of the factory in India. The chief and council's negotiations of collectivity and authority were performed in and through the consultation books. The knowledge order of accountancy included the moral and social order of the factory. And questions of power and discretion were battled out in the numbered paragraphs of the Company's letters. Collective decisions, mercantile civility, and legitimate authority did not precede writing and get reported or represented by it. Instead, they were made through the practices of writing.

Each of these forms of writing and accounting was important in that it shaped the Company's organizational form. New ways of writing minutes, accounts, and letters endeavored to draw a line between the public world of the Company's business and the private interests and failings of its employees.[146] Consultation books were to be public to guarantee that decision making be in the interests of the Company. The accounts were, like the "publique table," to be part of an ordered and idealized moral and social world, rather than shaped by what the directors saw as their employees' laziness, incompetence, or venal entanglements with Indian merchants. The Company's letters were also intended to ensure that its subordinates in India did not simply step out of their public role and pursue their private agendas to the detriment of the Company. Each traced out a line between the "public" and the "private." One key means of trying to stabilize and to guarantee all this, although it could never be finally guaranteed, was the construction and control of a particular space within the factories: the writing offices.

These offices were to be the restricted public space that could secure the integrity of the written documents through a combination of openness and closure. It was the "laying up" of the books in an office for authorized viewing that was the guarantee of their public nature, and the absence of such a space was a sure sign of mismanagement.[147] It also worked to connect the social order of the factory with orderly bookkeeping. For Puckle, one problem at Masulipatnam was that the "writers & Accomptants have always used to mannage their Acc^{ts} & do their writing worke in their respective Chambers, by reason whereof some have been negligent in their Duty, & many books & Papers of ye Hono^{ble} Comp^{as} concerns have been lost." His solution was that "[a] Roome in ye Factory house is prepared & ... all Bookes, Letters, Papers &c. are orderly disposed therein ... And ye Accomptants and writers

146. This was quite unlike the productive blending of public and private concerns evident in the letter books of the Atlantic trades; see David Hancock (ed.) (2002) *The Letters of William Freeman, London Merchant, 1678–1685*, London Record Society, vol. 36.

147. IOR E/3/86 f. 91 (31 December 1662); OIOC Eur Mss. E210/1/10 f. 29^r; and RFSG, *Diary, 1672–1678*, p. 160 (9 May 1675).

required to attend dayly & dispatch their worke therein." Each would have a desk, and "the Chiefe will dayly give directions & inspect y^r doings having a seat prepared for him."[148] Indeed, despite the cost, the Court of Directors supported this, ordering that each factory should have a writing office where "all our bookes and papers may be orderly kept & that none be carried out to any private house or lodging, but that all busines may be dispatched there."[149] This supported Master's previous attempt to establish offices at the fort and Hugli, and his tour of the subordinate factories included the repetition in each place of the same order:

> In every one of the Subordinate Factoryes there shall be a handsome con-
> venient roome, large, light and well scituated, near the Chief and Seconds
> lodgings, which shall be sett apart for the office, and never diverted from
> that use, In which roome shall be placed desks or tables to write upon,
> and presses with locks and keys, wherein the Registers of the letters, the
> Accompts, and all other writeings of the Factory shall be locked up and
> kept, which, upon the Remove of the Chief, are to be delivered over by a
> Roll or List to the succeeding Chiefs, that none may be imbezled.[150]

The aim was to construct a controlled space for writing and calculation that would seek to ensure the accessibility of the books, the orderly conduct of accountancy, the absence of the selfish interests of factory chiefs, and all that depended upon it. Understanding this specific and small-scale geography of writing and writing practices as an ordering of the relationships between power and knowledge in the making of global trade means recognizing the social and cultural relationships that lie right at the heart of the economic arrangements of mercantile capital. It is also the case that if the "logic" of capital was felt by those engaged in these forms of exchange as a "logic"—as an impersonal, inexorable, and determining force—then that was exactly the effect achieved by the separations, hierarchies, and controls instituted in the factories' writing offices as the sites of local practices of abstraction and standardization performed upon chains and compilations of inscriptions and reinscriptions.[151] There is therefore no history of capitalism that is only abstract and universal, standing apart from the institutional, social, and cultural forms, the different histories and geographies of capital, through

148. IOR G/26/12 p. 14.

149. IOR E/3/88 f. 250^r (23 June 1678).

150. Temple, *Diaries of Streynsham Master*, vol. 2 p. 333.

151. Theodore M. Porter (1995) *Trust in Numbers: The Pursuit of Objectivity in Science and Public Life* (Princeton University Press, Princeton, NJ).

which modes of production, consumption, and exchange are organized.[152] It was within these restricted public spaces, and only within them, that the English East India Company could turn their concerns into an objective and controlling profit-seeking force external to their servants' private interests, into the "logic" of capital.

Streynsham Master helped to make this world, but it was also part of his downfall. As well as reorganizing the factories, Master had indeed made a substantial fortune in private trade, which raised suspicions against him. When he finally lost the support of the Court of Directors and was removed as agent at Fort St. George, many claims were made that he had taken illegitimate advantage of his powerful position to extort funds for his own benefit. One claim was that he had sold four elephants belonging to the rich Indian merchant Pedda Venkatadri and pocketed the proceeds. Questioned on it by the council he claimed that his memory was so bad that he could not remember if there had been any transaction or not. When the final showdown came "his answers were all evasive & nothing but shifts, til a paper was shown to him under his own handwriting found in a desk in the Consultation room (wch he forgot to take away with him upon his removeall) mentioning the receipt of p^{as} 4930: for 4 Elephants he sould, wch soe abashed him he continued silent for a good while."[153] All that Master had done in using writing and accounting to construct a public sphere of Company business against which private interests would readily stand out had finally caught up with him.

---

152. Dipesh Chakrabarty (2000) *Provincializing Europe: Postcolonial Thought and Historical Difference* (Princeton University Press, Princeton, NJ).

153. RFSG, *Diary, 1681*, p. 60 (17 October 1681).

CHAPTER FOUR

The Discourse of Trade: Print, Politics, and the Company in England

Besides depending on the careful management and deployment of manuscript letters and account books, the English East India Company also had to involve itself in the world of print. Indeed, this chapter, which attends directly to the Company as it operated from London, argues that its fortunes, and at times its very existence, were a matter of the management of the geographies of the printed word. However, simple contrasts between the nature of script and print need to be treated with caution, as do arguments over the implications of print technology (see chapter 1). While it is possible to argue that in seventeenth-century England the printing press did make texts of certain sorts more widely available, and that the intention of going into print was to speak to a broader "public," any easy identifications between print and the public sphere, primarily associated with the work of Jürgen Habermas,[1] have been questioned by recent histories of print culture. These histories have queried the ease with which printed material became public, stressing the commercial and regulatory mechanisms through which it did so. They have also detailed the limited, changing, and contested nature of what that public might be, and how it was understood by writers, publishers, and readers.[2] This has differentiated and variegated both "print"

---

1. Jürgen Habermas (1989) *The Structural Transformation of the Public Sphere: An Inquiry into a Category of Bourgeois Society* (Polity Press, Cambridge).

2. Alexandra Halasz (1997) *The Marketplace of Print: Pamphlets and the Public Sphere in Early Modern England* (Cambridge University Press, Cambridge); Adrian Johns (1998) *The Nature of the Book: Print and Knowledge in the Making* (University of Chicago Press, Chicago); Michael Warner (1990) *The Letters of the Republic: Publication and the Public Sphere in Eighteenth-Century America* (Harvard University Press, Cambridge, Mass.); and David Zaret (1992) "Religion, science, and printing in the public spheres of seventeenth-century England," in Craig Calhoun (ed.) *Habermas and the Public Sphere* (MIT Press, Cambridge, Mass.) pp. 221–34.

and "public," and what follows argues that these are both issues of the geographies of print which require the careful reconstruction of the spaces of the printed word and its forms of collective address as they changed over time.

It is surely right to assert that "[t]he invention of printing effectively created a new social group with an interest in making knowledge public."[3] However, it is also the case that its arrival was met with a whole series of institutions and regulations that might serve the interests of those who wanted to see to it that certain forms of knowledge, particular ideas, or specific texts did not come out in print. States and religious authorities are the most obvious examples here. However, the mechanisms of restrictive laws in place in England before 1695 and their enforcement by the Stationers' Company and church officials, as well as appeals to the king and the Privy Council, could also be used by institutions such as the East India Company. Indeed, they were able to defend and promote their interests both by printing themselves and by trying to restrict the printing of others. They even found it possible to both distribute and act against publications in other countries. As this chapter will show, this combination of publication and restriction created a geography of print, not only in the sense of where particular books or pamphlets were available and where they were not, but in terms of the wide range of spaces—the king's levee, the Privy Council chamber, the assembly of the States General of the United Provinces, the printshops and booksellers of London, and the houses of Parliament—where publication in print might be halted, restricted, or promoted.

This, of course, has a history. This chapter shows that the early seventeenth-century English East India Company was often a reluctant entrant to the world of print. The printing press was one political resource among many, and a potentially risky one at that. Its use required careful political judgment and an understanding of the cultural meanings that print carried with it. The complex economy of influence that print was part of in the early seventeenth century, which also involved the artful deployment of speech and manuscripts, is explored here through detailed reconstructions of the Company's response to an attack printed in 1615, and the pamphlets that were part of conflicts with the Dutch in the 1620s. However, the explosion of printed material in the mid-seventeenth century changed matters. The flowering of the relationship between the press and the market in England, which powered the production of a huge diversity of printed material,

3. Peter Burke (2000) *A Social History of Knowledge: From Gutenberg to Diderot* (Polity Press, Cambridge) p. 176.

was conjoined with the establishment of a new relationship between print and politics.[4] By the 1670s it was impossible for the East India Company to seek to restrict the production of the printed material that challenged it. As a result, the later seventeenth century was marked by the increased use of print against print, and focused, in particular, on the specific practices of reading, writing, and speaking used in and around Parliament, where the Company's future was debated and decided upon.

As these material geographies of print shifted across the century, so did what might be called the imagined geographies of print. Just as it is important to specify the particular spaces that approximated the "public sphere," so is it necessary to identify the ways in which texts concerning the English East India Company constructed collective interests that they could address in their support for or attacks against such things as the exportation of bullion, the power of the Company directors, or the importation of Indian cloth. In different texts, and over time, the address shifted between notions of the "Common-wealth," paternalist and organic appeals to the monarch and his kingdom, and senses of the nation and the national interest. Through a discussion of the early mercantilist writings of the 1620s, it will be argued that the deployment of print, in dedications and in the layout of the printed page, was an important part of the construction of these collective entities, and of presenting their fortunes in terms of the balance of trade. Yet there was always the interpretative problem of whom to believe on these crucial questions of trade and politics. As in other spheres of knowledge, the printing press put rival assertions into much wider circulation than ever before.[5] And, perhaps more than elsewhere, there was always the sense that arguments professing to speak the truth, or at least to speak for the public, were actually guided by private interests. Once again the proliferation of print after 1640 was significant, and is addressed in the final section of the chapter. If there was a "crisis of knowledge" in the late seventeenth century, it was not only among philosophers, but also among those who sought to work out the relationships between private and public interests by critically reading the avalanche of texts on matters such as the East India trade.[6] We start, however, with one small book.

---

4. David Zaret (2000) *Origins of Democratic Culture: Printing, Petitions, and the Public Sphere in Early-Modern England* (Princeton University Press, Princeton, NJ); and Joad Raymond (2003) *Pamphlets and Pamphleteering in Early Modern Britain* (Cambridge University Press, Cambridge).

5. Elizabeth L. Eisenstein (1979) *The Printing Press as an Agent of Change: Communications and Cultural Transformations in Early-Modern Europe* (Cambridge University Press, Cambridge).

6. Burke, *A Social History of Knowledge*, p. 203.

The Trades Increase

On 12 February 1615, towards the end of the period of dealt with in chapter 2, the bookseller Walter Burre and the printer Nicholas Okes appeared at Stationers Hall in the City of London to register their rights to the copy of "a boke called *the trades Increase.*"[7] Licensed, according to regulations made by the Star Chamber in 1586, by Dr. Nidd (or Neede), one of the archbishop of Canterbury's chaplains, the work was printed on Okes's single press near Holborn bridge. It was quickly available for purchase at Burre's shop at the heart of London's print trades in St. Paul's Churchyard, alongside works of natural philosophy, theology, and travel writing published by the two stationers.[8] The bookseller, who had been in the business at least eighteen years, was at that time in trouble with the archbishop who oversaw the licensing process. Burre's printing of Sir Walter Raleigh's *History of the World* (1614), which its author had "hoped would please the king," had in fact provoked the opposite response. Judged dangerous for its "too free censuring of Princes," the archbishop, George Abbot, had ordered that Raleigh's *History* "should be suppressed and not suffered hereafter to be sould." All copies were to be removed from the printers' and booksellers' premises and taken into custody.[9] *The Trades Increase* would also cause a stir, coming to the notice of the king's Privy Council and requiring the attention of Archbishop Abbot. Yet in this case it was not royal power that was threatened, but the reputation of England's East India merchants.

Many of those who read *The Trades Increase* would not have known who had written it. Behind the initials "J.R." was Robert Kayll (or Keale), who, apart from his connection with this pamphlet, seems to have succeeded in remaining anonymous. But if the messenger was obscure, the message was clear. In fifty-six quarto pages Kayll set out a critique of the damage that the East India trade was doing to England, and the need to turn attention to reclaiming the herring fishery from the Dutch. He cited as his inspiration a treatise by Tobias Gentleman titled *England's Way to Win Wealth,* which had been published the year before, exhorting the king and

7. J.R. [Robert Kayll] (1615) *The Trades Increase* (London, printed by Nicholas Okes, and sold by Walter Burre); and Edward Arber (ed.) (1876) *A Transcript of the Registers of the Company of Stationers of London,* 5 vols. (London, privately printed) vol. 3 p. 563.

8. R. B. McKerrow (ed.) (1910) *A Dictionary of Printers and Booksellers in England, Scotland and Ireland, and of Foreign Printers of English Books, 1557–1640* (Bibliographical Society, London); and Johns, *The Nature of the Book.*

9. William A. Jackson (ed.) (1957) *Records of the Court of the Stationers' Company, 1602–1640* (Bibliographical Society, London) pp. 355–56.

Privy Council to support the herring fishery, and an older work by Robert
Hitchcock, which had promised solutions for all England's economic ills
through building four hundred fishing vessels manned by former beggars
and vagabonds.[10] Like them, Kayll emphasized the benefits that the Dutch
were gaining from the herring fishery and stressed that these could accrue to
England herself. The fishery would not only provide employment for fisher-
men (and be a suitable school in which landsmen could find their sealegs),
but would also provide work for a range of other trades.[11] Instead of paying
the Dutch for salted herring, England could capture the export markets from
them. Revenue, ships, and mariners would strengthen "this great *Machina*,
this goodly Engine of our Sea-state," and offer a stronger challenge to Dutch
mercantile and marine dominance. As he put it, "there are Herrings enough
to make us all rich."[12] Unlike Hitchcock and Gentleman, however, Kayll
did not restrict himself to fish. He sought to delineate the nature of England's trade with the rest of the world. This meant challenging the East India
Company.

Kayll mapped out England's trades and provided an account of their
health. The Dutch had forced England's merchants out of the Muscovy
trades, even while still trading English tin, lead, and coarse woollens. The
Mediterranean trades were "the worthiest in former remembrance, the
worst in present reputation." The Newfoundland trade was endangered
"by heathen and savage, as also by Pirates."[13] Trades with Spain, France,
Hamburg, and the Baltic were either not growing fast enough or were now
dominated by the Dutch. Even the Newcastle trade in coal was rife with
foreign shipping. Yet it was the East India trade that most threatened England's economic well-being. While it appeared to have increased trade, and
therefore shipping and the employment of mariners, things were not as they
seemed. Kayll sought to show that the East India Company was a drain on
the nation's resources.

First of all, shipping. The Company had bought ships, removing them
from other trades, but it had also built, he reported, more ships and bigger

---

10. Tobias Gentleman (1614) *Englands Way to Win Wealth, and to Employ Ships and Mariners*
(London, printed for Nathaniel Butter); and Robert Hitchcock (1580) *A Politique Platt for the
honour of the Prince, the greate profite of the publique state, relief of the poore, preservation of
the riche, reformation of Rogues and Idle persones, and the wealthe of thousands that knowes
not how to live* (London, printed by John Kyngston).

11. [Kayll], *Trades Increase*, pp. 50–51.

12. [Kayll], *Trades Increase*, pp. 1, 33.

13. [Kayll], *Trades Increase*, pp. 5, 13.

ships than ever before. Kayll counted twenty-one vessels from eighty tons
to eleven hundred tons, eight of which had been newly built and the rest
substantially rebuilt. Instead of adding "both strength and glory to the Kingdome by this your accession to the Navy," these ships stood as a loss. Four
had been cast away (including the *Ascension*, on which Alexander Sharpeigh
had come to grief), the others either remained to trade in the east or returned
so broken down as to be unserviceable for military use, and their timbers
unfit for building or repairing other ships. This trade had "devoure[d] our
timber" with no return. The trade had a similarly devastating effect on the
supply of mariners. Unlike the small vessels and short voyages of the Newcastle trade, the owners and masters of ships in the long-distance trades
were unwilling to take anyone but those who "can take their turne at the
helme, toppe and yard." The East India trade prevented landsmen from being turned into seamen. Moreover, it destroyed those sailors it did employ.
Kayll estimated that of the three thousand men that the Company had sent
out, over two thousand had been lost to "the dogged Starre of those Cly-
mates, [and] the stench of those Countries."[14] While Kayll did not dispute
the successes of these voyages—either in terms of profits for the Company
or by opening up the Indian trade and defeating the Portuguese at sea—he
accounted for them in the metric of England's timber, manpower, and naval
strength and found them wanting. As he put it of the ships, "they come
home so crazed and broken, so maimed and unmanned, that whereas they
went out strong, they returne most feeble; and whereas they were carried
forth with Christians, they are brought home with Heathen." He judged that
"amends cannot easily be made for so great a losse . . . for want of woods, &
spoile of shipping" and, more cuttingly, that "their commodities are at a far
deerer rate, being bought with so many men's lives."[15]

The greatest irony, and the sharpest dig at the East India Company, lay
in the pamphlet's title. This was not only because Kayll's diagnosis argued
that "trade's increase" was promoted by herring fishing and prevented by
the Company. It was also because the East India Company's greatest ship, a
custom-built vessel of over a thousand tons launched with great ceremony
by King James himself, was named the *Trades Increase*.[16] That this ship had
been, since 1613, a burnt-out and useless worm-ridden wreck half submerged

14. [Kayll], *Trades Increase*, pp. 15, 17, 25, 30.
15. [Kayll], *Trades Increase*, pp. 20, 27.
16. William Robert Scott (1910–12) *The Constitution and Finance of English, Scottish and Irish Joint-Stock Companies to 1720* (Cambridge University Press, Cambridge) vol. 2 p. 102.

in Bantam harbour, with its ordnance and gunpowder sold to the Dutch or the local king, only served to emphasize Kayll's critique:

> I heard a Ship-wright say on the losse of the *Trades Increase*, that if you ride forty miles from about *London*, you could not finde sufficient Timber to build such an other. It was a ship of eleven hundred Tunne: for beauty, burthen, strength, and sufficiency, surpassing all Marchants ships whatsoever. But alas! shee was but shewne, out of a cruell destiny shee was overtaken with an untimely death in her youth and strength; being devoured by those Iron wormes of that Country, that pierced her heart, and brake many a mans withall memorable in her misfortune, onely redounding to the Common-wealthes losse.[17]

What followed from this analysis of the interests of the commonwealth was a direct attack on the East India Company where it hurt most: their monopoly. Kayll argued that merchants as a whole, consumers, the king's customs, the navy, and mariners would benefit most from "a *freedom of Traffique* for all his Majesties subjects to al places." The end of the companies was argued for on the basis that "we are all *Britaines*, all subjects to one royall King, all combined together in one naturall league," and, as a result, the benefits of all trades should be open to all "fellow-subjects, and equall Citizens in this great Monarchie" rather than "the common-wealth being made private."[18] Indeed, the privatization of privilege, and of knowledge, threatened, for Kayll, the very basis of society and civility. Since "Society first beganne, and knowledge and civility by communication," if all possessions and knowledge had been kept private "there had not onely beene no civility, but no society."[19] Therefore, what began with herrings ended with an argument for the dissolution of the East India Company in the interests of the commonwealth, of civility, and of society itself. It was this public principle that Kayll had gone into print to proclaim.

The publication of *The Trades Increase* came at a bad time for the East India Company. By 1609 the Dutch and the English were both at peace with the Spanish. However, this did not prevent continuing conflicts over access

---

17. [Kayll], *Trades Increase*, p. 19. On the fate of the *Trades Increase*, and the fact that the Company in London did not know of it until June 1614 and were still receiving fresh reports in 1615, see W. Noel Sainsbury (ed.) (1862) *Calendar of State Papers. Colonial Series: East Indies, China and Japan, 1513–1616* (Longman, Green, Longman & Roberts, London) [hereafter *Cal. S.P. 1513–1616*] entries 730, 737, 739, 907.

18. [Kayll], *Trades Increase*, pp. 51–54.

19. [Kayll], *Trades Increase*, p. 55.

to the spice islands, and the fruits of the East India trade. The Dutch argued for their right to trade against the Spanish policy of exclusion, urging their claim via Hugo Grotius's famous *Mare Liberum*, published anonymously at Leiden in 1609.[20] The English East India Company protested against Dutch attempts to prevent them from trading, including the VOC's monopolistic contracts with spice growers. In turn, the Dutch protested against the insulting and hostile behavior of the English as they sought to break into the trade.[21] Eventually the merchants of the Dutch and English companies agreed that their disputes could be settled only via negotiation between them backed by the king and the States General on questions that were both matters of trade and matters of state. On the Dutch side there was strong support for the idea that amicable relations with the English, and the defeat of the Spanish, could come only with the merger of the two companies, albeit one that respected Dutch contracts and forced the English to pay a proportion of the charges of the trade, primarily the cost of forts and fighting ships. On the English side, any merger looked like a takeover. As one of the London correspondents of Ralph Winwood, English ambassador to the States General, put it in 1610, "we fear that in case of joyning, if it be on equall Terms, the Art and Industry of their People will wear out ours."[22]

The first conference on the matter brought Grotius and three prominent Dutch merchants to London in 1613 but produced no agreement. Indeed, the Dutch commissioners had not been empowered to deviate from their original position, and the discussions were most notable for Grotius's defense of the monopolistic contracts the Dutch had made with the spice growers against the English merchants' and lawyers' reading of his *Mare Liberum*.[23] For the second conference, the three English commissioners were Clement Edmondes, the clerk of the Privy Council, and two longstanding East India Company merchants, Robert Middleton, who had been involved in the 1613 conference, and Maurice Abbot, the archbishop of Canterbury's brother. They arrived in The Hague in January 1615, with the discussions set to take place at the house of the British ambassador, Sir Henry Wotton.

---

20. Hugo Grotius (1916) *The Freedom of the Seas, or the Right Which Belongs to the Dutch to Take Part in the East Indian Trade,* translated by Ralph van Deman Magoffin and edited by James Brown Scott (Oxford University Press, New York).

21. *Cal. S.P.* 1513–1616, 591, 606.

22. Edmund Sawyer (ed.) (1725) *Memorials of the Affairs of State in the Reigns of Q. Elizabeth and K. James I. Collected (chiefly) from the original papers of . . . Sir Ralph Winwood, Kt* (London, printed by W.B. for T. Ward) vol. 3 p. 239; and G. N. Clark and W. J. M. Eysinga (1951) *The Colonial Conferences between England and the Netherlands in 1613 and 1615,* pt. 2 (E. J. Brill, Leiden).

23. David Armitage (2000) *The Ideological Origins of the British Empire* (Cambridge University Press, Cambridge).

The diplomats were to make any treaties, the merchants were to work out the details of trade.[24]

However, by 1615 the dispute over trade in the east had been complicated by disputes over the Greenland whale fishery. Here, the English—and it was many of the same merchants involved—were determined to exclude the Dutch on the basis of prior discovery and effective occupation. Relations had also taken a turn for the worse in the east with the appointment of Jan Pieterszoon Coen as director general of the VOC, and the increasing support for his view that the English were the most dangerous enemy. As it stood, the Dutch were willing to negotiate to try to get the English to contribute towards the costs of trade rather than profiting from Dutch expenditure, and to maintain good relations with King James in Europe. They would, however, drive the English out of the spice islands by force if need be. The English were more divided. The king had previously shown himself to be in favor of the companies uniting, and the Dutch knew it.[25] It was even common talk among the Hollanders in Japan at the end of 1614 that it was "very likely the East India Companies of England and Holland will be united... to drive both Spaniards and Portugals out of these eastern parts of the world."[26] However, many in the English Company were against such a course, both on the grounds that the VOC had, after ten years, paid no dividends, and for fear of the "joint war" that would surely follow a joint stock. Yet they were also aware that the king could revoke the Company's charter at three years' notice. As the governor, Sir Thomas Smythe, put it, "he conceives it dangerous to cross His Majesty's intents, and impossible for the Company to join with the Hollanders."[27] Fortunately, the king's instructions to the commissioners did not insist on the companies uniting, requiring only that any discussions on the matter be referred back to him. The instructions also required that no agreement should be made that was prejudicial to the peace that James had made with the king of Spain. Since the Dutch line was that there were only three options in regard to the India trade—"either to leave it, or join stock with us, or undertake a vigorous war" against Spain— there was little chance of an agreement that would leave the English East India Company intact.[28] It was, therefore, within this context of global trade disputes and diplomatic negotiations over the future of the East India trades

---

24. *Cal. S.P. 1513–1616*, 831.

25. Clark and Eysinga, *The Colonial Conferences.*

26. *Cal. S.P. 1513–1616*, 823.

27. IOR B/5 *East India Company Court Minutes: December 1613 to 10th November 1615*, p. 331 (3 January 1615).

28. Clark and Eysinga, *The Colonial Conferences* p. 115; and *Cal. S.P. 1513–1616*, 957.

and of the Company itself that Kayll's *The Trades Increase* was composed and printed off by Nicholas Okes's pressmen.

The Company got wind of the pamphlet very quickly, and the directors certainly knew who had written it. Only four days after Okes and Burre had entered their title to the copy in the Stationers' Company register, Sir Thomas Smythe informed the Court of Committees that he had sent a copy of the book "divulgd and sett forth by one Mr Keale" to the archbishop of Canterbury.[29] The archbishop, in addition to being charged with the licensing of printing, and being the brother of one of the commissioners sent to the Netherlands, was a coadventurer with Smythe and Robert Middleton in the company formed in 1612 to exploit the northwest passage they believed had been found by Henry Hudson on a voyage sponsored by the East India Company in 1610. He also had financial interests in the East India Company itself.[30] More importantly, following the death of Robert Cecil, George Abbot had become one of the most important members of the king's Privy Council. Smythe was going straight to the top, and whereas Abbot treated most licensing work as a matter for his chaplains, he took a personal interest in this case.[31] In reply he sent Dr. Nidd, the chaplain who had been responsible for authorizing the printing, to deliver the archbishop's opinion, born of long experience no doubt, that suppressing the book (which was presumably what Smythe had requested) "will cause many men to seeke after yt the more earnestlie." He argued that the work "should rather be suffered to die," but offered to issue a warrant that "a stopp be made" if that was really what Smythe wanted. In discussion, it was made clear that what the Court of Committees desired was that an example be made of the author, rather than any action against the stationers that was in the power of the archbishop. Moreover, they wanted this recognized as a matter of state and not simply of their own private interest. As they argued, having the author "punisht . . . will quickly be bruited abroad and thereby discover the dislike the State hath to such pamphlets that shall taxe whatt the State hath approved."[32] As a result, they referred the book to two of their number to report on the possibility of having the author brought before the Star Chamber to answer for his text. The men chosen, Nicholas Leate and Robert Bell, had both been adventurers since the first East India Company voyage, had

29. IOR B/5 p. 370 (16 February 1615).

30. *Cal. S.P. 1513–1616*, 616, 786 (the court minutes for November 1614 noted "Doctor Gulston allowed to adventure 400l. in the joint stock for the Lord of Canterbury's sake").

31. Clark and Eysinga, *The Colonial Conferences* pp. 62–63; and Johns, *The Nature of the Book*, pp. 241–42.

32. IOR B/5 p. 370 (16 February 1615).

held positions of responsibility, and were actively involved in the northwest passage company in which the archbishop had invested. Leate had also done similar work for the Company the year before, when he was "requested to find out how some things have got abroad that should have been kept secret."[33]

Less than a week later Leate and Bell reported that the eminent lawyers whom they had consulted on the book "hath found some points very neare unto Treason, and all the rest very dangerous."[34] The Court of Committees referred them to the Company's solicitor for his opinion on the course that should be taken in pursuing a prosecution. However, at the same time as they were pursuing legal action, there were suggestions that they should also exploit the power of the press themselves. The governor reported that Sir Dudley Digges—another long-standing Company man, a key mover in the northwest passage company, and a former tutee of George Abbot's at University College, Oxford—was of the opinion that they should "have a booke sett forth in defence of the East India trade, and make knowne the benefitt that it brings to the Comon wealth."[35] This dual strategy was what was followed.

A few weeks later the need to act became more acute. Letters from the commissioners in the Netherlands were read before the Court of Committees, having also been seen by the king and his Privy Council. They revealed that *The Trades Increase* had traveled. In these letters Edmondes reported to Winwood that the differences between the Dutch and the English over the question of rights to trade remained unresolved, but that there was some possibility of moving ahead on the practicalities: how much the English should contribute, and what portion of the trade they would get for it. However, he feared that it would, as many had predicted, be a long business as the Dutch "dailie growe coulder in the matter."[36] Edmondes also noted that earlier that month he had seen a book published in England against the East India Company, which argued "that if it be thought fit to quit the trade we may bring home with us from hence 100,000*l* for our interest."[37] Kayll's pamphlet had appeared in the Netherlands; Edmondes read it in The Hague and understood its argument that England's development of the herring fishery

---

33. *Cal. S.P. 1513–1616*, 256, 257, 273, 616, quotation from 772. Robert Bell had procured the king's letters to the emperor of China and the king of Japan as sent out in 1614; *Cal. S.P. 1513–1616*, 702.

34. IOR B/5 p. 373 (22 February 1615).

35. IOR B/5 p. 373 (22 February 1615); and *Dictionary of National Biography*.

36. IOR B/5 p. 400 (29 March 1615).

37. *Cal. S.P. 1513–1616*, 947 (letters dated 16 and 21 March 1615 by the Dutch calendar).

might save the £100,000 that was spent on Dutch fish as offering a direct premium for giving up the East India trade (in which he was himself an investor) at the discussions in which he was taking part on behalf of a king who might warm to the idea.[38] Such interpretations did not help in driving a hard bargain and in preserving the Company and its trade.

Sir Thomas Smythe certainly drew direct connections between these costly stalled negotiations and what he called the "wronge and scandall the Companie is subject unto, by . . . envious eyes." He stressed that none of their detractors recognized the "greate hazard" the adventurers put themselves to, the "great burthens of charges" that fell upon them, and the profits that they brought to the king. Instead, he noted, there were those who were out to discredit the Company and its principal members. He cited the interloper Mr. Newman, who had "maliciouslie and disgracefullie" arrested the deputy governor, William Greenwell, in Southwark, and "Keale who had soe maliciouslie and unjustlie wrytten in publique disgrace of this Companie." The implication was clear. These actions substantially weakened the Company's position with both the king and the Dutch by denying that the merchants acted in the interests of the commonwealth. As Smythe concluded, the Dutch themselves would never behave this way. They were "soe jealous of there trade into the Indies as that they have given spetiall commandement for none to be soe bould to dare to make any invectives against the same."[39] He reassured the committees that he was determined to act against both Newman and Kayll, and it also became clear that Dudley Digges had been sharpening his quill in the Company's service.

As a result, Kayll appeared before the Privy Council at Whitehall in April 1615, just as the English commissioners were reaching the critical stage in their discussions with the Dutch.[40] If the intention was to give him a fright by having him answer for his book in front of the archbishop of Canterbury, the lord chancellor, the lord treasurer, the chancellor of the exchequer, several members of the nobility, and Ralph Winwood (to whom Edmondes had written about *The Trades Increase*), it did not initially seem to have worked. Kayll "ne[i]ther gave satisfaction to your Lordships nor demeaned himself as he ought," and they sent him for detention in the Fleet prison for an indefinite period. His release from custody nearly a fortnight later came only after the Privy Council had received a written petition

38. See [Kayll], *The Trades Increase*, p. 42 [misnumbering of p. 43].
39. IOR B/5 p. 400 (29 March 1615); and *Cal. S.P. 1513–1616*, 899, 928.
40. *Acts of the Privy Council of England, 1615–1616* (HMSO, London, 1925) p. 99; and *Cal. S.P. 1513–1616*, 966.

offering "his humble submission and acknowledgement of his offence," and
noting that their "lordships' most just and wise reprehensions" and his
"deserved punishment and imprisonment" had made him "truly and, as he
hopeth, sufficiently admonished both of his duty to the comonwealth and
your lordships."[41] Kayll had, to some extent, been forced to eat his words,
and an example had been made.

The complementary strategy—that of putting other words and other
interpretations into circulation—was pursued by an experienced author who
was able to negotiate the meanings and conventions of print to prosecute his
case. Dudley Digges had already published a treatise on military discipline
coauthored with his father Thomas, some panegyric verses for *Coryates
Crudities* (1611), and a short treatise on the circumference of the earth
written in support of the existence of a northwest passage, and providing
a justification for investment in the company formed to exploit it.[42] *The
Defence of Trade* (1615), Digges's response to *The Trades Increase*, took the
form of a letter to Sir Thomas Smythe and was printed by William Stansby.[43]

In the *Defence* Digges said he had no quarrel with support for the herring
fishery. What he regretted was the hijacking of Tobias Gentleman's scheme
by an anonymous invective against the East India trade. In countering it he
sought to turn around the charges of private interest and capture the prize
of benefit to the commonwealth. In doing so he used both substantive arguments and the cultural conventions of print to his advantage. With his
own name prominently on the title page, and his "love to the *East-India*
Trade" clearly stated, he set himself against the low literary strategies of
The Trades Increase.[44] Instead of revealing its author, he cast doubt on the
anonymous pamphlet and stated that his initial thought had been that the
Company should ignore this work "by some unknowne busie Person."[45]
Moreover, playing on anonymity allowed Digges to suggest darker forces at
work. He argued that the author did not understand what he had written,
and that "some *Ape* hath put the *Cats* foot in the fire, some cunning and

---

41. *Acts of the Privy Council*, p. 108.

42. Thomas Digges and Dudley Digges (1604) *Foure Paradoxes, or politique Discourses.
2 Concerning Militarie Discipline* [and] 2 *of the Worthinesse of Warre and Warriors* (London,
printed by H. Lownes for Clement Knight); Thomas Coryate (1611) *Coryates Crudities* (London,
printed by W.S.); and [Dudley Digges] (1612) *Of the Circumference of the Earth: Or, A Treatise of
the North-east* [sic] *Passage* (London, printed by W.W. for John Barnes).

43. Dudley Digges (1615) *The Defence of Trade* (London, printed by William Stansby for John
Barnes). It was registered with the Stationers' Company on 21 March 1615, again by Dr. Nidd; see
Arber, *Transcript*, vol. 3 p. 565.

44. Digges, *Defence of Trade*, unpaginated postscript.

45. Digges, *Defence of Trade*, p. 1.

malitious persons, for private ends, or *lewder* purposes infused the *Quicksilver* that set that running head awork." He suggested that it showed support for Spain against England's "best assured friends," the Dutch. As a result, he had decided, following Plutarch, to "make profitable use of enemies" by showing the ways in which the East India Company works to "advance the reputation and revenue of the *Common-wealth*."[46] That this could, indeed should, now be done as publicly as possible, and therefore in print, was to turn a problem into an opportunity:

> [T]his honourable enterprise, like *Hercules* yet in the Cradle, in the infancie hath beene assailed by Serpents slie aspersions, which *Envie* long since whispered in the eares of ignorance, of killing Mariners and carrying out the treasure of the land, in answere whereunto had the *East India Marchant* then but told a truth like *Martials, Bella, Dives, Puella, Fabulla,* hee might have beene (it may bee judged) neyther faire nor rich, nor chast, but only forward to commend himselfe: but now when as the poore Snake *Envie* growes to be a Monster, *Malice*, when the pratler late a creeping Worme is waxt a winged Goose, a setter forth in print of slanders. Now (me thinkes) you are, if not inforst, at least invited happily to shew the world the well deserving of the worthie Companie, whose innocence will shine more gloriously even to the eye obscur'd of him that dwels farthest from *London*, by passing through those vapours of an idle or corrupted braine these forst or forged imputations.[47]

Since sly whispers had now become printed slanders, they had taken a different form—the "creeping Worme" had grown into a "winged Goose"—and they demanded a different response. In replying the Company need not fear the censure of publicly commending itself. Indeed, being attacked in print gave it the perfect invitation to put its case without accusations of impropriety. However, doing so appropriately and convincingly also required care, at once epistemological, methodological, and stylistic. In contrast to the "fond reports of idle fellowes" upon which he claimed *The Trades Increase* was based, Digges set out a secure knowledge base. The substance of *The Defence of Trade* "was taken out of Custome-bookes, out of the *East-India* Companies bookes, out of Grocers, Warehouse-keepers, Marchants bookes, and conference with men of best experience."[48] He provided a

---

46. Digges, *Defence of Trade*, pp. 2, 4.
47. Digges, *Defence of Trade*, pp. 3–4.
48. Digges, *Defence of Trade*, p. 6 and unpaginated postscript.

detailed three-page list of East India Company shipping, and offered accurate figures in support of his claims about the state of the trades. Digges also took his opponent's words seriously. He stated that "as a ground-work of Integritie, first I would set downe what hee sayes even in his owne Apparell, Scarfe, and Feather too," and then quoted verbatim nine pages of Kayll's text. Finally, in presenting his case Digges imposed particular limits on writing style, stating that he would "apply my penne to satisfie an honest minde, rather than make him smart or carelesse Readers smile."[49]

On this basis, Digges seriously attended to the charges over timber, ships, and men, and in each argument sought to change the terms of the debate to the Company's advantage by emphasizing the productive use of resources rather than their destruction. Would the author of *The Trades Increase* prefer, he asked, that wood be used for "Beggers nests (that growing scurfe upon this Citie) new tenements, whose rotten rents make many Gentlemen before their time," or burned in fires or furnaces instead of coal? What "nobler use" was there for timber than to build ships "such as round about the World disperse the honour of the Crowne they serve, and then returne with wealth for King and Kingdome, and for those that set them foorth, in stead of Wood?"[50] His detailed accounting of the Company's twenty-four ships played down the losses reported by "this Vulture that thus followes wreckes and dead mens bodies" in relation to the losses of the Dutch, the Portuguese, or even in the Newcastle trade.[51] Likewise, Digges rewrote the loss of the *Trades Increase* and the other three wrecks as unfortunate accidents, or stemming from the bad actions of individuals (particularly the sinking of the *Ascension*), rather than misguided policy.

As for mariners' lives, Digges disputed Kayll's calculations and argued that many of the men brought death on themselves "through their owne abusing of themselves, with the hot drinkes and most infectious women of those Countries."[52] More significantly, Digges argued that working bodies "are the Bodies of our King, and of our Countrie," contrasting the wealth of well-populated Jacobean England with the "poore naked King" Powhatan and uninhabited Virginia. As he put it, "Living bodies, unimploi'd, are nothing." Employing them is the only way to "inrich us, or our Masters, or serve the King, or good the Common-wealth." Citing Tacitus, he argued that it is better that mariners die as "good men," that is men "such as doe good," by

49. Digges, *Defence of Trade*, pp. 5–6.
50. Digges, *Defence of Trade*, pp. 27, 28.
51. Digges, *Defence of Trade*, p. 23.
52. Digges, *Defence of Trade*, p. 37.

sailing merchant ships to the East Indies, than that they die as poor men on the gallows at Tyburn, or "be forced to turne Sea-robbers" and meet their end on the executioner's steps at Wapping.[53] Finally, Digges answered objections that the Company only brought back unnecessary wares, claiming that these employed many people in preparing and undertaking the voyages and that they were part of trades that profited the commonwealth both by exchanging goods (as well as bullion) for cheaper spices and by re-exporting goods from the East Indies to Europe. In all these ways the Company's work served king and commonwealth, and Digges's argument was prosecuted without ever straying onto the dangerous ground of explicitly defending its monopoly.

Therefore, Digges combined nascent mercantilist arguments and exploited the meanings of print to challenge the author of *The Trades Increase*. He countered anonymity and slander with, as he had it, open authorship and the true facts of the case. Yet, despite himself, he also drew on comic and polemic genres to simultaneously undermine the importance of both pamphlets, and of such disputes in general, by ending with the wish that they be seen as "Inke-wasting toies," that *The Trades Increase* be used "to packe up fish" and the *Defence of Trade* "to wrappe up spice," and that both "should have seene no other light then the fire."[54] Taken together with the prosecution of Kayll, it is apparent that the Company's interests lay as much in curtailing debate in print over their actions and contributions to the commonwealth as in promoting it. This was particularly evident when such debate might jeopardize the continuing existence of the Company, just what was at stake in the conferences with the Dutch and communications with the king. In responding to *The Trades Increase*, however, through both prosecution and printing, the Company showed both an acute awareness of the politics of print in its deployment of the regulatory mechanisms of licensing and law in their own interests and, via Dudley Digges, a careful appreciation of the cultural meanings and resources of print that might be useful in arguing that the Company worked in the collective interest. As will be argued below, both these sets of maneuvers were increasingly important in the 1620s.

In the end, the Anglo-Dutch conference produced no agreement.[55] The Dutch still wanted a joint stock, but the English could not countenance an "offensive" war with Spain. In the months that followed the East India

53. Digges, *Defence of Trade*, pp. 32–33.
54. Digges, *Defence of Trade*, unpaginated postscript.
55. The final positions are well summarized in Huntington Library Mss. EL 9606.

Company had to present its case against a further Dutch proposal for the union of the two companies, which the king had judged "a worthie project."[56] Despite this, Edmondes, Middleton, and Maurice Abbot were able to argue for pursuing an economic rather than a military offensive against the Spaniard, "by underselling him in all parts of Christendom."[57] Further conflict in the east eventually produced an uneasy accord in 1619, ending competition between the companies without combining them.[58] When this accord, and Anglo-Dutch relations, collapsed, it did so amid a flurry of pamphlets that revealed both the Company's careful use of printing and the ways that the politics of print were beginning to change.

In and Out of Print: The Amboina Pamphlets

The issue over which the accord finally collapsed was the execution by the Dutch governor on the island of Amboina of ten English merchants and factors, nine Japanese soldiers, and a Portuguese. They were convicted of being involved in a conspiracy to overthrow the Dutch company's fort and seize power, but almost every aspect of their torture, trial, and execution was disputed between the two companies. These different accounts played into the companies' attempts to generate support from their respective governments within the ever-changing politics of foreign policy in the early seventeenth century.[59] The use of printing was a crucial part of that process, but its deployment had to be negotiated by the Company within and across a range of different spaces and with close attention to the potential political effects of the production of printed material.

The English East India Company was quick to turn this treachery into text. After the six survivors' evidence had been presented to an admiralty court, they instructed John Skinner to produce a narration of "that most cruel and bloody butchery of their men in Amboyna" based on their testimony.[60] The aim was, as they put it, "the setting down some relation of the truth of that proceeding, for the suppressing of such rumours as are spread

---

56. IOR B/5 p. 438 (4 July 1615).

57. *Cal. S.P. 1513–1616*, 1014.

58. Clark and Eysinga, *The Colonial Conferences*. For a lively version of these events, see John Keay (1991) *The Honourable Company: A History of the English East India Company* (HarperCollins, London) pp. 24–51.

59. Simon Adams (1983) "Spain or Netherlands? the dilemmas of early Stuart foreign policy," in Howard Tomlinson (ed.) *Before the English Civil War: Essays on Early Stuart Politics and Government* (Macmillan Press, London) pp. 79–101.

60. W. Noel Sainsbury (ed.) (1878) *Calendar of State Papers: Colonial Series—East Indies, China and Japan, 1622–1624* (Longman & Co., London) [hereafter *Cal. S.P. 1622–1624*], entry 516.

among the vulgar in justification of the Dutch."[61] Yet the Company, through Robert Barlow, its representative in Holland, was aware that printing it could jeopardize their interests in a situation where the king, the increasingly influential duke of Buckingham, Prince Charles, and the Privy Council were not eager to alienate the government of the United Provinces as the plans for an Anglo-Spanish alliance unraveled.[62] So the manuscript, backed by the Company's threat to end the East India trade, was performed and distributed in other ways in pursuit of their claim to the king for justice, restitution, and the dissolution of the Anglo-Dutch treaty that only he could deliver. A version was read before the Company's General Court. It was sent to Dudley Carleton (the English ambassador), Robert Barlow, and Edward Misselden, the Company's commissioner in the Netherlands, who had it translated into Dutch in "the sharpest style the translation would bear, the hardest expression being not bad enough for the subject." It was also presented, along with a petition, as "a book" to the king in his bedchamber at Wanstead.[63] The initial responses to these communications were gratifying. The king promised action, and, as a result, the narration was read before the Privy Council, "whereat sundry of the greatest shed tears" and offered support for the Company's claims. The king set the States General a deadline of 12 August 1624 by which to answer the charges, and backed it with threatened action against VOC shipping in the Narrow Seas.[64] Meanwhile, the Company maintained its refusal to resume operations "until they may see some action either from the States or the King," and continued to distribute manuscript copies of the narration, gratified that they "begin to be much asked after especially by the knights and burgesses of parliament."[65]

The first foray into print came from the Dutch. This was a short pamphlet without an author's or printer's name, and therefore a "libel" in Dutch law, which, as Barlow put it, "would insinuate to the States the upright carriage of the business, and the foulness of the fact in ours."[66] It appeared in large numbers in the Netherlands, and Barlow, who sent copies to London, reported that the VOC's active partners (the Bewindhebbers) were busy distributing it among their friends. He also suspected, despite VOC denials, that it was written by Jacques Boreel, chief of the VOC chamber at

61. *Cal. S.P. 1622–1624*, 481.

62. *Cal. S.P. 1622–1624*, 516; and Adams, "Spain or Netherlands?"

63. *Cal. S.P. 1622–1624*, 496, 503, 526. It was so gruesome that one secretary gave up the translation, since he "felt his body not able to bear it."

64. *Cal. S.P. 1622–1624*, 534.

65. *Cal. S.P. 1622–1624*, 544.

66. *Cal. S.P. 1622–1624*, 551, 537.

Middelburg, "for there were divers of them looked at him when I spake it."[67] Going into print could, however, be a tactical political mistake when the relationships between the companies and their governments were as finely balanced as they were in the mid-1620s. Dudley Carleton took advantage of the pamphlet's existence to set out the English case as dramatically as possible at the assembly of the States General. He argued that having been dealing in private with various deputies on the matter of Amboina, he now came to them in full assembly, having been "driven or rather dragged by the hair to do so by the libel" that had been published. This allowed him to present the English "narration," to dispute the facts of the case, and to suggest that the governor of Amboina had been supported in his actions by the VOC in Amsterdam, on the basis that the pamphlet itself could have come from no other source (its false imprint at The Hague not fooling him one bit). He demanded justice against the author and printer, and action in the wider case before the king's deadline.[68]

The anonymous Dutch publishers' strategy had indeed backfired. The pamphlet brought the whole Amboina business out into the open and gave Carleton the upper hand. It openly contradicted the States General's request to the English Company that nothing be done "that may tend to the exaggerating of the proceedings passed in Amboyna," which arrived by letter in London the same day as the printed pamphlet.[69] The wedge that this drove between the VOC and the States General was exacerbated as the Bewindhebbers, maintaining their denial, tried to suggest that the pamphlet might have been produced by the government's own clerks. The States proclaimed it a libel on 11 August 1624, stressing their determination to punish its author, printers, sellers, and distributors and offering a four hundred–guilder reward for the discovery of those who had written or printed it. Placards to this effect were fixed to the pillars of the burse and to all of Amsterdam's street corners, much to the embarrassment of the Bewindhebbers, "for now many men's mouths are open and speak very largely concerning all their miscarriages toward their fellow adventurers."[70] Barlow certainly sought to press home the English Company's advantage by sending men out to find the printers, thought to be in Zeeland, and through them to locate the author. However, he had no success in his attempts to confirm Boreel's authorship.[71]

67. *Cal. S. P. 1622–1624*, 537, 538.
68. *Cal. S. P. 1622–1624*, 548.
69. *Cal. S. P. 1622–1624*, 561 I and II.
70. *Cal. S. P. 1622–1624*, 551, 555 I, 576, 583, 588.
71. *Cal. S. P. 1622–1624*, 563, 576, 593.

The East India Company in London planned a response to the Dutch publication as soon as it was distributed in England. This rejoinder was to be carefully written, "and so much the less haste made that it may be the more advisedly done and pinch home." Moreover, in contrast to the libelous anonymity of the Dutch pamphlet the reply was to be officially endorsed by the Company.[72] Once again, however, the question arose as to whether it was a good time to go into print. Would it propel the States General and the king into action, or would it sour relations between them to the detriment of the Company? Barlow advised the Company to hold back and later argued that any English Company publication "would have received the same disgrace the Dutch pamphlet has had."[73] Yet, in the face of a lack of real action from the States General and the king—the 12 August deadline came and went without any resolution—it was unclear what the best strategy was. Should they continue to lobby the king, or should they apply pressure via publicity and Parliament? As Maurice Abbot put it, "Some of the Company advise to beseech his Majesty to put it to the judges of the kingdom, but many of the gentry themselves to rely upon the House of Parliament."[74] This was a complicated situation. While Parliament had no constitutional role in decisions over foreign affairs, it was crucial in providing the fiscal resources for warfare. There was, therefore, some need for the king and his ministers to engage in discussion in Parliament over foreign policy, but there was also, as Simon Adams puts it, "the fear of being forced into a disliked foreign policy by popular pressure," particularly as the support for alliances with the Spanish or the Dutch became increasingly factional and divisive issues in Stuart politics.[75] Thus, the English Company proceeded cautiously, and in manuscript, but made ready to print. They presented a translation of the Dutch libel (and the States General's proclamation) to the king and asked for the stay of Dutch shipping in the seas between them to be put into effect. The Company also received permission from the king to print its official response to the Dutch work. This was granted on condition that the pamphlet be vetted by the commission James had established to consider the Amboina affair to ensure that "it contained no bitterness against the States."[76]

---

72. *Cal. S. P. 1622–1624*, 561 and 554.

73. *Cal. S. P. 1622–1624*, 593, 553, 559, 563.

74. *Cal. S. P. 1622–1624*, 589. Carleton also warned the prince of Orange that "his Majesty will be driven to other resolutions such as the resentment of the nation assembled in Parliament shall require" (602).

75. Adams, "Spain or Netherlands?" p. 82.

76. *Cal. S. P. 1622–1624*, 607, 566, 575, 594, 609.

What finally pushed the Company into print was a renewed need to re-
fute the Dutch claims, combined with a sense that increased pressure had
to be applied to the king and Privy Council. Not only was the Dutch libel
still around, the placards having never been proclaimed in Zeeland, but an
English translation was published at Middleburg and Flushing. Carleton detected the hand of Boreel's son, an *advocaat* of the Company, whose brother
was secretary to the States of Zeeland. Again, the pamphlet was read as confirming official VOC support for the governor of Amboina's actions.[77] It demanded a response. As Abbot wrote to Carleton, "which how to leave unan-
swered the Company know not, without reproach to the nation."[78] Closer
to home, the Company needed to keep the pressure on the Privy Council,
which had—much influenced by the direct and harrowing testimony of the
six survivors (who had been kept on hand for such purposes by the governor and committees) and the Company's narration of their statements—
supported action against Dutch shipping, but had not endorsed the ending
of the treaty and the separation of the companies. Going into print thus
answered the Dutch in the name of the nation, not the Company, and was
justified in its censure of the Dutch actions on the basis "that the truth may
appear and that those innocent souls that have suffered in their persons, may
not suffer a second time in their reputations."[79] The pamphlet that appeared
was titled *A True Relation of the Unjust, Cruell, and Barbarous Proceedings
against the English at Amboyna in the East-Indies*, and contained versions
of the original narration, a translation of the Dutch libel, and a response
to it.[80]

The Company recommended that a higher than average run of two thousand copies be printed in English and that the prime readership should be
targeted by ordering that "some few copies be given to some principal persons of the nobility, and if they be well taken, to adventure to put abroad
the rest, which if they shall do, the benefit will pay for the printing." In addition, they recommended that a thousand copies be printed in Dutch. The
production details were to be negotiated with the printers by Thomas Mun,
already an experienced author (see below).[81] The work was carefully done.

77. *Cal. S. P. 1622–1624*, 602, 614. In fact, it turned out that the translation was by an
Englishman, John Winge, who thought he was doing the Company a service (622).

78. *Cal. S. P. 1622–1624*, 614.

79. *Cal. S. P. 1622–1624*, 620, 623.

80. *A True Relation of the Unjust, Cruell, and Barbarous Proceedings against the English
at Amboyna in the East-Indies, by the Neatherlandish Governour and Councel there* (London,
printed by H. Lownes for Nathanael Newberry, 1624).

81. *Cal. S. P. 1622–1624*, 636, 643.

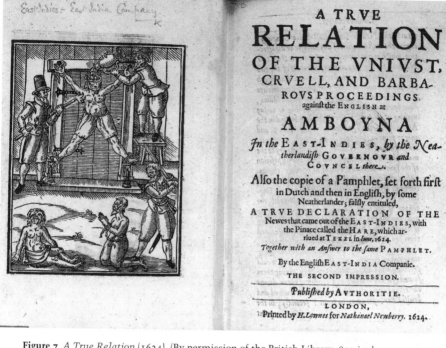

Figure 7 *A True Relation* (1624). (By permission of the British Library, 802.i.5.)

The pamphlet's title page exhibited more expensive dual-color printing and, for the second edition, a woodcut that depicted the tortures and executions (Figure 7). There was no author named. Although the "Epistle to the Reader" was drafted by John Skinner, it was presented as a collective statement, and in the main body of the text the survivors' own testimony was to speak as directly as possible. The pamphlet established its veracity and credibility through reluctant authorship and official truth telling. Facing the epistle, which stressed that the Dutch pamphlet had meant the English merchants were "inforced, contrary to their desire and custome, to have recourse also to the Presse," was the Company's coat of arms, "that it may not be taken for a libel...[and] in token that they avow them to be true."[82]

The Company's political intent was made clear when the pamphlets were distributed. Dudley Carleton received ten English and forty Dutch copies, along with expressions of disquiet at the continued lack of satisfactory action by the king or the States General, and notification that the Company had published this material "lest the world should think that they

82. *A True Relation*, sig. A2ᵛ; and *Cal. S. P. 1622–1624*, 639.

can forget the bloody injuries done to their people."[83] The Company sought to stir up political opinion. Every committee was to have five or six copies "for themselves and their friends, and that the Lords of the Council and the principal nobility residing in and about London be each presented with one in the fairest binding." Further effort was made to ensure that they were "dispersed in all parts of England," and in the Netherlands, although it was reported that Boreel was trying to suppress it by buying up as many copies as he could.[84] The Company also used it to continue to lobby the king. An audience with James at Newmarket involved direct testimony from the six survivors "to report to his Majesty the truth of their cruelties," and received his acknowledgment of the printed work. It produced the "assurance that he will not leave the Company unsatisfied."[85]

These proved to be empty words. In the face of inaction by the States General and the king in the early months of 1625, pressure was brought to bear in a wider and more public realm. This was done both by the Company and by those who supported it either for ideological reasons or because it would line their pockets. Print was an increasingly important medium here, alongside other forms of popular representation. The licenser of the Company's *True Relation*, Dr. Thomas Worrall, who had a reputation as something of a soft touch, inquired whether they wanted him to permit a "printed piece of the several tortures in effigy of our men at Amboyna." Archbishop Abbot was also keen to speak to his brother about these "printed models of the tortured English in Amboyna [that] had been brought over hither by the porter of the Archduchess' Ambassador's house and printed on the other side."[86] Alongside these images was a printed ballad "to the tune of Braggendary" entitled *Newes out of East India*, and a play on Dutch wrongdoings around the world that was about to be performed in London. Somewhat more decorously, Dr. Thomas Meriall presented the Company with twenty-four small vellum-bound copies of a sermon preached before the king, and the accompanying epistle in "bitter English" on the subject of the "abhored cruelties of the Dutch," with a "loving dedication to the

83. *Cal. S. P. 1622–1624*, 659, 682.

84. *Cal. S. P. 1622–1624*, 660, 708, 723.

85. *Cal. S. P. 1622–1624*, 688; and W. Noel Sainsbury (ed.) (1884) *Calendar of State Papers: Colonial Series—East Indies, China and Persia, 1625–1629* (Longman & Co., London) [hereafter *Cal. S.P. 1625–1629*] entry 10.

86. *Cal. S.P. 1625–1629*, 43, 59. This is discussed in Pamela Neville-Sington (1997) "The primary Purchas bibliography," in L. E. Pennington (ed.) *The Purchas Handbook: Studies of the Life, Times and Writings of Samuel Purchas, 1577–1626* (Hakluyt Society, London) vol. 2 pp. 532–33.

Company." He was given £10 for them.[87] Finally, the Company commissioned Mr. Greenbury to paint a large canvas of the tortures, having previously contemplated getting the same scenes "cut out in brass."[88]

All of these works were directly political, but it was not clear if they were also politically acceptable. In the changing foreign relations of 1624 and 1625, and in relation to expressions of anti-Dutch sentiment, "what was considered acceptable or politic one day was not necessarily so the next."[89] This is well illustrated by the publishing history of Samuel Purchas's compilation of English voyages, *Hakluytus Posthumus*, caught as it was between the demands of king and Company. On the eve of publication in 1624, it seems that Purchas felt concerned that he had produced a dangerously anti-Dutch work. Too late for revisions to the text or for extensive reprinting, the copies of the work intended for the four influential dedicatees—Prince Charles, the duke of Buckingham, John Williams (the lord keeper of the great seal), and George Abbot—and for other important friends and patrons, were subtly amended. Two leaves were reprinted to omit an indiscreet remark against the Dutch, corrections were made by hand, and cancel slips with the bland phrase "English voyages to the East Indies" were printed to be pasted over more offensive headlines such as "The Hollanders base usage of the English" and "Dutch crueltie. Their lying devices." Needless to say, this process led not simply to two versions of Purchas's text but to many versions as these corrections were applied differently to different copies. However, because of Purchas's close relations with the English East India Company, the arrival of the news from Amboina just as his book was being completed raised further confusions (and he did include an abridged version of the Amboina pamphlet). Purchas now became unsure whether he had produced too *pro*-Dutch a text for some of his readers. His presentation of the work to the Company in January 1625 brought him a reward of £100 and was accompanied by an epistle "wherein the generall injuries of the Dutch in the Indies was set downe." He wanted to have this printed and bound into the three copies that the Court of Directors had ordered. However, aware of their own liabilities for the contents of the works that they produced, the printer, William Stansby once again, could not be persuaded to print it, and the bookbinder, "who hath taken advise thereupon, & is told it may be dangerous," refused to insert it between the covers. Despite its

87. *Newes out of East India* (London, printed for F. Coules, 1624); and *Cal. S.P. 1625–1629,* 71, 59.

88. *Cal. S. P. 1625–1629,* 105.

89. Neville-Sington, "The primary Purchas bibliography," p. 529. The remainder of this paragraph is based upon the evidence presented by Neville-Sington.

author's intentions, this more anti-Dutch version of the text could not be constructed.

It seems that the printer and bookbinder were politically astute. The sermon, play, pictures, and painting drew complaints to the Privy Council from the Dutch ambassadors in London that "they may be greatly endangered by the fury of the people" on Shrove Tuesday, also known as Apprentices' Day.[90] The Company countered that they had nothing to do with anything but the painting, which was simply to be kept in their own house as "a perpetual memory of that most bloody and treacherous villainy." More fault, they argued, lay with the Dutch in London, whose preachers had not spoken out against the Amboina outrage. The Privy Council—who told the Dutch "that books, pictures and plays are not the revenge that his Majesty intends"—the king, and the Dutch were, the Company surmised, really concerned that "this large picture is prepared for the view of the approaching Parliament," to generate support among the nobility and gentry. Indeed, it was ordered by the Privy Council to be shut away "because it should not provoke them as it hath many who heretofore have taken view thereof." The authorities also called in the book, forbade the play, and ordered that a watch of eight hundred men be ready on Shrove Tuesday "to see the city be kept quiet."[91] It is clear that with both print and painting the Company and its supporters were entering politics and trying to shape public opinion in ways that had not been possible before. However, there were still significant limitations here. First, the Company had a very restricted sense of whose opinion mattered. Its aim was to directly influence Parliament, and it complained that Greenbury had shown the picture to too many people when it was at his house, and that was the reason why not only it but the other pictures, texts, and performances had been "quashed."[92] Second, it was still not necessarily Parliament that mattered in the making of foreign policy.[93] As it turned out, the Company had to relinquish its painting to the duke of Buckingham, who, in his attempts to destroy support for a Spanish alliance and to promote the idea of a war with Spain, had no need of depictions of Englishmen tortured by the Dutch.[94]

In the end, therefore, the Company's appeals to the king had failed. Although the Anglo-Dutch alliance in the east was dissolved there was no

90. *Cal. S. P. 1625–1629*, 61, 64.
91. *Cal. S. P. 1625–1629*, 61, 65, 70, 71.
92. *Cal. S. P. 1625–1629*, 105.
93. Simon Adams (1985) "Foreign policy and the parliaments of 1621 and 1624," in Kevin Sharpe (ed.) *Faction and Parliament: Essays on Early Stuart History* (Methuen, London) pp. 139–71.
94. *Cal. S. P. 1625–1629*, 73.

action taken against the Dutch, and the English Company was subsequently forced out of the trade in the Indonesian archipelago and had to turn its attention to India (see chapter 3). The king's government had been able to suppress or ignore attempts by the Company and its allies to shape policy through speech, manuscript, print, painting, and performance. In part this reflected the Company's own avoidance of print in favor of considering the king and his advisors as the key audience to be addressed in more private ways. Yet, by 1625 the *True Relation*, the Company's carefully crafted foray into print, was still in circulation. This pamphlet, albeit one for which the king's approval was sought, represented a shift in the Company's relationship to politics since its use of the printing press had been directed to those in Parliament, a body of political opinion that could put at least some pressure on the monarch. This shift in the political geography of print would become increasingly important later in the century.

In its response to both *The Trades Increase* and the Dutch pamphlets over the incident on Amboina, the English East India Company did eventually go into print. However, because printing was not yet central to politics, doing so was part of a careful strategy that paid close attention to the economy of influence within which the Company operated. This involved action in a wide range of different spaces. The Company sought to deploy the testimony of the survivors in person, script, and, finally, in print, in order to sway different audiences or readerships: the king at Wanstead, the Privy Council in its chamber, and the knights and burgesses in Parliament. Indeed, as the Dutch Company became aware, there were potential costs in going into print, and such public pronouncements might be turned against their publishers, as Carleton succeeded in doing before the States General. As with *The Trades Increase*, awareness of the regulatory apparatus, this time in the Netherlands, was a key part of the Company's strategy. There was, therefore, a particular geography of print, and one that always involved the relationships between print, script, and speech within particular practices of politics and publicity. Finally, as both Dudley Digges and John Skinner were aware, going into print meant addressing a collective interest. The Company's endorsement of the *True Relation* in 1624 in the name of a nation wronged by the Dutch needs to be understood in relation to the changing forms of collective address in other writings on trade in the 1620s.

King, Kingdom, Nation, and the Balance of Trade

As has been shown, going into print in the first half of the seventeenth century meant negotiating the relationships between monarchical power, parliamentary opinion, and different versions and definitions of private interests and public benefits. Going into print required a justification for publication that, in turn, was a matter of establishing a political position from which to write and publish. In the 1620s, just prior to the concerns over Amboina, discussion of the East India Company was part of a wider debate over the contemporary commercial crisis, and merchants attached to the Company and those critical of it engaged in both print and manuscript in an attempt to shape the state's policies on currency and trade. As Mary Poovey has argued, these early seventeenth-century debates were formative in constructing trade "as a semiautonomous, law-abiding domain," and in positioning merchants as the source of particular expertise about it.[95] These textual interventions were not simply a response to new economic relationships, but an active part of conceptualizing and shaping this new domain.[96] An important element of this was the way in which different authors, with different relationships to the East India Company, used the resources of print as part of this contested construction of the politics of trade. At stake was the way in which the interests of merchants were to be understood in relation to collective interests differentially defined in terms of monarch, kingdom, commonwealth, and nation.

These debates over the politics of trade were of crucial importance to the East India Company. Its merchants depended upon the state to secure their position and profitability within international trade by granting privileges such as the restrictions on other merchants entering its trade and the ability to export bullion (see chapter 2). In justifying its claim to these privileges, the Company also depended upon demonstrating the autonomy of the world of commerce and its own pivotal role within it. None of this was fixed in the 1620s. The worlds of commerce and politics were being constructed in these debates as new ways were found to understand, write, and publish about trade. The fortunes of the English East India Company depended on the outcome of these discussions.

During the early seventeenth-century commercial crisis King James I established a series of committees and a trade commission to investigate the shortage of currency and the outflow of bullion. Their deliberations and their manuscript memoranda formed the basis for the printed books and

95. Mary Poovey (1998) *A History of the Modern Fact: Problems of Knowledge in the Sciences of Wealth and Society* (University of Chicago Press, Chicago) p. 67; and Andrea Finkelstein (2000) *Harmony and Balance: An Intellectual History of Seventeenth-Century English Economic Thought* (University of Michigan Press, Ann Arbor).

96. Joyce O. Appleby (1978) *Economic Thought and Ideology in Seventeenth-Century England* (Princeton University Press, Princeton, NJ).

pamphlets that carried the debate to a wider audience.[97] Shifting from script to print, however, meant constructing justifications for writing and publishing for that wider audience. These were necessarily political positions as they attempted to secure relationships between the state, commerce, and specialist expertise in the debate over trade. Their presentation of different versions of the public realm can be seen by considering the work of Gerard de Malynes, and of two men we have already encountered: Edward Misselden and Thomas Mun.

Malynes was a merchant and an assay master of the mint. He had published on economic matters since 1601, and was one of the authors of the first committee report for King James in 1622 and a series of subsequent memoranda. He understood England's economic problem to be one of exchange rates, believing that the conspiratorial machinations of European bankers had systematically undervalued the English currency in relation to its intrinsic precious metal content. Silver coin was, therefore, flowing out of the country. His solution was direct state intervention, a royal proclamation that all exchanges should be conducted at par. The king would enforce the rule that metallic content was value.

This absolutist political economy was given public form by his use of print and it justified his public pronouncements. Malynes fulsomely dedicated his main printed work, the *Lex Mercatoria* (1622), to the king, and pointed out in his epistle to the "Courteous Reader" that merchants are "the meanes and instruments" to make kingdoms and commonwealths flourish "to the Glorie, Illustration, and Benefit of their Monarchies and States."[98] He also dedicated pamphlets on the commercial crisis to the king and to Prince Charles.[99] Doing so put the monarch, and his law, right at the heart of questions of trade and commerce:

> The state of the Monarchie must needes be the Supreamest thing under the cope of Heaven, when Kings are not only Gods Lieutenants upon earth and sit upon his throne; but also are called Gods, by God himselfe, in regard of their *Transcendent Preheminences and Prerogatives*, whereby they maintain *Religion* and *Justice*, which are the onely

---

97. Barry E. Supple (1964) *Commercial Crisis and Change in England, 1600–1642: A Study in the Instability of a Mercantile Economy* (Cambridge University Press, Cambridge).

98. Gerard de Malynes (1622) *Lex Mercatoria, or The Ancient Law-Merchant* (London, printed by Adam Islip) sig. [A5ᵛ].

99. Roger Chartier (1995) "Princely patronage and the economy of dedication," in *Forms and Meanings: Texts, Performances, and Audiences from Codex to Computer* (University of Pennsylvania Press, Philadelphia) pp. 25–42.

true supporters and fundamental stayes of all Kingdomes and Common-weales, so naturally united and conjoyned, that where both of them are not, properly there can be neither. These high *Attributes* cause their Lawes to be sacred, and consequently religiously to be observed; whereby *Justice* is administered, which is *Distributive* and *Commutative*. The *Commutative* part includeth *Traffick*, which is the sole peaceable in-strument to enrich kingdomes and common-weales, by the meanes of *Equalitie* and *Equitie*, performed especially by the *Law-Merchant* by reason of her stabilitie.[100]

For Malynes, quoting James I himself, it was the monarch's law that held mercantile practice together and enabled it to support and enrich kingdoms. Indeed, Malynes was suspicious of merchants' profit seeking.[101] He argued that "Kings and Princes were to sit at the sterne of Trade," and should treat merchants' advice cautiously since "they doe onely study for private benefit." Thus, in constructing a model of society, Malynes deployed con-ventional patriarchal and organic depictions of the monarch as the "father of the great familie of the common-wealth" and as the head, and therefore "the seate of judgement," in relation to the body of the people. He argued that monarchy—"a common-wealth where one sole prince hath the absolute government"—was the ideal form of state, "when . . . one person (imitating nature) dothe governe (as the head) all the parts and members of the bodie, for the general safeguard and weal publicke." This safeguarding was to be by royal proclamation over exchange rates. Without this, he argued, the East India Company's trade would remain "unprofitable for England" because bullion could only be imported into the country at a loss and would not "increase treasure."[102]

One of Malynes's bitterest opponents was Edward Misselden, a merchant whom we have already seen at work as the English East India Company's commissioner in the Netherlands. Although Misselden was highly critical of Malynes's arguments and of his presentation of them, they shared a com-mon scholasticism and Erasmian humanism that shaped their evidence, argumentative styles, and forms of rhetoric and logic.[103] It also shaped how these arguments were presented in print. Misselden's texts adopted the con-ventions of the humanist printed page. They were furnished with marginal

100. Malynes, *Lex Mercatoria*, sig. A3[r–v].
101. Finkelstein, *Harmony and Balance*.
102. Gerard de Malynes (1623) *The Center of the Circle of Commerce* (London, printed by William Jones) pp. 45, 70, 102, 114; and Malynes, *Lex Mercatoria*, pp. 62, 63.
103. Poovey, *History of the Modern Fact*.

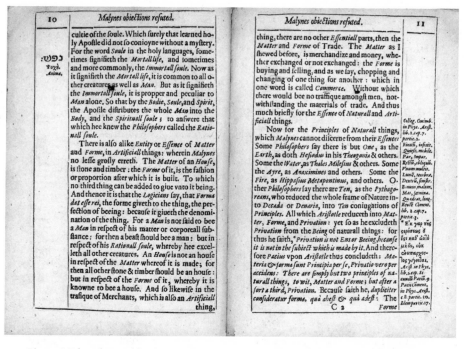

Figure 8 Edward Misselden (1623) *The Circle of Commerce.* (By permission of the British Library, 1029.b.2.)

notes in a range of ancient languages—Latin, Hebrew, and Greek—and with references to ancient sources and commentaries by scholars such as Ramus and Bodin (Figure 8).[104] These authorities underpinned a political economy based on the centrality of the balance of trade, which was different from Malynes's concentration on exchange rates, but also contained a notion of the monarch's crucial role in managing this balance.[105] Thus, monarchical understanding—the king "as the *intellectual part* of this *Microcosme*"— was to be combined with mercantile expertise in "the *Parliamentation* and *Consultation* of all the *Parts* together about these *Causes* and *Remedies*."[106] This was to be undertaken, as by Misselden in publishing his works, in the service of a wider public.

104. Evelyn B. Tribble (1993) *Margins and Marginality: The Printed Page in Early Modern England* (University Press of Virginia, Charlottesville).

105. Finkelstein, *Harmony and Balance.*

106. Edward Misselden (1622) *Free Trade. Or, the Meanes to Make Trade Flourish* (London, printed by John Legatt, for Simon Waterson) pp. 4–5.

Misselden dedicated his pamphlet *Free Trade* (1622) to Prince Charles and his *The Circle of Commerce* (1623) to the lord treasurer. Through these dedications, and in his arguments, Misselden constructed "the kingdom" as the public realm with which he was concerned and for which he published for both "the honour of the King, and service of the publique."[107] In his dedication to Charles he noted that the king "hath carried a quick Eie, over the Commerce of this Kingdome: because it hath relation both to the Revenue of the Crowne, and the Common-wealth of all His Kingdomes."[108] This was echoed in his final evocation of the balance of trade in *The Circle of Commerce* as a great glass globe, like the one made by "Sapor King of Persia" to view "the motions and revolutions of the Starres, rising and falling under his feet: as if he that was a mortall man, would seeme Immortall." As he put it, "surely if a King would desire to behold from his throne, the various revolutions of Commerce, within and without his Kingdome, he may behold them all at once in . . . this Globe of glasse, *The Balance of Trade*." This "balance"—an accounting of imports and exports—would be constructed for the monarch through the knowledge and expertise of merchants like Misselden, but it would be the king, as "The *Royall Merchant, the Regall Father of that great family of a Kingdome*," who would make the decisions, including those over the East India Company. More fundamentally, for the Company the decision over which trades should be "governed"—and therefore "invested with part of the KINGS *Honour*"—should be on the basis of their "*Publique Utility*."[109]

Many of these arguments were explored more thoroughly by Thomas Mun, the East India Company merchant who would later be employed to see the *True Relation* through the press. In many ways his writings worked from a set of premises similar to those of Malynes and Missenden, but there were also significant shifts in writing style, alterations in the uses of print, and changes in the terms of debate that would become increasingly influential in subsequent discussions of trade.[110] Mun's style, evident in his *A Discourse of Trade from England Unto the East Indies* (1621), ran contrary to the agonistic humanism of Malynes and Misselden. While the layout of his printed page remained the same as Misselden's, Mun used a plain and declarative mode of writing quite different from the philosophical

---

107. Edward Misselden (1623) *The Circle of Commerce. Or the Balance of Trade, in Defence of Free Trade* (London, printed by John Dawson for Nicholas Bourne) p. 2.

108. Misselden, *Free Trade*, sig. A3^r.

109. Misselden, *Circle of Commerce*, pp. 130–31, 142; Misselden, *Free Trade*, pp. 53, 67.

110. Poovey, *History of the Modern Fact*, highlights the changes while Finkelstein, *Harmony and Balance*, concentrates on the continuities.

arguments of the humanists. He also eschewed lengthy marginal citations of other sources in favor of summaries of his own arguments (compare Figure 8 and Figure 9). In this intervention into the ongoing debate over the East India Company, Mun clearly declared his authorial position and his intention to make his views public on the basis that "the clamorous complaints against the same, are growne so loude and generall, that (my selfe being one of the Society) it hath much troubled my private meditations, to conceive the means or true grounds of this confusion." His address, and his pamphlet came without any prefatory dedications whatsoever, was to a broad anonymous audience:

> That so these mis-understandings and errours may bee made known unto the whole body of this Kingdome, which at this present time is most worthily represented in those noble assemblies of the high Court of *Parliament;* where I hope the worth of this rich Trade, shall be effectually inquired, and so in the end obtaine the credit of an honourable approbation.[111]

What he presented was a defense that included many of the issues that Dudley Digges had addressed in 1615: the exhaustion of timber supplies, the mortality of mariners, and the lack of provision for their bereaved families. Mun answered these in familiar terms. What was new was his attempt to make a coherent case for the Company in relation to the accusation that its exportation of bullion damaged the kingdom. In doing so he foregrounded the mechanisms of supply and demand rather than the royal regulation of trade. Proceeding through the comparison of lists of prices, he argued that, while remaining well within the limits on bullion exportation, the Company was lowering prices for a wider range of East Indian goods, and importing increased levels of silver via the re-export trade to Europe. Moreover, his accounting of exports and imports from and into "the Kingdomes stocke" allowed him to differentiate this "balance" from the profits of merchants. He reasoned that mercantile profits were dependent upon the costs of moving merchandise—customs, wages, insurance, victuals, shipping—that played no part in increasing or decreasing the kingdom's balance of trade. While at this stage Mun offered a variety of reasons for the outflow of bullion, some of them agreeing with Malynes, his principal concern was still the relationship between imports and exports and his main recommendations for "us all in

111. Thomas Mun (1621) *A Discourse of Trade From England Unto the East Indies,* 2nd ed. (London, Printed by Nicholas Okes for John Piper) p. 4.

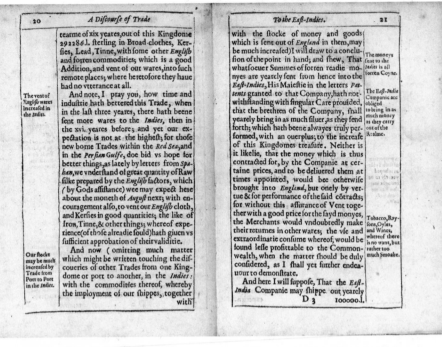

Figure 9 Thomas Mun (1621) *A Discourse of Trade*. (By permission of the British Library, 1029.c.28.)

generall, and every man in his particular" lay with encouraging exports "by industrie, and encrease of Arts" and decreasing imports by curbing "our common excesses of food and rayment."[112]

Mun's arguments were sharpened in the debates of 1622 and 1623. He was an important member of the committee (which also included Robert Bell and John Skinner) that criticized and countered Malynes's recommendations and responses. These manuscript memoranda formed the basis for his *England's Treasure by Forraign Trade*, where he argued that the only significant issue was the balance of trade.[113] That was what determined exchange rates and supplies of treasure, since where imports "overbalanced" exports, the difference would have to be paid in bullion. This concentration on the balance of trade emphasized the importance of merchants, who Mun called *"The Steward of the Kingdoms Stock,"* and the necessity of taking their advice and expertise seriously. In what appears to be a criticism of Malynes for emphasizing mercantile untrustworthiness, Mun argued that "[i]t

112. Mun, *Discourse of Trade*, pp. 44, 55, 56.
113. Supple, *Commercial Crisis and Change*.

is ... an act beyond rashness in some, who do dis-enable their [merchants']
Counsel and judgment (even in books printed) making them uncapable of
those ways and means which do either enrich or empoverish a Common-
weal, when in truth this is only effected by the mystery of their trade."[114]
While there is some ambiguity here as to whether the printed books are
works of mercantile counsel or those that disavow it, there is a clear sense
that this rejection of merchants' books or of merchants in books, with the
associated claims to veracity and the attention of a broader audience, is even
more rash than other forms of disapprobation. For Mun, more so than for
Malynes, there was a stronger differentiation between manuscript memoranda for a royal trade commission and printed works on trade. For Malynes,
both forms addressed the central figure of an absolutist monarch. For Mun,
they addressed a complex terrain of public and private interests in different
ways.

Malynes identified king, commonwealth, and merchants as a single,
albeit differentiated, body (or family) unified by the God-given power of
the monarch. Mun disaggregated their interests, but sought to bring them
together in a conscious and chosen recognition of their potential mutual
benefits and responsibilities. The commonwealth is interested in the balance of trade, merchants are interested in their profits, and monarchs are
interested in taxation. Gains in any one of these could coincide with losses
in the others, although they could also, with careful attention, reinforce one
another. Thus, merchants' "Commerce with other Nations" was "a work of
no less *Reputation* than *Trust*, which ought to be performed with great skill
and conscience, that so the private gain may ever accompany the publique
good."[115] This meant emphasizing "the plainness of our dealing" instead
of the *"Conjuring"* that Malynes argued characterized the mechanisms for
honoring merchants' bills of exchange. Indeed, Mun accused Malynes, and
his style of writing, of this bad faith. Malynes was guilty of "cunning delu-
sions which might deceive the unskilful Reader" and of having "disguised
his own knowledge with Sophistry to further some private ends by hurting
the publick good."[116] In contrast, Mun's texts would be plain and trustworthy, a fact as evident in their typographical avoidance of the trappings of
scholastic erudition as in their plain language and appeal to the authority of
number.[117]

114. Thomas Mun (1664) *England's Treasure by Forraign Trade* (London, printed by J. G. for
Thomas Clark) pp. 1, 3.
115. Mun, *England's Treasure*, p. 1.
116. Mun, *England's Treasure*, pp. 45, 48.
117. Poovey, *History of the Modern Fact*.

As a result, Mun's printed pamphlets addressed themselves not to the monarch but to a wider audience, whose interests constituted the semi-autonomous realm of commerce and could be reconciled only through political engagement rather than royal fiat. He emphasized the importance of parliaments, which he called "an excellent policie of Government, to keep a sweet concord between a King and his Subjects, by restraining the Insolence of Nobility, and redressing the injuries of the Commons, without engaging a Prince to adhere to either party, but indifferently to favour both." He also reemphasized the problems of idleness and overconsumption that determined the balance of trade, "the general leprosie of our Piping, Potting, Feasting, Fashions, and misspending of our time in Idleness and Pleasure (contrary to the Law of God, and the use of other Nations)." For Mun, the level of bullion was not so much a matter for kings, but for merchants, producers, and consumers, since "so much Treasure will only be brought in or carried out of a Commonwealth, as the Forraign Trade doth over or under balance in value. And this must come to pass by a Necessity beyond all resistance." While these arguments were less specifically tied to a defense of the East India Company, they could clearly be used to support it if it could, as Mun believed, be shown to be profitable to the nation. Indeed, he referred to the East India trade as "this great and noble business, so much importing the Publique wealth, Strength and Happiness."[118]

These were significant debates in the construction of the realms of trade and politics, and the relationships between them. The contributions to them were important in construing and highlighting different and competing versions of the relationship between public and private interests. In doing so the authors presented different versions of collective address, emphasized through, and in part secured by, the politics of dedication and the forms of knowledge implied in their typographic and stylistic conventions. In particular, Malynes's absolutist and humanist political economy contrasts to Mun's moralized "balance of trade," where the responsibility for the "Common-wealth" lies in the hands of every producer and consumer in the nation. Although Mary Poovey has argued that it is Mun's work that foreshadows important developments in the understanding of trade and wealth, there was no secure way for readers to decide that he was right and others were wrong.[119] Moreover, it is significant that his pamphlet *England's Treasure by Forraign Trade* was not published until 1664. By that time

118. Mun, *England's Treasure*, pp. 11, 66–67, 73, 87.
119. Poovey, *History of the Modern Fact.*

the conditions within which debate in print over the East India trade were conducted had changed dramatically.

Print and Company Politics in the Late Seventeenth Century

England's mid-seventeenth-century explosion of printing meant that more publications were issued between 1640 and 1660 than in the previous century and a half. This was primarily due to the role of print in the religious and political conflicts of the 1640s, particularly the ways in which printed material became the mechanism for invoking and expressing public opinion. Thus, after 1650 the prevalence of print became associated with coffeehouse discussion, newspaper circulation, and political and religious dissent. While there were continuities with scribal cultures, it can be argued that print gave new form to political debate and established a relationship between itself and Parliament based on appeals to public interest and public opinion.[120]

There was also a proliferation of printed material concerning the English East India Company. Pamphlets, petitions, broadsides, and single-sheet tracts, as well as books and items in the newspapers, were published by, for, and against the Company in increasing numbers as the seventeenth century wore on. Many of the concerns were the same as in previous decades, particularly the question of the beneficial or detrimental effect of the East India trade. Others were more specific. The organization of the East India Company as a limited joint stock, and the restrictions that placed on other merchants, was questioned not only by interlopers at sea but also at law, in front of the king, and in Parliament. Legal challenges over the Company's monopoly in the 1680s became debates over whether it was the king or Parliament that had the power to restrict trade. These conflicts were replayed in the 1690s over the vexed question of the renewal of the Company's charter. These concerns, and the value of the trade to the nation, fed directly into debates over the bill to restrict the importation of cloth from India, and the division of the trade between the Old Company and the New Company between 1698 and 1709, which, incidentally, pitted Streynsham Master against William Langhorn once again. In each case, these differences—differences that could mean the end of the East India Company, or a particular incarnation of it—were fought out in print, through print, with print, and over

120. Zaret, *Origins of Democratic Culture*; Joseph Monteyne (2000) *The Space of Print and Printed Spaces in Restoration London, 1669–1685,* unpublished Ph.D. thesis, University of British Columbia; and Markman Ellis (2004) *The Coffee-House: A Cultural History* (Wiedenfeld & Nicholson, London).

print, in ways that sought to make and shape the debate both inside and outside the houses of Parliament. Print, it will be argued, became part of the politics of the East India Company much more directly than ever before.

One somewhat idiosyncratic example can serve as an entry point. *England's Almanack* (1700) was a single sheet printed only on one side, and crammed with textual, numerical, and visual information and arguments that the East India trade should be restricted to preserve the woollen manufactures (Figure 10). First, its form raises the question of how printed materials like this were used at home, in the coffeehouse, or even in Parliament. Second, its content drew upon, reproduced, and mimicked other textual forms—the almanac, the newspaper, shipping lists—quoting from and referring to other printed texts as it did so. Finally, the *Almanack* sought to use the power of print to establish the credibility of its arguments, by presenting the facts of the case in tables of shipping and bullion, in stories from the news, and in close readings of the authors it opposed. In short, documents like this one were part of a flourishing, differentiated, and vibrant world of print. This, it will be argued, shaped the nature of the political conflicts of which they were a vital part.

The Uses of Print

The expansion of printing meant that printed texts were anything but homogeneous. Printed materials were produced and used differently in different contexts, and their use must continue to be understood in relation to script, speech, and gesture, and in relation to the places and purposes that those supporting or challenging the East India Company intended them for. While print was sometimes used as a mechanism through which Company business was organized, it proliferated most readily in the political arena, and particularly where the Company's future and the future of the East India trade were debated in Parliament.[121] That was the context that mattered most. It affected when the Company itself went into print. For example, in 1681 the Company published a work on the East India trade by its governor Josiah Child (Figure 11), the man to whom Robert Boyle had had to make solicitations regarding the vacancies for chaplains (see preface). This treatise, published under the pseudonym "Philopatris" and arguing that the Company pursued "the most national of foreign trades," appeared as part of debates among the adventurers, in the law courts, and in Parliament about

---

121. For example, printed notifications of the dates for adventurers to pay subscriptions, in BL, *Petitions,* Cup.645.b.11.(24**).

whether the trade should be opened to other merchants. Debates where, as Child saw it, "the Company was assaulted by the private designs of particular men."[122] Indeed, one of the crucial issues at stake in the 1683 test case, the prosecution of Thomas Sandys (or Sands) for interloping on the Company's privileges, was whether it was the king or Parliament that had the right to place restrictions on trade. Although Judge Jeffreys ruled, unsurprisingly, that any usurpation of the king's prerogative in this respect was tantamount to being a rebellious act, the late seventeenth-century debate over the East India trade increasingly shifted, although never wholly, from the monarch's court and the law courts to the houses of Parliament.[123]

Much printed material relating to the East India Company was, therefore, produced directly for a parliamentary audience. One pro-Company pamphlet complained that "their Enemies have endeavoured of late to impose, not only upon the unthinking Vulgar, but upon the Members of the Honourable House of Commons, by putting Papers in their hands; wherein besides the scurrility of the Language sutable to the Authors, there are many things notoriously untrue."[124] For example, thirteen articles detailing "Abuses and unlawful Practices" by the Company were "delivered to the Members of the Honourable House of Commons," and, in turn, answers to each were "given in Print" and "humbly submitted to the Judgement of those Honourable Members of that House."[125] This meant matching the form to the readership. As one author noted in producing a summary of his work, "there are many Persons will read one side of Paper, who will not read so much as is before writ of this Subject."[126] The presses on both sides therefore issued forth many single sheets, printed on one or both sides, with

---

122. Philopatris [Josiah Child] (1681) *A Treatise Concerning the East India Trade* (London, printed by J.R. for the Honourable the East-India Company) p. 41.

123. *The Argument of the Lord Chief Justice of the Court of King's Bench concerning the Great Case of Monopolies, between the East-India Company, Plaintiff, and Thomas Sandys, Defendant* (London, printed and sold by Randall Taylor, [1689]); and *A Journal of Several Remarkable Passages, before The Honourable House of Commons, and the ... Privy Council: Relating to the East India Trade* [1693?].

124. *A Modest and Just Apology for; or Defence of the Present East-India-Company* (London, 1690) p. 1.

125. *A Brief Account of the Great Oppressions and Injuries which the Managers of the East-India Company have acted on the Lives, Liberties, and Estates of their Fellow-Subjects* ... [1691?] [BL Cup.645.b.11.(25)]; *Some Equitable Consideration, Respecting the Present Controversie between the Present East India Company, and the New Subscribers or Petitioners Against Them* [1691?] [BL Cup. 645. b.11. (24***)] p. 1; and *A Reply on Behalf of the Present East-India Company, to a Paper of Complaints, commonly called, The Thirteen Articles* ... [1691] [BL Cup.645.b.11. (26)].

126. *To Preserve the East-India Trade* (London, printed by Freeman Collins, 1695) p. 4.

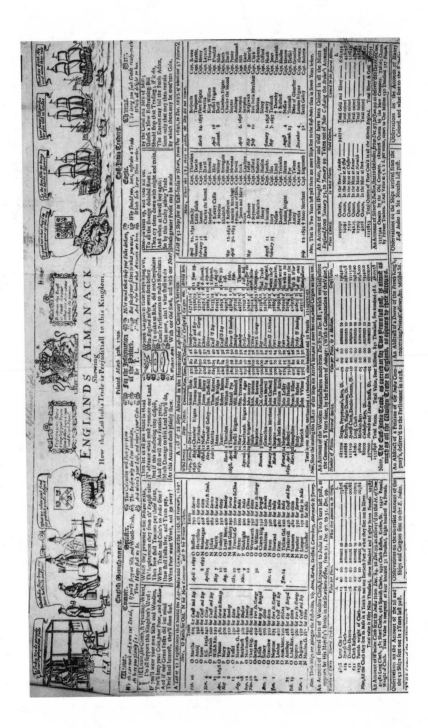

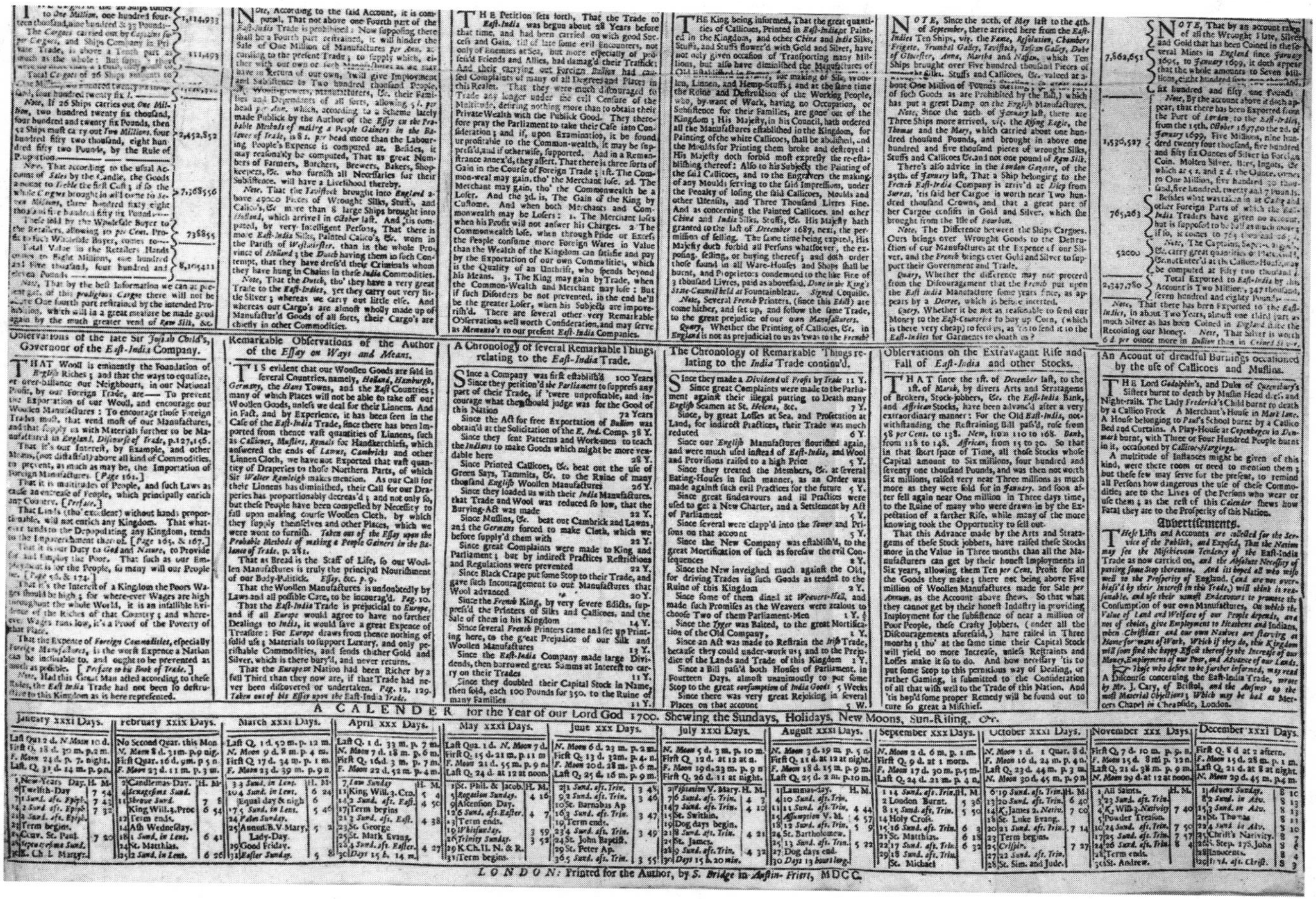

Figure 10 *England's Almanack* (1700). (By permission of the British Library, 816.m.11.(92).)

S.ʳ Josiah Child. Bar.

Published as the Act directs April 11798, by W. Richardson Castle Street Leicester Square

Figure 11 Sir Josiah Child. (By permission of the British Library, Oriental and India Office Collection, P1461.)

arguments, evidence, or suggested clauses and amendments. These were ready for use. Some were printed so that they could be folded to a pocket size, with the title still visible. They were intended to be portable, pocketable aids to action in specific circumstances. They did not stand alone as printed texts, but were to be mobilized within Parliament's distinctive oral culture of speech making and debate. They were performative texts meant to be consulted, brandished in anger, and waved in agreement, and if they were dropped on the floor of the House after the debate then no matter.[127]

Longer works were also aimed at a parliamentary audience, including accounts of materials presented before Parliament and the proceedings of both houses.[128] These were the products of scribal cultures that involved print, but did not simply rely upon it. One particular instance is the legal tradition, most evident in the Sandys case, of the close reading of manuscript and printed sources to determine in various ways their credibility as legal authorities in arguments for and against the restriction of trade.[129] More generally, there was a recognition that print served particular purposes. While it is necessary to be cautious about statements of a reluctant entrance into print, Josiah Child noted in 1668 that he had written his *Brief Observations Concerning Trade and Interest of Money* "not then intending to publish it, but onely to communicate it to some honourable and ingenious Friends of the present *Parliament*, who were pleased to take Copies of it for their more deliberate consideration and digestion of the Principles therein asserted." Even when it was printed, one reason was so that copies might better be distributed to members of the house.[130] Indeed, Parliament was full of readers and writers whose deliberations on the India trade mixed print, script, and talk. For example, it is not clear whether the East India papers of Francis North, lord keeper of the great seal, which his brother found among "Vast quantitys of materialls, viz., notes, memorandums, reasonings, etc" all in

127. There are many examples in *Tracts Relating to Various Trading Companies* [BL 816.m.11]; and *Petitions* [BL Cup.645.b.11].

128. For example, *Reasons Humbly Offered Against Grafting or Splicing, and for Dissolving this Present East-India Company* (London, 1690); *The Answer of the East-India Company, to Two Printed Papers of Mr. Samuel White* (no date); and *A Journal of Several Remarkable Passages*.

129. *Argument of the Lord Chief Justice*. Sandys's counsel, John Polloxfen, dismissed textual authorities on various grounds including being printed during the interregnum and "not licensed by any Judge or Person whatsoever." *The Argument of a Learned Counsel, upon an Action of the Case Brought by the East-India-Company Against Mr. Thomas Sands, an Interloper* (London, printed for B. Aylmer, 1696) p. 54. Others highlighted the misleading differences between manuscript and printed laws. See, for example, W.A. (1690) *An Apology for the East-India Company* (London, printed for the author).

130. Josiah Child (1668) *Brief Observations Concerning Trade and Interest of Money* (London, printed for Elizabeth Calvert and Henry Mortlock) p. 17.

Francis's hand, "and those some no better than incoherent scrapps of paper, and entrys in little books, and crouded into odd corners," were destined for the press or ever reached it. One, clearly prompted by his reading, was "An essay of the East India Company shewing how it is degenerated into a monopoly, and not fitt to be continued without regulations opposed to the arrogant disputes and pamflets of Sir Josia Child." There were also "upon a loos paper some further doubts and considerations concerning the East India Company." Rough notes, published materials, and drafts that might find their way into print jostled together as part of deliberations on the future of the Company. These writings, drawn together in a gentleman's closet or library, were, as Roger North put it, the "thoughts and deliberation his Lordship had with himself, while this great matter was depending."[131]

All of this meant, of course, that printed material, and with it debate and discussion, could not be restricted to Parliament. Again, the Company's supporters complained of what their rivals were doing: "What great indefatigable industry hath bin imployed? What Arts and Devices made use of, to blast the Reputation of the present *East-India*-Company, is notoriously evident to all, who either give themselves the trouble of listning to those Calumnies, dayly inculcated in all noted Coffee-Houses against them, or to the reading of those Prints exposed publickly, and delivered *gratis* in the said Coffee-Houses, to all such as will but accept them."[132] The late seventeenth-century connections between print, politics, and the coffee-house are well established. What they meant for the Company was that news, and the relationship between coffeehouse talk and print, became a constant commentary on its actions.[133] Yet the proliferation of print, and of places in which it might be read, discussed, and argued over, meant that the Company could not respond as it had done in the early seventeenth century. For example, in 1698 the Company's attempts to find out the source of a story in the *Post Boy* newspaper that an East India Company ship had turned pirate and taken a ship belonging to the Secretary of the Mughal emperor were halted when they were told by its author that "it was the common discourse of the Coffee House."[134] All they could do was demand he print a

131. Mary Chan (ed.) (1995) *The Life of The Lord Keeper North by Roger North* (Edwin Mellen Press, Lampeter) pp. 474–76, 484.

132. *Modest and Just Apology*, p. 1.

133. Stephen Pincus (1995) "'Coffee politicians does create': coffeehouses and restoration political culture," *Journal of Modern History*, 67, pp. 807–34; for a sense of jokes between insiders put into print, see *An Elegy on the Death of the Old East-India Company* (London, printed for the author, 1699).

134. IOR B/41 *East India Company Court Minutes: 24th April 1695 to 26th April 1699*, p. 350 (25 November 1698). I am grateful to Philip Stern for this reference.

retraction. Indeed, the only response to the proliferation of print was more print.

Printing and Reprinting

One characteristic of printing, and of printed pamphlets in particular, was the recycling of materials as pamphlets absorbed each other and responded to other texts. As with the religious and political pamphlets and the printed petitions of the 1640s, the rhetorical strategies adopted in the printed materials generated around the East India Company in the late seventeenth century depended on the presumption of the reliable reproduction of other texts to which they replied.[135] The Company's supporters and detractors quoted extensively from each other—and especially, as *England's Almanack* shows, from the influential works of Josiah Child, Charles Davenant, and John Cary—as they challenged particular positions and constructed others. They reprinted older pamphlets, including those on Amboina, as part of ongoing debates about economics and politics, which deployed them in changing contexts.[136] Finally, pamphlets included printed transcriptions and translations of other documents, and parts of documents, including statutes, charters, proclamations, a declaration of the French king against imported cloth, deliberations between the Dutch and English companies, and portions of Company letters.[137]

These were, as the preface to a translated reprint of an "Unjust and Slanderous" Dutch treatise put it, presented for "the Judgement of the unbiassed Reader."[138] Readers of works on the Company in the 1680s and 1690s were presented with these textual sources and invited to draw their own conclusions. Or, more precisely, they were invited to confirm that the evidence matched the interpretation offered by the author. So, the manuscript commissions granted by Elizabeth I and James I to the generals on the Company's early voyages (see chapter 2) gained a new lease of life in printed form eighty years later. In 1683 and 1684 the Company's governor on St. Helena used a

135. Raymond, *Pamphlets and Pamphleteering*; and Zaret, *Origins of Democratic Culture*.

136. *The Emblem of Ingratitude* (London, printed for William Home, 1672) and *Insignia Bataviæ; or, the Dutch Trophies Display'd* (London, printed for Thomas Pyke, 1688) were both reprints of *A True Relation* (1624). *A Collection of Papers Relating to the East India Trade: Wherein are shewn the Disadvantages to a Nation, by conjoining any Trade to a Corporation with a Joint-Stock* (London, printed for J. Walthoe, 1730) speaks for itself.

137. *Reasons Humbly Offered*, pp. 33–36; *An Impartial Vindication of the English East-India-Company* (London, printed by J. Richardson for Samuel Tidmarsh, 1688); and *Proposals for Setling the East-India Trade* (London, printed and sold by E. Whitlock, 1696) p. 14.

138. *Impartial Vindication*, title page.

commission of martial law to put down violently a series of what he saw as rebellions among the island's smallholders as they tried to resist attempts to tie them into a plantation economy. After this act of "Arbitrary Government" and its presumption of "a Despotick Power and Sovereign Authority" came to light and was condemned as illegal in Parliament, it was seized upon by the Company's critics, particularly after 1688, who used it to damn the Company for its associations with James II, and to call for a reorganization of the trade.[139] The Company's defenders stressed the need for martial law in the government of their colonies. The abstractor of Josiah Child's work referred readers to *"Purchas* his *Pilgrims,* the first part, printed in the Year 1625. wherein by the course of the *History* he will find that *Martial Law* is more necessary in *India,* than Bread is to the support of Mans Life; and the *East-India Company* had constantly *Commissions* from the *Crown* for that purpose in the blessed time of *Queen Elizabeth,* and during all the peaceable Reign of King *James* the first."[140] Another supporter went further, arguing that while the Company's actions were not "strictly justifiable," they were "fairly excusable" on the basis that "their present Charter, and the Kings particular Commission, and the Customs and Usages of their Predecessors, sufficiently attested by several Original Commissions of Martial Law lying by them, seemed to warrant them." In support of this, the author reprinted Elizabeth's commission for James Lancaster and referred to another from James I.[141]

Any such text was, however, open to different readings. At the end of the Company's ill-advised war with the Mughal empire in the late 1680s, its supporters and detractors reproduced and debated the *farmān* that the emperor had granted. George White, an old adversary of the Company, reprinted the text and encouraged readers to "see the Honour of his Mother Country expos'd to this Contempt and Ignominy, and our common Name stigmatiz'd with such opprobrious Epithetes."[142] In *News from the East Indies* (1691) "An Exact Copy of the *Phirmaund* for the Bay of *Bengale"* was presented as "the Instrument that stands upon Record in the Court of that Great Prince, to the Eternal Infamy of the English Nation."[143] The Company's supporters

---

139. *A Brief Abstract of the Great Oppressions and Injuries which the late Managers of the East-India-Company Have Acted on the Lives, Liberties and Estates of their Fellow-Subjects* (1698?), pp. 1, 4.

140. *A Discourse Concerning Trade, and that in particular of the East-Indies* (London, printed and sold by Andrew Sowle, 1689) p. 11.

141. W.A., *An Apology,* p. 7.

142. George White (1691) *An Account of the Trade to the East-Indies* (London) p. 8.

143. *News from the East-Indies* (1691) [BL 816.m.11(77) recto].

offered other interpretations. It was argued that these critics had misunderstood the *farmān*'s language. Far from being "a base and dishonourable Peace," it was "well known to all, who have observed the stile of the Eastern Princes, That it is no other, than what the said *Mogol*, and others of the Great Eastern Monarchs generally use, in their Treaties, they make with their Neighbouring Princes."[144] As for the terms of the *farmān*, Josiah Child himself argued "that if a Blank had been delivered to the Company in *England*, to write down their own Terms, They would not have desired more than is granted by the said Articles." In support of this he reprinted, as a postscript, the letter from the Mughal governor of Surat to John Child, stressing that the *farmān* is "much for your Profit."[145] Yet none of this could determine the readings that were made as contending parties made their "appeal to the reason of every impartial Man."[146] Indeed, the contexts in which those readings were made, and the other knowledge that was brought to bear, shaped interpretations of these texts. With the East India Trade at a very low ebb, and with the Company under attack, readers were preparing their responses to Child's bullish interpretation of the *farmān* even before it appeared. As the anonymous letter writer who kept John Ellis, the secretary to the commissioner of the king's revenue in Ireland, abreast of metropolitan developments reported in 1688, "Sir Josiah Child is preparing a Relation of his late success against the Great Mogoll, yet it is thought his rhetoric will scarce longer gain belief."[147] Indeed, part of the issue here was Child's own reputation, and the problems of his credibility as both a merchant and a commentator on the politics of trade.

Print and Authority

Josiah Child's problem was an acute version of that faced by all authors, booksellers, and readers when it came to constructing, publishing, and judging accounts of trade. There was a general recognition, voiced by Thomas Mun but traceable to classical sources, that the gain of any particular merchant was not necessarily the same as the gain of the kingdom or nation as a whole. As Child (as "Philopatris") himself argued, citing Cicero and Bodin, "Trading Merchants," while involved in business, "are not always the best

144. *Reply on Behalf of the Present East-India Company*, p. 3.

145. [Josiah Child] (1689) *A Supplement... to a former Treatise, concerning the East-India Trade* (London) pp. 9, 14.

146. *Modest and Just Apology*, p. 23.

147. George J. W. A. Ellis (ed.) (1829) *The Ellis Correspondence*, 2 vols. (Henry Colburn, London) vol. 2 pp. 103–4.

judges of Trade, as it relates to the profit and power of a Kingdom," since their eyes are so fixed "upon what makes for their particular Gain or Loss." The same went for shopkeepers, clothiers, and manufacturers "until they leave off their Trades, and being Rich, by the purchase of Lands, become of the same common Interest with most of their Countrey-men."[148] Yet while this might be accepted in general, there was no consensus over the particular situations in which private and public interests clashed. Indeed, this was the very basis of the seventeenth-century disputes in print and Parliament over the English East India Company.

In debates over the costs and benefits of the East India trade, over the form that an East India Company should take, over the granting of charters, and over the bill to prohibit the importation of printed calicos, each side claimed to be speaking for the public, while decrying the often hidden private interests of their adversaries. For example, those opposed to the East India Company as a joint stock argued that its monopolizing tendencies, and the gathering of powers within it into a few private hands, could not be in the national interest, as the Company exported bullion, distorted prices, and excluded other traders. Opening up the trade would be in the national interest.[149] Those arguing for the East India Company countered that freeing up the trade would mean that individuals seeking their own short-term profits would be reluctant to invest in the forts, factories, and diplomacy necessary for commerce in the Indies. Doing so would lose the nation's most profitable trade to the Dutch, who actively supported their own East India Company.[150] Thus, King James II's 1685 proclamation against interlopers referred to "several ill disposed Persons preferring their particular Gains before the Profit and Reputation of the Nation."[151]

This debate took its most extreme and personalized form in the reaction to the printed works of Josiah Child himself. The merchant's rapidly amassed personal fortune, his dominance of the East India Company, and the rumors of his influence at the Stuart court shaped impressions in print, script, and person. John Ellis's correspondent reported "a great grumbling in the City against a certain great East India merchant, whose first name rhymes with Goliah." George White's anti-Company pamphlet also

---

148. Philopatris [Child], *Treatise Concerning the East India Trade*, pp. 1, 2.

149. [John Polloxfen] (1697) *A Discourse of Trade, Coyn, and Paper Credit* (London, printed for Brabazon Aylmer); and White, *An Account of the Trade*.

150. Josiah Child (1693) *A New Discourse of Trade* (London, printed and sold by John Everingham) pp. 80–81; *Some Considerations Offered touching the East-India Affairs* [1698], pp. 1–2; and *A Journal of Several Remarkable Passages*, p. 15.

151. *Proclamation: James II. East India Company* (1 April 1685).

presented him as the "GREAT GOLIAH," one who "raises his *Own Family* on the BLOOD and RUIN of his Neighbours." John Evelyn, who visited Child's newly acquired Wanstead estate to see the tree planting, described him, as was noted in the preface, as "most sordidly avaricious &c:" and as one of the "over growne & suddainly monied men."[152] His power within the Company was such that opponents raised the accusation that its whole administration was in the hands of ten men, led, if not controlled, by Child, "Whence a *Particular* Interest is carried on under the name of a *Publick*."[153] From this sprang accusations of stock jobbing and corruption (see chapter 5), which shaped how his works were read. As one critic of his *Discourse on Trade* wrote, it "shewed the World how difficult a Thing it is for some Men to be impartial, where private Interest comes to stand in Competition with the publick."[154]

What this meant was that the question of trade and how to organize it, which was a matter for the king and Parliament as well as for merchants themselves, was always a problem of knowledge. Epistemological questions about how trade was to be known about could not be separated from questions of its profitability (personally or publicly), or from political questions of how trade was to be regulated. Positions or decisions on matters of "public interest" could not be adopted without entering into disputes over how that interest could be defined—and therefore questions such as whether bullion was just another commodity or the only significant element of national wealth—and how a nation's balance of trade could be effectively calculated.[155] In all of this the key problem was that of merchants' knowledge and how to evaluate it. Merchants had established themselves as the experts on trade, but they were suspect in that they lived from it too.[156]

The problem was, it seems, exacerbated when the discussion appeared in print, with all its associations with the expression and formation of public opinion. As one anti-Company pamphleteer wrote, as if to a member of

152. Ellis, *Ellis Correspondence*, vol. 2 pp. 119–20; George White (1689) *A Letter to Mr Nathaniel Tenche*...(London) p. 8; and John Evelyn (1955) *The Diary of John Evelyn*, E. S. de Beer (ed.) (Clarendon Press, Oxford) vol. 4: *Kalendarium, 1673–1689*, p. 306.

153. *An Arrest on the East India Privateer* [1681] p. 3.

154. *A Discourse concerning the East-India Trade* (1693), in Lord John Somers (1751) *A Third Collection of Scarce and Valuable Tracts, on the most Interesting and Entertaining Subjects* (London, F. Cogan) p. 201.

155. Examples are the disputes between Davenant and Polloxfen (John Polloxfen [1697] *England and East-India Inconsistent in their Manufactures* [London]) and Josiah Child's questioning of Thomas Mun ("Concerning the ballance of trade," in *A New Discourse of Trade*, chap. 9).

156. Poovey, *History of the Modern Fact*; and Steven Shapin (1994) *A Social History of Truth: Civility and Science in Seventeenth-Century England* (University of Chicago Press, Chicago).

Parliament, "In fine, Sir, 'Tis believ'd the Liberty given the Press was, That you might constantly receive full and free Information in all things relating to Publick Affairs, and the Good of the Kingdom; and that, to the best of my *Knowledge* and *Judgment*, I have given in this Case; and I bless GOD I have not writ this for Bread, or any other Interest, but as before said: And I thought that the making it more Publick than for your single Satisfaction, may not be Unacceptable or Useless."[157] As a result, the Bristol merchant John Cary was delighted when both Edmund Bohun, author of a well-known geographical dictionary and licenser of the press, and the philosopher John Locke, to whom he had sent his *Essay on Trade*, concurred that it was "Written with so disinterested an aire that no man can possibly tell where your trade lyes by it." Cary asked Bohun whether he might print this testimonial as part of the second edition of the work, and informed Locke, who had suggested the essay might be longer, that "[t]here are some things I leave in the dark, as thinking them more proper to be Spoken in the house of Commons, then made the subject of a discourse, they might bring me under the name of a projector, wch I carefully endeavour to avoid." Cary, who understood the House of Commons to be unduly governed by those "who too often examine things by the Touch Stone of their own Interests," was not unusual in seeking to shape his published works to the broader audience for print in ways that negotiated the epistemological problems of mercantile knowledge.[158]

This might be done in various ways. There were explicit statements of disinterestedness or of a concern only with the public interest. There were also expressions of the willingness to be corrected, examined, or answered so that the debate might be extended and the truth discovered.[159] These came from all sides. However, as Child recognized, such appeals could not "hinder Men from thinking their own way," and the debates were characterized by the close reading of others' texts to reveal false notions or evidence of self-interest.[160] Authorship might be manipulated to the same ends. Anonymous or pseudonymous authors could evade some charges of interest. Booksellers claimed that manuscripts had fallen into their hands accidentally or "without any Attestation by the Author," and that after due

---

157. A. N. (1699) *England's Advocate, Europe's Monitor: Being an Intreaty for Help in Behalf of the English Silk-Weavers and Silk-Throwsters* (London, printed for George Larkin) p. 56.

158. *Letters and Papers of John Cary* BL Add. Mss. 5540 ff. 57^r, 72^v.

159. For example, *Free, Regulated Trade, Particularly to India, the Interest of England* (1691?) p. 4; and [Polloxfen], *A Discourse of Trade*, pp. 166–67.

160. Philopatris [Child], *Treatise Concerning the East India Trade*, p. 27; and *Answer to a Late Tract Entituled, An Essay on the East-India Trade* (London, printed for Tho. Cockerill, 1697) p. 44.

deliberation and consultation they were publishing them "for the Common good."[161] It was also a question of style. Plainness of expression and the plain facts of the case ruled the rhetorical roost once again where merchants were concerned (see chapter 3), and Charles Davenant in particular was criticized for his "extraordinary Computations and Rhetorical flourishes" and "his Superfine Spun Linse-woollse Discourse." This questioned his grasp of mercantile matters, as a man of "Letters and Theory," whose knowledge came from reading and not from practice. It also accused him of trying to foist false notions on the public by confusing them with "fine Words."[162] Finally, pamphleteers might adopt forms that allowed the expression of "private" opinions and partisan arguments in the public sphere. There was a proliferation of printed letters—from and to barristers, MPs, persons of quality, and friends in the country—and of dialogues between contending parties, both of which allowed polarized positions to be expressed and, as it were, overheard.[163]

This negotiation of mercantile knowledge and interests in print also shaped the individuals and collectives imagined and addressed in these texts. The arguments might be presented to "any unprejudiced Person, whose Judgement is not biassed by his own private Interest."[164] However, as Child's earlier comments on merchants and gentlemen show, this could take a particular form. John Locke reminded John Cary that it was high time someone should "awaken & inform" the "country Gent" of his stake in the world of trade. Other pamphlets addressed themselves directly to "Country Gentlemen" as representatives of the national interest and preached the gospel of the free-born Englishman.[165] Indeed, it was the nation as much as the

---

161. Raymond, *Pamphlets and Pampleteering*; [Polloxfen], *A Discourse of Trade*, sig. A2ʳ; and *An Answer to Two Letters, Concerning the East-India Company* (1676) p. 1.

162. *An Answer to the Most Material Objections that have been Raised Against Restraining the East-India Trade* [BL 816.m.11(91)] p. 4]; *A True Relation of the Rise and Progress of the East India Company* [BL 816.m.11(73)] p. 2; *Reasons humbly offered on behalf of the Bill for restraining the wearing of East-India and Persia wrought Silks* [1696?] [BL Cup.645.b.11.(21)]; and *Answer to a Late Tract*, p. 35.

163. Raymond, *Pamphlets and Pamphleteering*. For example, *Two Letters Concerning the East India Company* (London, 1676); *A Letter to a Friend Concerning the East-India Trade* (London, printed and sold by E. Whitlock, 1696); *A Letter from a Lawyer of the Inner Temple to His Friend in the Country, Concerning the East India Stock* (London, 1698); *A Dialogue Between a Director of the New East-India Company, and one of the Committee for preparing By-Laws for the said Company* (London, printed for Andrew Bell, 1699); and *A Dialogue Between Jest, an East-India Stock-Jobber, and Earnest, an Honest Merchant* (London, 1708).

164. *Discourse concerning the East-India Trade*, p. 216.

165. BL Add. Mss. 5540 f. 70ʳ; *Some Reflections on a Pamphlet, Intituled England and East-India Inconsistent in their Manufactures* (London, 1697) p. 18; and *Two Letters*, p. 9.

kingdom whose collective interests were at stake. This was most explicit in arguments in the 1690s for broadening the Company to make "room for the *whole Body* of the Nation, according to their *Birthright*." There were proposals that the Company might be run by "men of considerable Fortunes, that come not in merely to raise themselves an Estate"; that it might be overseen by a council appointed by Parliament; and that shares in the Company should be proportioned to land ownership so that "the Landed-men of the Nation may have the whole Benefit of this great *East-India* Trade."[166] This, of course, never occurred. However, when the directors of the Company managed to acquire their new charter in 1693 by deploying their influence at court, its justification on the basis of being "for the Honour and Profit of the Nation" serves to demonstrate both the necessity and the mutability of such forms of collective address in the debates in print and Parliament in the late seventeenth century.[167]

Conclusion

By the 1690s, therefore, the uses of print were actively shaping the politics of the East India Company in England. In large part, the form that parliamentary politics took was as a part of the larger debate in print and through print over issues such as trade, the royal prerogative, and the rights of the free-born Englishman. Parliamentary debates and deliberations over the Company and the East India trade were conducted by politicians who had at their fingertips printed materials that could support their speech making and voting. Those who took the time to do so could gather up in their libraries or closets published texts, copies of documents, notes, and memoranda through which arguments might be constructed for the printing press, oration, or conversation. Indeed, these conversations continued in houses and coffeehouses, supported by broadsides, pamphlets, and newspapers that countered, confirmed, and extended each others' arguments. Among these texts were faithful, or presumed to be faithful, copies of other texts: opponents' treatises, legal documents, royal charters and commissions, and proclamations of foreign emperors and kings. These were presented to the public, at least to an imagined reading public, for their deliberation and decision. What they made of them was shaped by their expectations of print as

---

166. *An Essay Towards a Scheme or Model for Erecting a National East-India Joynt Stock or Company* (London, printed for the author, 1691) pp. 5, 14; and *The East-India Trade* (1693) [BL 816.m.11(81)] p. 3.

167. *Reasons Proposed for the Encouragement of all People to Under-Write to the New Subscriptions, Appointed to be Made to the late East-India Companyes Stock* (1693) p. 1.

both the medium of veracious textual reproduction and the realm of endless debate in the interests of the public. Printing implied a broad anonymous audience. It implied a voice that spoke in the public interest, highlighting the need to detect private interests and raising most acutely the problem of mercantile knowledge and the need to construct collective forms of address that could be mobilized to advantage within the politics of trade. Overall, therefore, print mattered because the English East India Company depended upon privileges that could be removed by political authorities for whom print had become the medium of politics. The Company and its opponents engaged fully in print and in the formation of parliamentary and public opinion because the outcomes were significant. The power of the press meant that print had to be countered with print.

In part this had always been the case. The English East India Company had, of course, countered *The Trades Increase* with *The Defence of Trade*, and the Dutch pamphleteering over Amboina with the *True Relation*. Thomas Mun had also challenged Gerard de Malynes with the Company's interests at heart. However, as this chapter has shown, these engagements with the printing press have to be set within a politics of influence that changed over time. Engaging with printing in the early seventeenth century was as much about seeking to restrict, curtail, and prevent the use of print by deploying diplomatic and legal power and playing the regulatory system as about pressing one's own case. Understanding Company politics in England has to involve a geography of printing that can seek out what went on in a whole range of spaces from the king's bedchamber to the street corners of Amsterdam. This means that print must always be considered alongside script and speech, which work in other spaces and travel by somewhat different circuits. It is also possible to trace historical changes here. There is a shift, albeit an incomplete one, from the dominance of the monarch and his law to the need to influence parliamentary debate. This was both a product of print and something to which those making use of the press had to respond. It is signaled, again in incomplete ways, in the shift of forms of collective address adopted within what we might see as East India Company print culture from Common-wealth, to king and kingdom, and to nation.

Print, therefore, could change what the Company was, and it could change who considered it part of their concerns. One way in which it could do so was via the power of print to "destroy the *Credit* of the *Company*."[168] This was understood not merely in terms of its political credibility, but in terms of the ways in which the value of the Company's stocks depended

168. *An Answer to Two Letters, Concerning the East-India Company*, p. 1.

upon what was known about it and its trade, and what people believed that implied for the future. On the one hand, the Company's supporters could argue that there were malign forces at work "since a great deal of Industry has been used thro' some late weekly Papers, to depreciate the Credit of the Court of Directors, and by that means to strike at our very Trade to the *East-Indies*."[169] On the other hand, as the next chapter will show, the Company's detractors believed that the greatest manipulators of information were within the Company itself.

169. *A Letter to the Chairman of the East-India Company* (London, printed for J. Roberts, 1727) pp. 3–4.

CHAPTER FIVE

Stock Jobbing: Print and Prices on Exchange Alley

The sorts of printed materials discussed in chapter 4 certainly found an active readership in seventeenth-century England. In the summer of 1682 Robert Boyle received a series of letters from the depths of rural Somerset. They were sent by the Reverend John Beale, an active member of the Royal Society and a long-term enthusiast for agricultural improvement who had been part of the Hartlib Circle in the 1650s.[1] Beale told Boyle about the works on religion, nature, and trade that he had been reading and what he had made of them. Beale approved of John Tillotson's sermons, and particularly of Dr. Parker, the archdeacon of Canterbury, and his "demonstration of the Divine Authority of the law of Nature, and of the Christian Religion: in 2 parts, printed for Mr Royston & Mr Chiswell 1681." He told Boyle that "I doe not think there was a fitter Treatise written since the Apostles age, for our Nation, in our circumstances; Since Pride, & Prodigality, Luxury, and all excesses doe abound amongst Us, to the utmost perill of the City, & of this whole Monarchy, as Mr Mun the Merchant like a wise and pious Christian hath demonstrated fully, in his 3 & 19th Chapter [of *England's Treasure by Forraign Trade*], highly approved by the wisest of our Statesmen, & therefore Licensed in Whitehall."[2] Indeed, he saw Parker's book as a defense against

---

1. Mayling Stubbs (1982) "John Beale, philosophical gardener of Herefordshire. Part 1: Prelude to the Royal Society (1608–1663)," *Annals of Science*, 39, pp. 463–89; and Mayling Stubbs (1989) "John Beale, philosophical gardener of Herefordshire. Part 2: The improvement of agriculture and trade in the Royal Society (1663–1683)," *Annals of Science*, 46, pp. 323–63.

2. Robert Boyle (2001) *The Correspondence of Robert Boyle*, Michael Hunter, Antonio Clericuzio, and Lawrence M. Principe (eds.) (Chatto and Pickering, London), vol. 5: *1678–1683*, John Beale to Robert Boyle, 1 July 1682, p. 306. Beale is referring to Samuel Parker (1681) *A Demonstration of the Divine Authority of the Law of Nature, and of the Christian Religion* (London, printed by M. Flesher for R. Royston and R. Chiswell).

the attacks on the Royal Society by "Hobbians & Stubbians, Atheists, & Scoff-
ers, Theatricall Buffoons, Blasphemers, & the Burlesque & Travesty which
explode all Religion, graces, & virtues... which are the strength & honour
of our nation." He argued that the volume should be "mentioned with
just applause" in the Society's weekly memorials "that the Kingdome may
know, that the Royal Society doe heartily espouse true & solid virtues together with serious Christianity."[3] As he had done throughout his career,
Beale sought to use printed communications to promote the conjunction of
science and religion in the national interest.

Beale's reading of natural philosophy, theology, and the work of Thomas
Mun, coupled with his understanding of the importance of print, meant
that when he had turned to the recently published serial *A Collection of
Letters for the Improvement of Husbandry and Trade* by John Houghton, a
fellow member of the Royal Society, he was horrified. This was all the more
surprising because Beale and Houghton were otherwise engaged in a common project. Houghton's *Collection* can be seen as pursuing the idea for the
promotion in print of agricultural and commercial reform that Beale had
impressed upon Henry Oldenburg, the editor of the *Philosophical Transactions*, in the 1670s. Both Houghton and Beale were Tories and strong
supporters of the monarchy.[4] They had corresponded, with Houghton acting
as messenger between Beale and Boyle, and the younger man had presented
contributions from Beale before the Royal Society.[5] Indeed, the first number
of Houghton's *Collection* had been dedicated to John Beale, stressing the
author's obligations to him and his reputation as "a great favourer of any
thing that tends to the improvement of your country."[6] Yet, as Beale read it,
the fifteenth line of the fifty-second page of the fifth number of Houghton's
Collection, published on 27 April 1682, was an "ugly business." On reading
what he called "*a damnable* sentence, contrary to all the Histories, that I
have seen, and especially the whole tenour of the holy Scriptures," Beale
wrote to Houghton "with as much affection as I could express, & with the
best Arguments of Reason, & Divinity, that I could suggest," including those

3. *Correspondence of Robert Boyle*, vol. 5: *1678–1683*, John Beale to Robert Boyle, 26 June
1682, pp. 299–300.

4. John Houghton (ed.) (1681–1683) *A Collection of Letters for the Improvement of Husbandry
and Trade* (printed for John Lawrence, London); James R. Jacob (1980) "Restoration ideologies and
the Royal Society," *History of Science*, 18, pp. 25–38; and Stubbs, "John Beale [part 2]."

5. *Correspondence of Robert Boyle*, vol. 5, John Beale to Robert Boyle, 16 February 1681;
vol. 6, appendix, John Beale to John Houghton, 1 February 1681; and Thomas Birch (1756–57) *The
History of the Royal Society of London* (printed for A. Millar, London) vol. 4 p. 69.

6. Houghton, *Collection*, 8 September 1681 p. 2.

of Mun, asking him to issue a retraction. He also wrote to Boyle, complaining that Houghton's publication would damage the reputation of the Royal Society itself, since "they will say the Virtuosi doe now publiquely declare against all fundamentall virtues, & commend Prodigallity, Pride, Vanity, and luxury, *to bee for the good of the Nation.*"[7] It is readers such as John Beale that demonstrate that the debate over trade, and over the East India Company, was not simply a matter for merchants and politicians, but also for that group of men who mixed theology, experiment, and ideas of improvement to construct late seventeenth-century natural philosophy.

John Houghton's great offence was to write "That those who are guilty of *Prodigality, Pride, Vanity,* and *Luxury,* do cause more Wealth to the Kingdom than Loss to their own Estates." Consumption, he reasoned, drove trade, industry, and improvement. Consuming domestic produce would employ "many Idle People." Where rents were increased to fund landlords' spending, tenants would be induced by necessity to try new projects and increase their industry. Consuming foreign imports was also "a great Advantage, as well as a Security to the Nation, for 'twill increase our Seamen, and by Consequence our Naval Strength. How shall we expect Trade to other Places if we consume none of their Products." More specifically, he argued that the Turkey and East India trades had been almost wholly supported by "Our *Prodigality* in wearing of Silk . . . and it hath Increast us such a Manufacture, that in Time may spoil the *Silk-Work* of all *Europe.*"[8] In contrast, Beale's more careful evaluation of the balance of trade, and his more austere Protestantism, was bolstered by Mun's arguments for individuals to restrict their consumption (see chapter 4). In the case of the East India and Turkey trades Beale relied upon Mun's reasons for the benefits of long-distance imports to the nation, but also pointed towards "the fullness of Theologicall demonstrations, shewing, how such excesses in Pride, & Luxury (according to the proportions of the Increaseing, & abounding commerce) have been the ruine, and precipitancy of all the Great Empories, and of all the Monarchys, that have ever-yet been, and that it shall ever be so." For Beale, therefore, there was an economic theology, supported by Mun's "most excellent Booke," which prescribed appropriate forms of production, exchange, and consumption:

> We see that god hath ben gracious & bountifull to us above many other
> Nations in preserving & blessing us with peace and plenty, with a universall

---

7. *Correspondence of Robert Boyle,* vol.5, John Beale to Robert Boyle, 8 July 1682, pp. 307–8.
8. Houghton, *Collection,* 27 April 1682, pp. 52–53.

Commerce, and with a great advancement in usefull arts and Sciences, in Manufactures, Mechanicals, and all sorts of good Culture (of which I have much to say hereafter): and now we Should be notoriously ingratefull, if we should abuse gods graces, mercys, and bounty, with pride, luxury, and wantonness, whereas we are the more bounden to give god the glory by our true humbleness, & Sutable industry.[9]

Thus, he was outraged by Houghton's insistence that unrestricted imports might increase trade, that increased consumption could be beneficial, and that there were no sharp moral lines to be drawn between forms of economic activity.[10] Both men supported the East India trade and the project of developing national strength through agricultural and commercial improvements, yet they could read Thomas Mun's treatise quite differently. Beale reported that Houghton had written back to him about Mun's chapters three and nineteen, saying "that hee had read both those Chapters, & could find matter enough there in which hee [Mun] contradicted himself, and Justified Mr Houghtons sentiments &c."[11]

As the Reverend Beale's outrage as a reader shows, John Houghton's *Collection* was, inevitably, part of the contentious debate about trade discussed in the previous chapter. However, there was more to it than that. Beale's letters—particularly in his fears for the reputation of the "Virtuosi" of the Royal Society of which he, Boyle, and Houghton were all members—signal that this was a concern about the ways in which economic activity (including the fortunes of the East India Company), knowledge, and "improvement" were connected, and of the role of print in making these into issues of public concern and national interest, as well as matters of the fortunes and reputations of private individuals. While Beale's critique operated mainly on the terrain of consumption and luxury, there was another aspect of Houghton's publishing project, particularly as it developed in the 1690s, that also raised important questions about new economic relationships, the making of reliable knowledge, and the morality of profit seeking. This concerned the buying and selling of stocks on the stock market, and the role of John Houghton's *Collection* as the first place where stock prices (and, prominently, those of the East India Company) were regularly published in print for an extensive and nonspecialist readership.

9. *Correspondence of Robert Boyle*, vol. 5, John Beale to Robert Boyle, 8 July 1682, pp. 308, 310–11.

10. Joyce O. Appleby (1978) *Economic Thought and Ideology in Seventeenth-Century England* (Princeton University Press, Princeton, NJ).

11. *Correspondence of Robert Boyle*, vol. 5 John Beale to Robert Boyle, 8 July 1682, p. 308.

The printed publication of the prices of stocks in newspapers and periodicals for a broader audience has subsequently become utterly commonplace, a simple and familiar listing of words and numbers. This chapter seeks to set out the context for their first publication in London in the late seventeenth century. In part this is simply to demonstrate the developing role of print in another crucial arena of the Company's operation. As a joint-stock company, the English East India Company increasingly raised capital, and had its ownership and control decided, through the workings of the stock market. One important way in which the Company was known in England, and particularly in London, was through these listings of figures. Much of the significance of the Company was understood and thought about in terms of the price and availability of its stock. Moreover, through the new forms of public debt financing developed during the financial revolution of the 1690s, joint-stock companies, and their shareholders, became large-scale public creditors. Stock prices were a political issue too. This chapter shows how these developments fueled a concern over the manipulation of stock prices, what was called "stock-jobbing." For both political and economic reasons, it was argued, prices might be driven up and down by rumors, hopes, fears, and artful management. It was against this background that John Houghton sought to produce a printed listing of stock prices. The chapter, therefore, explains the distinctive features of his project. Doing so takes us away from the direct workings of the East India Company, which the rest of the book deals with, to other spaces in London within which the Company was of concern, and which in turn were important to the Company. By focusing on Houghton rather than directly on dealing in Company stocks, it demonstrates that he drew upon the knowledge-making practices of the Royal Society, the mechanisms of print production and dissemination, and the protocols of political arithmetic—the resources of a particular culture of information, knowledge, and fact—to offer a specific political vision of the changing relationships between knowledge, economy, and society in the 1690s. Houghton, it is argued, used print to try to stake a claim over what people's relationship to the stock market, and to institutions like the East India Company that were traded on it, should be at a time when these new political and economic relationships were first being forged.

Stock-jobbing: The East India Company on Exchange Alley

English joint-stock companies had existed since 1553, with the founding of the Muscovy Company and the Africa Company. However, it was not until the late seventeenth century that there developed an active market in London for buying and selling these stocks as speculative investments.[12] The growth of this market was due to intertwined economic and political processes that created new roles for investment via the stock market and made the question of when stock was sold and at what price a crucial issue.

By the 1690s the English East India Company was one of many ventures that used the joint-stock form to raise sufficient working capital and to spread risk. As has already been seen in chapter 2, on the Company's early voyages the stock (or capital for the venture) was subscribed only for the duration of a voyage. When the ships returned the investors took their share of the cargoes, or the proceeds from their sale, or suffered the losses. It was not until 1657 that a permanent joint stock was established. This created investments in the Company that did not have a limited life span. Moreover, possession of stock gave rights (albeit depending on how much was owned and other entry criteria) to vote in the elections to appoint the Company's governor and directors.[13] The importance of stockholding was magnified by the Company's practice of raising capital by issuing bonds—short-term loans at interest—rather than increasing the amount of stock available.[14] While the stocks could always be bought and sold, there was not an active market until late in the century. In the early 1660s, an average of only 44 trades in Company stock occurred each year. By the late 1680s this had risen to over 650 trades a year. This increased activity reflected two processes. First, the stock became concentrated in the hands of a few large purchasers drawn from the ranks of merchants, bankers, and traders. This was a direct result of the struggles between the Company and its enemies in the 1680s and 1690s, which were fought out through stock ownership as well as in Parliament and print. By 1691 there were eight investors who had over £10,000 worth of stock each (including Josiah Child, who had £50,000 worth) and accounted for more than a quarter of the whole Company's value. Their dominance was to increase in subsequent years.[15] Second, there was also a diversification of those who held stocks, which meant that overall the capital was passing out of the hands of the broader body of merchants. Between 1675 and 1691

12. William R. Scott (1910–12) *The Constitution and Finance of English, Scottish, and Irish Joint-Stock Companies to 1720*, 3 vols. (Cambridge University Press, Cambridge).

13. Scott, *Constitution and Finance*, vol. 2 pp. 128–79.

14. P. G. M. Dickson (1967) *The Financial Revolution in England: A Study in the Development of Public Credit, 1688–1756* (Macmillan, London).

15. Bruce G. Carruthers (1996) *City of Capital: Politics and Markets in the English Financial Revolution* (Princeton University Press, Princeton, NJ); and K. G. Davies (1952) "Joint-stock investment in the later seventeenth century," *Economic History Review*, 4, pp. 283–301.

the number concerned in the stock fell from 547 holdings to 467 holdings, but the number of women stockholders doubled to 56 (12 percent).[16] There were also 67 dead stockholders, and a third who had also been stockholders in 1675. As K. G. Davies put it, "[a]ll of these signs suggest the approach of a gilt-edge."[17]

By the late seventeenth century, East India Company stocks were fully negotiable and assignable commodities—a remarkably liquid and secure form of investment compared to buying land or putting money directly into a merchant ship. Buyers and sellers could, of course, simply agree to trade. However, to facilitate this meeting of supply and demand a London stock market emerged rapidly in the late 1680s and early 1690s.[18] This market became concentrated in Exchange Alley, a couple of short and turning passages cut between Cornhill and Lombard Street in the City of London that were developed as what Samuel Pepys called a "little town" by goldsmith-bankers Edward Backwell and Charles Everard before and after the Great Fire (see Figure 12). Just south of the Royal Exchange, and a short walk from East India House on Leadenhall Street (Figure 13), the alley increasingly became the haunt of money dealers as the exchange itself became overcrowded. The brokers who worked from there and dealt increasingly in Company stocks could be found particularly in Jonathan's and Garraway's coffeehouses, but also in Baker's and Elmer's.[19] Coffeehouses, first established in London in the 1650s, had long been used for auctions of all sorts of goods, from beaver skins to art, and it was unsurprising that some became known as places to buy and sell stocks. The trade developed through specialist brokers buying and selling for their clients on commission and, increasingly, stock-jobbers who bought and sold on their own account. Before 1689 there were fewer than twenty joint-stock companies in England to trade in. However, war with France left overseas traders seeking alternative investments, and by 1695 there were around 150 ventures, and a corresponding increase in the number of transactions in companies old and new to be entered in the transfer

16. Davies, "Joint-stock investment"; and Carruthers, *City of Capital*. Carruthers shows that women made up 14 percent of Company shareholders in 1712, but tended to have less share value and be less active in trading than men.

17. Davies, "Joint-stock investment," p. 300.

18. Carruthers, *City of Capital*; and Larry Neal (1990) *The Rise of Financial Capitalism: International Capital Markets in the Age of Reason* (Cambridge University Press, Cambridge).

19. [Daniel Defoe] (1719) *The Anatomy of Exchange-Alley. Or a System of Stock-Jobbing* (London, printed by E. Smith); Samuel Pepys quoted in John Biddulph Martin (1892) "*The Grasshopper" in Lombard Street* (Leadenhall Press, London), p. 185; and Kenneth Rogers (1935) *Old London: Cornhill, Threadneedle Street and Lombard Street, Old Houses and Signs* (Whitefriars Press, London and Tonbridge).

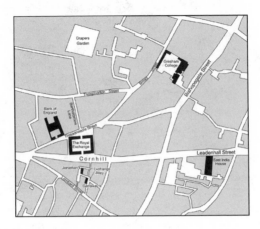

Figure 12 Stock-jobbers' London.

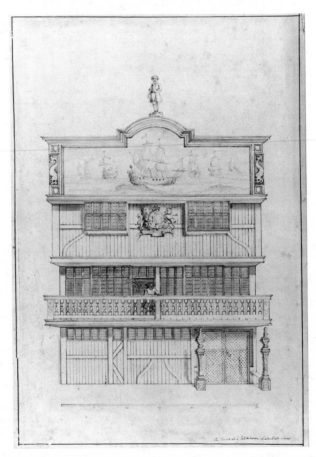

Figure 13 East India House, c. 1711. Pen and ink sketch by George Vertue. (By permission of the British Library, Oriental and India Office Collection WD1341.)

books and stock ledgers.[20] Stocks were sold in Exchange Alley coffeehouses by open outcry. Those with company stocks, lottery tickets, and annuities to sell cried out their wares and called for deals. Bargains were struck on the trading floor in a snowstorm of paper and a cacophony of mysterious words and numbers—tontines, actions, puts, and refuses—which kept this a matter strictly for those in the know.[21]

The increased importance of stock trading to the East India Company, and the greater prominence of Exchange Alley, meant that the debates over the organization of the East India trade in the late seventeenth century were debates over the working of the stock market (see chapter 4). In part this was a question of access. Company supporters argued that there were no restrictions on who could enter the trade, as anyone could buy stock on the open market. Their opponents countered that this was no freedom at all, since it was dependent upon the willingness of current stockholders to sell, and their ability to manage the prices at which they did so.[22] This raised other questions of control, and the term "stock-jobbing" was used to identify the manipulation of market prices in favor of particular interests. Thus, Josiah Child and his circle were accused of concealing the true value of the stock, giving them the opportunity to "ingross so great a part thereof," enabling them not only to pursue "their own *private Designs*" but to manipulate stock prices to their advantage: "that fraudulent and mischievious trade of *Stock-Jobbing.*"[23] Indeed, by the 1690s it was a common accusation against the Company's directors that trading in East India Company stock had become detached from its grounding in a trade in real commodities, producing illusions and monstrosities:

'Tis by this means, that our noble *Trade* has degenerated into *Trick*, and instead of employing a Stock in honest and generous Adventures abroad, according to the laudable Practice of Merchants; There is lately set up

20. Scott, *Constitution and Finance*, vol. 1 pp. 326–51.

21. Brian Cowan (2005) *The Social Life of Coffee: The Emergence of the British Coffeehouse* (Yale University Press, New Haven), discusses coffeehouse auctions; Markman Ellis (2004) *The Coffee-House: A Cultural History* (Wiedenfeld & Nicholson, London) describes stock trading.

22. See, for example, Philopatris [Josiah Child] (1681) *A Treatise Concerning the East India Trade* (London, printed by J.R. for the Honourable the East-India Company) pp. 14–15; and *An Arrest on the East India Privateer* [1681], pp. 5–6.

23. *A Brief Account of the Great Oppressions and Injuries which the Managers of the East-India Company have acted on the Lives, Liberties, and Estates of their Fellow-Subjects...*[1691?] [BL Cup.645.b.11.(25)]. See also *Some Remarks upon the Present State of the East India Company's Affairs* (London, 1690); and John Polloxfen (1697) *England and East-India Inconsistent in their Manufactures* (London).

a new Society of *Artificers*, who blow the Price of Stock up and down, as best suits their Design of enriching themselves by the ruin of others; and this *Legerdemain* is manag'd by a strange sort of *Insects* call'd *Stock-Jobbers*, who devour men on our *Exchange*, as the *Locusts* of old did the Herbage of *Egypt*.[24]

Not only were such market manipulations seen as responsible for the economic crisis of 1695 to 1697, but anti-Company pamphleteers argued strongly that "[t]hat pernicious and infamous Trick of Stock-jobbing... had its Original from this Company."[25] As a result, those who campaigned for an end to the Company's monopoly in favor of a free or regulated trade did so at least partly on the grounds that it would also see an end to stock market trickery.[26]

Buying and selling stock was, however, not simply about the management of particular trades and the control of their profits. Their involvement in the innovations of the English financial revolution of the late seventeenth century meant that the major joint-stock companies were political institutions as well as economic ones. Put simply, the changes in the organization of public credit in the 1690s aimed to shift towards forms of long-term borrowing by the state from domestic sources at low rates of interest. Initially, there was a series of less than satisfactory initial experiments with tontines (collective annuities funded by excise duties), life annuities, and increasingly elaborate lotteries. However, the greatest source of loans to the state before 1720 was the major joint-stock companies. The Bank of England was established in 1694 specifically to manage government debt. Investors subscribed £1.2 million, which was then loaned to the government at 8 percent interest, with certain customs and excise revenues earmarked by Parliament to fund the interest payments (plus an annual £4,000 management fee). In 1697 problems with further short-term government debts were resolved by allowing their holders to exchange them for Bank stock, and an additional sum of more than £1 million was subscribed. It was also this sort of conversion of debt to stocks that floated the South Sea Company in 1711 with a £9 million loan to the government at 9 percent interest.[27]

24. George White (1691) *An Account of the Trade to the East-Indies* (London) p. 5.

25. *Proposals for Setling the East-India Trade* (Printed and sold by E. Whitlock, London, 1696) p. 4; and Huw V. Bowen (1993) "'The pests of human society': stockbrokers, jobbers and speculators in mid-eighteenth-century Britain," *History*, 78, pp. 38–53.

26. *A Regulated Company more National than a Joint-Stock in the* EAST-INDIA TRADE [BL 816.m.11(71)] verso; and [Henry Martyn] (1701) *Considerations upon the East-India Trade* (printed for A. and J. Churchill, London).

27. Carruthers, *City of Capital*; and Dickson, *The Financial Revolution*.

In each case these were matters of party politics too. The Bank was a Whig project, the South Sea Company a Tory initiative. Each operated as part of the ongoing conflicts between the parties over religion, foreign policy, the independence of Parliament, the power of the monarch, and the sources, control, and uses of public finance. The East India Company was not created specifically to deal in the national debt, but it had a long history of making loans and gifts to a succession of monarchs (see chapter 2). While never simply controlled or supported by one political party or the other, its loans to James II aligned it publicly, and problematically after 1689, with the Tory party, and it was challenged by Whig interests in the 1690s via the formation of the Scottish East India Company in 1695 and, more successfully, the New East India Company in 1698. Indeed, the formation of the New Company sparked a remarkable bidding war over which group of merchants could lend most money to the government. The Old Company's offer of £700,000 at 4 percent was rejected in favor of the New Company's £2 million at 8 percent. However, the subscription of £380,000 to the New Company's stock by the Old Company made it the largest shareholder and ensured its continuation. It also underpinned the eventual merger of the two companies and the ongoing involvement of the United Company in party political rivalries during the eighteenth century.[28] Overall, this meant that owning a share of a major joint-stock company was to hold a slice of the national debt. Therefore, the buying and selling of stocks was also a trade in government borrowing, and part of the political process of public finance as well as a matter of private interest. This meant that stock prices responded not only to the conditions of trade but also to the political climate. Under William and Mary, for example, Jacobite scares made stock prices fall, and good news from the wars with France saw them rise. Indeed, the effects of speculation and rumor were so great that price movements in East India Company stock in the eighteenth century were not related to changes in the Company's profitability.[29]

This combination of economic and political interests, and the responsiveness of the markets to rumor and manipulation, also underpinned accusations of stock-jobbing. Two Tory pamphlets from 1701 attributed to Daniel Defoe explicitly connected the stock market and politics. In *The*

---

28. Carruthers, *City of Capital*; Philip Lawson (1987) *The East India Company: A History* (Longman, London); and Lucy S. Sutherland (1952) *The East India Company in Eighteenth-Century Politics* (Clarendon Press, Oxford).

29. Carruthers, *City of Capital*; and Philip Mirowski (1981) "The rise (and retreat) of a market: English joint-stock shares in the eighteenth century," *Journal of Economic History*, 41, pp. 559–77.

Villainy of Stock-Jobbers Detected, Defoe demonstrated their ability to manipulate the value of Old Company shares so that "the Price shall dance
attendance on their designs, and rise and fall as they please, without any
regard to the Intrinsick worth of the Stock."[30] This, he argued, damaged
public credit, weakened the government, and threatened the nation. The
argument was continued in *The Free-Holders Plea Against Stock-Jobbing
Elections of Parliament Men*, which questioned the government's wisdom
in the handling of the rivalry between the Old and New East India Companies, strongly suggesting that the only motivation was the securing of
a loan to the state, regardless of the consequences for the East India trade
and the country. Indeed, this conflict, and the machinations of the money
men, was now threatening to undermine the electoral process as the rival
companies sought to secure members of Parliament to their cause. This was
"*Parliament-Jobbing*, a New Trade":

> For as this Stock-Jobbing in its own Nature, is only a new Invented sort
> of *Deceptio Visus*, a *Legerde-main* in Trade; so mix'd with Trick and
> Cheat, that twou'd puzzle a good *Logician* to make it out by *Syllogism*:
> So nothing can be more Fatal in *England* to our present Constitution;
> and which in time may be so to our Liberty and Religion, than to have
> the Interests of Elections Jobb'd upon Exchange for Mony, and Transfer'd
> like *East-India* Stock, for those who bid most.[31]

He charged that elections, and therefore the parliaments that would make
decisions over war and peace, were being bought and sold at Jonathan's
and Garraway's, replacing the representatives of country freeholders with
the particular interests of tradesmen, stock-jobbers, and placemen. This political and economic critique of stock-jobbing, therefore, expressed concern
over the volatility of a market that was detached from notions of "Intrinsick
worth" and full of "Giddy People, fancying every thing about them to be in
Motion, when all the stir is only in and from their own Imaginations."[32] Under these conditions, people responded to rumor, fantasy, and trickery, and
"unskilful and unwary Men are entic'd away, from certain Profit to pursue

30. [Daniel Defoe] (1701) *The Villainy of Stock-Jobbers Detected, and the Causes of the Late
Run Upon the Bank and Bankers Discovered and Considered* (London) p. 6.
31. [Daniel Defoe] (1701) *The Free-Holders Plea Against Stock-Jobbing Elections of Parliament Men* 2nd ed. (London) pp. 9–10.
32. *Some Considerations upon the late* ACT *of the Parliament of* SCOTLAND, *for Constituting
an* INDIAN COMPANY (1695) [BL 816.m.11(84)] p. 4.

uncertain Hopes."[33] However, over and above these private interests were the implications for government credit and the independence of Parliament. This was also a matter of public knowledge about the stock market and of stock prices in a situation where the value of these intangible commodities was highly dependent upon news of war, trade, and politics. As one advocate of dismantling the East India Company put it, "Stocks in the Warehouses of private Merchants rise and fall, and no Man knows it but themselves.... Companies are frighted by wars and rumours of Wars; the Joint-stocks fall, and every one must hear it: And this engages the private Interest of some, the Fear of others, to disturb the publick Resolutions."[34] Where there was no guarantee that the Company would "put out an Advertisement in some publick Print, to hinder such an abominable Cheat," stock-jobbers could prosper from spreading rumor and misinformation.[35] There was, therefore, both a demand for reliable information about the stock market and a problem of trust and credibility to be overcome by those who provided it.

Stock Prices in Print

Printed newspapers and the stock exchange both have their origins in Amsterdam. However, although the English forms of news publication and stock broking were heavily influenced by these Dutch forerunners, it seems that it was in London that the first regularly printed stock price lists appeared.[36] Commodity price currents had been published in the capital from at least the 1630s, but it was with James Whiston's *The Merchant's Remembrancer* in the early 1680s that stock market information was first made available in print. Whiston listed the prices of hundreds of commodities from all over the world available for purchase in London. His innovation was to devise a system of symbols that indicated for each of them, including East India Company and Royal Africa Company stocks, whether the market was standing, rising, falling, highest, or lowest. This way of presenting stock values, the earliest example of which is from 4 July 1681, was an attempt to deal with the problems of presenting this information in a printed list: the inevitable delay between taking the temperature of the market and publication, and the problems of expressing "something as tenuous as a 'prevailing market'

33. [Martyn], *Considerations*, p. 30.
34. [Martyn], *Considerations*, pp. 30–31.
35. T. S. (1697) *Reasons Humbly Offered for the Passing a Bill for the Hindering the Home Consumption of East-India Silks, Bengals &c.* (London, printed by J. Bradford) p. 14.
36. Neal, *The Rise of Financial Capitalism*.

in rigid tabular form."[37] It seems that Whiston's publication was issued, in both English and French editions, in competition with Robert Woolley's *Prix Courant des Merchandises a Londres*. Woolley soon hit back, and by his French edition of 25 October 1681 he was giving the prices of stocks for the East India, Royal Africa, and Hudson's Bay companies.[38] Woolley continued to publish until 1696 and Whiston until 1709, by which time John Castaing's *The Course of the Exchange* was on its way to becoming the definitive guide to stock prices, fighting off later competition from stockbroker John Freke.[39]

It is difficult to assess accurately the readership and uses of these printed listings. Information on the state of the market and the ways in which prices had changed over time was necessary for profitable speculation in stocks. However, the most active traders, seeking the most privileged and up-to-date information, would find it face-to-face with brokers and other dealers gossiping and exchanging news around the tables of the Exchange Alley coffeehouses, or in the shouted market making of the open outcry. Beyond that, they might pay high premiums for regularly updated information by manuscript letter from privileged sources. Therefore, the audience for a weekly or biweekly listing of prices available in print to all who paid a few pennies was likely to be the occasional investor, or the speculative trader who wanted to cross-check word of mouth, specialist, and confidential information against a more standardized source that would, collected over time, build a longitudinal picture of the market.[40]

The printed lists offered, issue upon issue, a standard tabular format by which prices might be easily compared and changes noted. The seemingly definitive and objective statement of the state of the market was no doubt intended to encourage readers to take up the services of the brokers who published them. Whiston's first list noted for the stocks that "[t]he Auth. J.W. will buy or sell," and John Freke advertised that he could easily be

37. *The Merchants Remembrancer* (London) 4 July 1681; J. M. Price (1954) "Notes on some London price-currents, 1667–1715," *Economic History Review*, 7:2, p. 245; and Michael Harris (1997) "Exchanging information: print and business at the Royal Exchange in the late seventeenth century," in Ann Saunders (ed.) *The Royal Exchange* (London Topographical Society, London) pp. 188–97.

38. Prices were listed in Woolley's English edition of *The Prices of Merchandise in London* by at least 3 January 1682.

39. Price, "Notes on some London price-currents." John Castaing's *The Course of the Exchange, &c* (London), started c. 1697, and there are collections of John Freke's *Freke's Prices of Stocks, &c.* (London) for 1715 to 1722.

40. Larry Neal (1988) "The rise of a financial press: London and Amsterdam, 1681–1810," *Business History*, 30, pp. 163–78.

found in his office in Exchange Alley.[41] However, these statements of price belie some critical conditions and decisions. A single daily price for each stock made sense only under specific institutional conditions. It was conceivable only when the increased volume of trades and, more crucially, the active engagement of brokers in those trades made something that might be thought of as a single price out of the many meetings of buyers and sellers.[42] Even under those conditions, the price to state for a particular day was not necessarily an easy decision, especially in a volatile market. That Castaing's and Freke's lists quoted quite different prices for South Sea stock in the bubble year of 1720 may have been due to quoting for different times of the day—Castaing stated that he gave "Prices of stocks from Twelve to Two of the Clock"—or it may have come down to the relationship between politics and the stock market. Freke was a "hot Whig" with associations with the South Sea Company and may have had particular hopes and expectations for its performance, which shaped his quotation. It is notable that his listing lasted no longer than the South Sea Company did, going out of circulation in 1722.[43] A broader politics can also be detected. James Whiston is best known for his promotion, through printed pamphlets in the 1680s and 1690s, of merchant influence in parliamentary decisions over mercantile matters via "a standing council for the regulation of trade."[44] His publication of commercial knowledge in print might therefore be seen as both an enactment of the importance of trade to the nation, a barometer of its condition, and a public positioning of what might otherwise have been exclusive mercantile knowledge. Yet there was little to appeal to a broader readership in Whiston's close-set columns of commodity names and prices. It was, as already noted, the apothecary, coffee and chocolate dealer, and virtuoso John Houghton who first made stock prices available in print to a broad and nonspecialist (at least in mercantile terms) audience when he began publishing them in his *Collection of Letters for the Improvement of Husbandry and Trade* in 1692.

Science, Print, and Circulation: Houghton's *Collection*

John Houghton's *Collection* was one of a range of mostly unsuccessful periodical publications on natural knowledge that appeared in the wake of the

---

41. *The Merchants Remembrancer*, 4 July 1681; and *Freke's Prices of Stocks, &c.*
42. Davies, "Joint-stock investment."
43. Carruthers, *City of Capital*, p. 173; and Neal, *The Rise of Financial Capitalism*, p. 32.
44. Perry Gauci (2001) *The Politics of Trade: The Overseas Merchant in State and Society, 1660–1720* (Oxford University Press, Oxford) p. 185.

foundation, disappearance, and reappearance of the *Philosophical Transactions* in the 1670s and 1680s.[45] It was initially published in September 1681 and for nearly three years carried a series of essays on agricultural improvements such as marling or the growing of parsley; on trades like malting and brick making; and on the development of industries, banks, and branches of commerce. This was pursued through original contributions, reprinted essays, and reviews of books Houghton wished to bring to his readers' attention. After folding in 1684, it reappeared in March 1692 with a new prospectus and the intention of providing useful commercial intelligence alongside the more didactic articles.[46] Within a month it had moved to twice a week, on Wednesdays for the agricultural prices and Saturdays for the bills of entry into the Custom House in London. However, success was by no means guaranteed, and publication ceased at the end of June 1692, with Houghton noting that the conditions were not right for his periodical since "the Generality" were "at this time, minding Political News, and ways how to preserve what they have, rather than Methods how to improve."[47] Yet he was back six months later, reporting that his difficulty in financing the publication had been resolved by "a Contribution of a Guinea a Year from some Gentlemen," and that, in anticipation of more subscriptions, he would continue to publish the letters for a penny each.[48] While Houghton's problems with subscribers were never satisfactorily resolved, his finances were, no doubt, assisted by the addition of a sheet of advertisements in summer 1693; publication continued until 1703.[49]

John Houghton only began publishing stock prices with the *Collection's* relaunch in 1692. Figure 14 shows both sides of the single sheet for 30 March 1692, with its "DESIGN" for the publication on the front and its tabulation of information on the back. Both sides show in their different ways how Houghton's publication of stock prices was embedded within a broader project of the communication of agricultural and commercial knowledge.[50] The design promised more "Accounts or Discourses" on subjects such as

45. Adrian Johns (1998) *The Nature of the Book: Print and Knowledge in the Making* (University of Chicago Press, Chicago). Another, more successful example was the *Athenian Mercury*; see Helen Berry (2003) *Gender, Society and Print Culture in Late-Stuart England: The Cultural World of the Athenian Mercury* (Ashgate, Aldershot).

46. The last number is dated 16 June 1683, but the sequence suggests that it was from 1684.

47. Houghton, *Collection*, 20–27 June 1692.

48. Houghton, *Collection*, 20 January 1693.

49. Houghton, *Collection*, 21 July 1693.

50. Nastasha Glaisyer (2000) "Readers, correspondents and communities: John Houghton's A Collection for Improvement of Husbandry and Trade (1692–1703)," in Alexandra Shepard and Phil Withington (eds.) *Communities in Early Modern England: Networks, Place, Rhetoric* (Manchester

the types of earth and their cultivation; the histories of trades such as malting, tanning, and glassmaking; the arts of fishing and fowling; instruments
of husbandry; and the trading strength and policy of other nations. In short,
"all useful things fit for the *Understanding* of a plain Man."[51] Indeed, it
appears that Houghton had been thinking about his intended readership,
the type of information that they might need, and the sort of uses that
they might put it to. Between an earlier manuscript draft of his proposal,
probably sent to Hans Sloane, the printed version that advertised the forthcoming *Collection*, and the *Collection* itself, Houghton had imagined an
ideal and active readership.[52] Unlike the manuscript draft, the printed versions described this audience of plain men, and identified them by their
use of specific information for instrumental purposes. The *Collection* was
to be

for the Advantage of *Tenant, Landlord, Corn-Merchant, Mealman, Baker,
Brewer, Feeder of Cattel, Farmer, Maulster, Grazier, Seller* and *Buyer of
Coals, Hop-Merchant, Soap-Boyler, Tallow-Chandler, Wool-Merchant,*
their *Customers*; &c... *weekly* to give an account from the most *prin-
cipal* places of the Kingdom, of the *Prizes* of *Wheat, Rye, Mault, Oats,
Pease, Hay, Coals, Hops, Tallow* and *Wool:* To which I add, the Price, at
London, of *Beef, Pork,* and *Mutton.*

Also, for whom it might concern, it was design'd to give *weekly* from
hence an Account of the value of *Actions* of the *East-India, Guinea,
Hudsons-Bay, Linnen,* and *Paper Companies.*

Also a *weekly* Account of the several parts of the *Bills of Mortality*
of *London.*

Also a *weekly* Account of the *Shipping* into and out of the *Port of
London,* with the place *witherto* bound and from *whence.*[53]

This was, as the second page of Figure 14 shows, digested and presented as
a printed table of figures. Stock values were only a small part of this tabulation, the bulk of it being taken up by domestic agricultural prices. The
initial listing of "Actions," Houghton adopting the French term for stocks,
covered only eight companies, and was headed by the East India Company.

University Press, Manchester) pp. 235–51; and Margaret R. Hunt (1996) *The Middling Sort: Commerce, Gender, and the Family in England, 1680–1780* (University of California Press, Berkeley).
    51. Houghton, *Collection,* 30 March 1692.
    52. BL Sloane Mss. 2903 ff. 167–68, "A proposall for Improvement of Husbandry and Trade"
[1691?]; and John Houghton (1691) *A Proposal for the Improvement of Husbandry and Trade.*
    53. Houghton, *Collection,* 30 March 1692.

A COLLECTION for Improvement of
Husbandry and Trade.

Wednesday, March 30. 1692.

The DESIGN,
The Price of Corn, &c. *in many Counties of the Kingdom.* Price *of* Actions *in Companies.* Price *of*
Flesh. *A List of* London Ships *in and out.* An Abstract *of the* Bills of Mortality.

IN *November* last I dispers'd a *Proposal* for Improvement of *Husbandry* and *Trade*, wherein is mention'd, That it was intended to lay out for a large *Correspondence*: And for the Advantage of *Tenant, Landlord, Corn-Merchant, Mealman, Baker, Brewer, Feeder of Cattel, Farmer, Maulster, Grazier, Seller* and *Buyer of Coals, Hop-Merchant, Soap-Boyler* , *Tallow-Chandler* , *Wool-Merchant* , their *Customers*; &c. For these, I say, it was intended, *weekly* to give an account from most of the *principal* places of the Kingdom, of the *Prizes* of *Wheat, Rye, Mault, Oats, Pease, Hay, Coals, Hops, Tallow* and *Wool* : To which I add, the Price, at *London*, of *Beef, Pork,* and *Mutton.*

Also, for whom it might concern, it was design'd to give *weekly* from hence an Account of the value of *Actions* of the *East-India, Guinea, Hudsons-Bay, Linnen,* and *Paper Companies.*

Also a *weekly* Account of the several parts of the *Bills of Mortality* of London.

Also a *weekly* Account of the *Shipping* into and out of the *Port of London*, with the place *whitherto* bound and from *whence.*

And besides these one or more Accounts or Discourses *weekly* upon some of the following Heads; such as are,

A full Catalogue of all forts of *Earths* that may be procured, to anatomize and distinguish them into their *Classes*, and to discover what is the proper *Product* of each; what the best fort of *Compost* or *Manure* for each, and how to be *cultivated* ; as also the Arts of *Dreining* and *Flooding Lands*, and such like.

Ample and exact Histories of Trades, as *Maulting, Brewing, Baking, Tanning, Dying, Potting, Glass-making,* and many others.

An Account of Goods *imported* and *exported* from the *Custom-House* of *London*, to draw the *particulars* into *gross Sums*, in order to examin the *Expence* and *foreign Vent* of each Commodity.

The Description and Structure of all Instruments for *Husbandry* and of *Carriages*, with the *Deficiencies* of the fame, and Proposals for their *Amendment* ; together with the Arts of *Conserving* and *Amending* the *Roads* or *High ways.*

The Arts of *Fishing, Fowling, Hunting,* and *Destroying of Vermin.*

An Account of the *Rivers* of *England*, of their *Fountains, Bridges, Tides,* and *Exits*; how far they are *Navigable*, and with what *Vessels.*

An exact Comparison of all sorts of *Weights* and *Measures* used at this time in all parts of the World.

The comparative Strength and Weight of *Woods, Cordage, Metals,* and other *Materials* in order to *Building* and *Stowage.*

Experiments on the *Vegetation* of *Plants*, to examin what *Mixtures* do promote or hinder it.

Epitomes of Books of *Husbandry* and *Trade*, and

Collections of what else is to be met withal useful for the Advancement of *Husbandry.*

As good an Account as may be had of the *Trade, Strength* and *Policy* of other Nations.

In short, all useful things fit for the *Understanding* of a plain Man.

For which I did then propose a *quarterly* Payment, to all who were willing to have them sent to them.

The whole *Proposal* I stick to, only am *advis'd*, instead of the quarterly Payment, to *print* on such Paper you here see, and let them be fold each for *Two Pence*; which all that like may have after the fame manner they have *Gazettes.*

I have taken good Advice in this Affair, and design nothing but what may tend to the *Real Advantage* and *Honour* of this my Country, letting it know what *Necessary Useful Matter* I can learn ; as also *explaining* much Matter already divulg'd, but in such *Terms* as t nerality understand not.

And thanks to my good *Friends*, to whom I am much obliged, and shall always gratefully acknowledge, my self bound, I have the *Testification* following;

November 11. 1691.

THESE *are to testifie our Knowledg of the approv'd Abilities and Industry of Mr.* JOHN HOUGHTON, *Citizen of* London, *and Fellow of the* ROYAL SOCIETY *there, in the Discovery and Collection of Matters worthy Observation, and more particularly such as relate to the Improvement of* Husbandry *and* Trade. *Towards his furtherance wherein, and in his laudable Inclination already experienc'd , and now further design'd, to the communicating the Effects thereof to the Publick; We do hereby most willingly give him this* Testimony *of our Knowledg and Esteem, in order to the recommending him to the Notice, Assistance, and Encouragement of all Gentlemen and others, desirous of promoting the Endeavours of a Person so qualify'd and dispos'd to the Service of his Country.*

Robert Southwell.	Thomas Gale.	William Hewer.
Thomas Meres.	John Scott.	Henry Whitler.
John Hoskins.	Robert Plot.	Alexander Pitfield.
Peter Pett.	Daniel Coxe.	Richard Waller.
Anthony Deane.	Nehemiah Grew.	Edward Haynes.
John Evelin.	Edward Tyson.	Thomas Langham.
Thomas Henshaw.	Frederick Slare.	Francis Lodwick.
Abraham Hill.	Robert Pitt.	Edmund Hally.
Samuel Pepys.	Hans Sloane.	
John Creed.	Hugh Chamberlen.	

These things consider'd, such like may be expected, at least *Once a Week*, from

England*'s hearty Well-Wisher,*

John Houghton, *F.R.S.*

Fror

Figure 14 John Houghton's *Collection*, 30 March 1692. (By permission of the British Library, 522.m.11(2).)

	Wheat per Bushel	Rye per Bushel	Malt per Quarter	Oats per Quarter	G.Pea per Bushel	W.Pea per Bushel	Coals Chaldr.	Hops per Hundred	Hay p.Load	Tallo p. Hund.	Wool per Tod
	High pr. Low	High pr. Low	High price. Low	High price. Low	High pr. Low	High pr. Low	s. D.	High price. Low	s. D.	s. D.	High price. Low
	s. D. s. D.	s. D. s. D.	s. D. s. D.	s. D. s. D.	s. D. s. D.	s. D. s. D.	s. D.	s. D. s. D.	s. D.	s. D.	s. D.
Brecon	4: 6	2:10	20: 0	8: 0	1:10	3: 6	10: 0	30: 0	40: 0	28: 0	11:8
Colchester	4: 0	2: 0		10: 0	2: 6	2: 8	25: 0	30: 0			
Derby	3: 6 3: 0	2: 6 2: 4	16: 0	8: 0	2: 0	3: 6	5: 10	56: 0	20: 0	20: 0	14: 0
Dorchester	4: 6 3: 0			11: 4 10: 8	2: 8				28: 0	18: 6	16: 0
Farnham	5: 3 4: 0	3: 0	18: 8	13: 6 11: 0	2: 6	3: 0		36: 0 30: 0			
Hertford	5: 6 4: 1	3: 0 2: 8	17: 6 15: 0	14: 0 12: 0	2: 5 2: 3	2: 8 2: 6	46: 8	36: 0 34: 0	30: 0	23: 4	18: 0 14: 0
Ipswich	3: 9	1: 9	16: 0	12: 6							
Leicester	3: 6 3: 2	2: 2 2: 0	14: 4 13: 4	12: 0 11: 4	1:11 1: 9	4: 0 3: 8	13: 4	74: 8 65: 4	30: 0	21: 4	21: 6 14: 0
London	5: 3 5: 0		21: 0 19: 0	12: 0 10: 0	2: 8 2: 6	3: 6 3: 0	28: 0	33: 0 28: 0	43: 0	28: 0	
Regis	4: 0 3: 0	1:10 1: 9	14: 0 13: 0				22: 0		25: 0	28: 6	
...	5: 6 4: 8		10: 8	2: 2 1:10	2: 6						
...stle	3: 6 2: 9	2: 2 1: 9	20: 0 13: 0	10: 0 6: 0			10: 6	36: 0	26: 8	20: 0	11: 0 5: 6
Norwich	3: 6 2: 9	1: 9 1: 7	14: 0 13: 0	10: 0 9: 0	1: 9 1:7½	2:1½ 1:10½	22: 0	60: 0 50: 0	30: 0	28: 0	13: 6
Pembroke	3: 8	2: 2	19: 4	6: 0	1: 8	3: 4	13: 0	37: 0	40: 0	28: 0	17: 6 12: 0
Reading	5: 6 4: 6		15: 0 12: 0	2: 8 2: 3			36: 0	40: 0 30: 0	40: 0	30: 0	14: 0 9: 0

Yet within only a few weeks this had expanded to thirteen ventures, which
he soon began to publish in alphabetical order. By 1694 the list had expanded
to cover fifty companies (see Figure 15), although numbers subsequently de-
creased with the end of the boom in the formation of joint-stock companies
in 1696.[54]

It was within a particular culture of information, knowledge, and fact
that Houghton's *Collection* presented prices from the stock market in print
and explained the workings of trade and finance. Unlike the price currents
of James Whiston or Robert Woolley, where stock prices were simply listed
among hundreds of other commodities, the context was not the purely mercantile one of metropolitan overseas traders and their buying and selling.
Houghton's understanding of knowledge and the uses of print was derived
instead from the practices of the Royal Society and its ongoing discussions
about the improvement of agriculture and trade. Also, unlike the later more
specialist stock price listings of John Castaing and John Freke, Houghton's
Collection was not simply centered on the workings of Exchange Alley and
its specialist brokers and dealers. Indeed, Houghton's project should be seen
as one that sought, for political reasons, to use print to construct a real and
imagined geography of knowledge about improvement and trade, including
the trade in stocks, that cut against that of Exchange Alley but without
simply denigrating it as the home of stock-jobbery. For Houghton, the publication of stock prices, including those of the East India Company, and the
effort to explain the workings of the new stock market drew upon the ways
of making knowledge provided by the new natural philosophy, and upon a
geography of knowledge made possible by print. It was an attempt to realize
a specific political vision of the relationships between knowledge, economy,
and society.

Houghton sought to present information that his readers could accept
as reliable and credible. Key to this was his role as a member of the Royal
Society and the positioning of his *Collection* within their forms of knowledge making. In his preface to the first number, in 1681, Houghton told
his readers that he was honored to be a member of the Royal Society and,
more particularly, of its Committee for Agriculture.[55] Indeed, he had been
a fellow of the Society since the year before, having been nominated by
Robert Hooke, and was actively involved in the early 1680s in nominating

54. Dickson, *The Financial Revolution*; and Scott, *Constitution and Finance*, vol. 1 pp.
352–54.
55. This was founded as the Georgical Committee, at least partly on the suggestion of John
Beale; see Stubbs, "John Beale [part 2]."

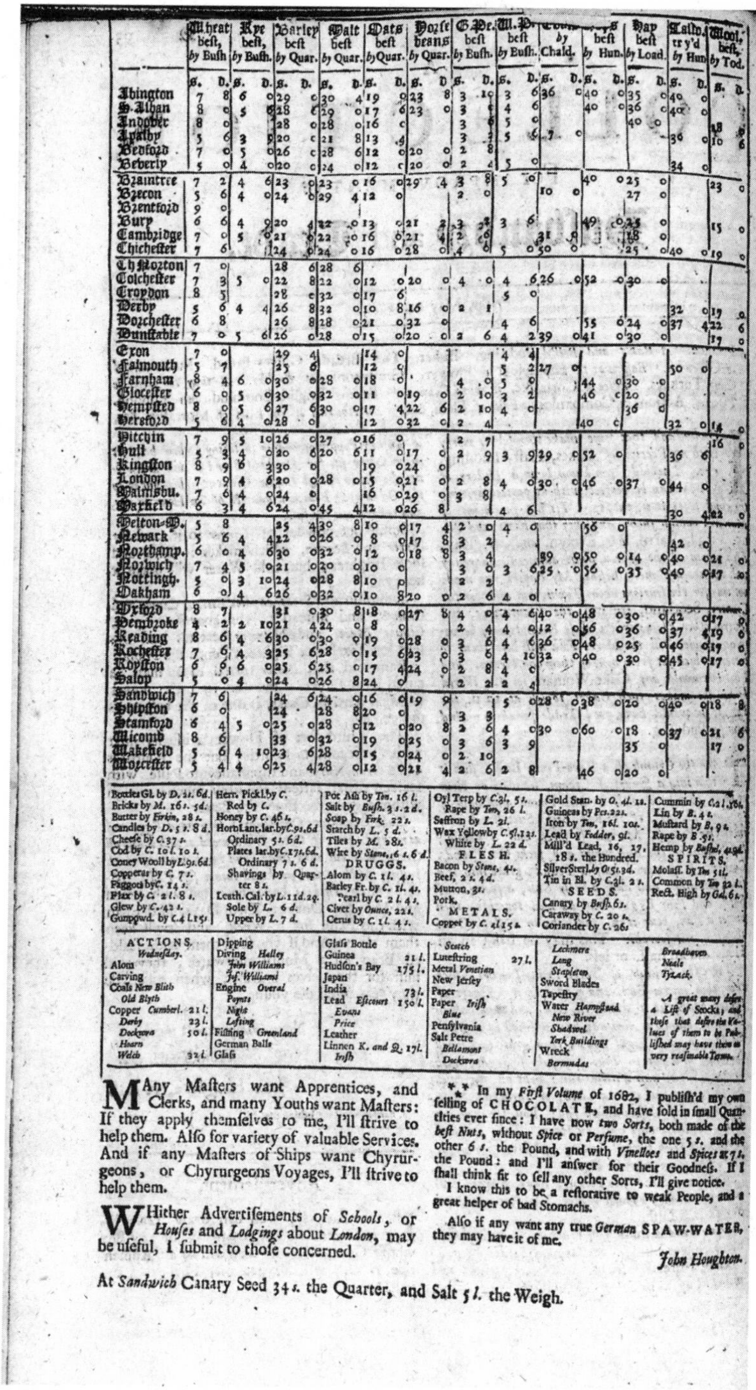

The column headings of the table read:

	Wheat best by Bush.	Rye best by Bush.	Barley best by Quar.	Malt best by Quar.	Oats best by Quar.	Horse beans by Quar.	G.Pe. best by Bush.	All.P. best by Bush.	...Chald.	...by Hun.	Hay best by Load	Tallo tt y'd by Hun	Wool best by Tod
Ibington													
S. Alban													
Andover													
Aylsby													
Bedford													
Beverly													
Braintree													
Brecon													
Brentford													
Bury													
Cambridge													
Chichester													
Chi Norton													
Colchester													
Croydon													
Derby													
Dorchester													
Dunstable													
Exon													
Falmouth													
Farnham													
Glocester													
Hempsted													
Hereford													
Hitchin													
Hull													
Kingston													
London													
Malmsbu.													
Maxfield													
Melton-M.													
Newark													
Northamp.													
Norwich													
Notting.													
Oakham													
Oxford													
Pembroke													
Reading													
Rochester													
Royston													
Salop													
Sandwich													
Shipston													
Stamford													
Wicomb													
Wakefield													
Worcester													

Bottles Gl. by D. 2s. 6d.
Bricks by M. 16s. 3d.
Butter by Firk. 28s.
Candles by D. 5s. 8d.
Cheefe by C. 37s.
Cod by C. 10l. 10s.
Coney Wooll by L.9s.6d.
Copperas by C. 7s.
Faggood by C. 14s.
Flax by G. 2l. 8s.
Glew by C. 43s.
Gunpowd. by C.4 Lrs.

Herr. Pickl.by C.
Red by C.
Honey by C.46s.
Hort.Lanc.lar.by C.9l.6d
Ordinary 7s. 6d.
Plates lar.by C.17s.6d.
Ordinary 7s. 6d.
Shavings by Quar-ter 8s.
Leath. Cal. by L. 11d.2q.
Sole by L. 6d.
Upper by L. 7d.

Pot Afh by Ton. 16l.
Salt by Wey. 3.1.2d.
Soap by Fork. 22s.
Starch by C. 3d.
Tiles by M. 28s.
Wire by Stone, 16s. 6d.
DRUGGS.
Alom by C. 2l. 4s.
Barley Fr. by C. 1l. 4s.
Tranfly C. 2l. 4s.
Civet by Ounce, 22s.
Cerus by C. 1l. 4s.

Oyl Terp by C.3l. 5s.
Rape by Tun, 26 l.
Saffron by L. 2l.
Wax Yellow by C.5l.12s.
White by L. 2s. 4d.
FLESH.
Bacon by Stone, 4s.
Beef, 2s. 4d.
Mutton, 3s.
Pork,
METALS.
Copper by C.4l.5s.

Gold Stan. by O. 4l. 1s.
Guineas by Pcs.22s.
Iron by Tun, 16l. 10s.
Lead by Fodder, 9l.
Mill'd Lead, 16, 17.
18s. the hundred.
Silver Sterl. by O. 5s.3d.
Tin in Bl. by C.3l. 2s.
SEEDS.
Canary by Bufh.6s.
Caraway by C. 20 s.
Coriander by C. 26s.

Cummin by C.2l.18s.
Lin by B. 5s.
Muftard by B. 9s.
Hemp by Bufhel, 4s.9d.
SPIRITS.
Molaff. by Tun 35l.
Common by T. 4s.
Rect. High by Gal. 6s.

ACTIONS Wednesday.

Alom	Dipping	Glass Bottle
Carving	Diving Halley	Guinea
Coals New Blith	John Williams	Hudfon's Bay 175l.
Old Blyth	Jof. Williams	Japan
Copper Cumberl. 21l.	Engine Overal	India
Derby 23l.	Fopts	Lead Efticont 150l.
Deckoys 50l.	Negin	Evoans
Hearn	Lefuing	Price
Welch 32l.	Fishing Greenland	Leather
	German Balls	Linnen K. and Q. 17l.
	Glass	

Scotch	Lichmere	Broadhaven
Luteftring 27l.	Long	Neale
Metal Venetian	Stapleton	Tynelt
New Jerfey	Sword Blades	
Paper	Tapeftry	A great many defire
Paper Irifh	Water Hampftead	a Lift of Stocks, and
Blue	New River	thofe that defire to be Va-
Penfylvania	Shadwel	lues of them to be Pub-
Salt Petre	York Buildings	lifhed may have them in
Bellamont	Wreck	very reafonable Terms.
Deckoys	Bormudas	

MAny Mafters want Apprentices, and Clerks, and many Youths want Mafters: If they apply themfelves to me, I'll ftrive to help them. Alfo for variety of valuable Services. And if any Mafters of Ships want Chyrurgeons, or Chyrurgeons Voyages, I'll ftrive to help them.

WHither Advertifements of *Schools*, or *Houfes* and *Lodgings* about *London*, may be ufeful, I fubmit to thofe concerned.

At *Sandwich* Canary Seed 34s. the Quarter, and Salt 5l. the Weigh.

*** In my *Firft Volume* of 1682, I publifh'd my own felling of CHOCOLATE, and have fold in fmall Quantities ever fince: I have now two Sorts, both made of the beft Nuts, without Spice or Perfume, the one 5s. and the other 6s. the Pound, and with Vinelloes and Spices 7s. the Pound: and I'll anfwer for their Goodnefs. If I fhall think fit to fell any other Sorts, I'll give notice. I know this to be a reftorative to weak People, and a great helper of bad Stomachs.

Alfo if any want any true *German* SPAW-WATER, they may have it of me.

John Houghton.

Figure 15 John Houghton's *Collection*, 11 May 1694. (By permission of the British Library, 522.m.11(2).)

other fellows (including William Penn and a number of merchants and tradesmen), pursuing those who had not paid their subscriptions, and auditing the Society's accounts.[56] Houghton also made a variety of contributions at the Society's meetings in the 1680s. Some of these were directly connected to agricultural issues, such as his demonstration of a new plow or his intention to investigate the price of corn in London.[57] Others exhibited an interest in seeds, plants, and other materials (including various metals, stones, and glass beads), and uses for them or works contrived from them.[58] Many contributions deployed his mercantile connections with Asia, Africa, and the Americas, including his presentation of exotic plants, "two bodies, supposed to be the pizzles of a shark," and "a pair of Indostan shoes, an arrow, and some writing upon a palm leaf."[59] His contributions, including the presentation of "a monstrous chicken hatched with four legs at Godalmin in Surrey," were part of the characteristic activities of the broad community of the curious in the recently formed European scientific societies as they forged a new natural philosophy based on the examination of "facts."[60]

Houghton's *Collection* reflected his involvement with the Royal Society. In many ways it was a continuation of Francis Bacon's and Samuel Hartlib's ideas for a series of histories of the trades, which had been taken up by the Society and in particular by Robert Hooke, Robert Boyle, William Petty, and John Evelyn.[61] Moreover, much of the content of the *Collection* came from fellows of the Society. For example, Evelyn on French bread; John Beale

---

56. Michael Hunter (1982) *The Royal Society and Its Fellows, 1660–1700: The Morphology of an Early Scientific Institution*, British Society for the History of Science Monograph 4 (British Society for the History of Science, Chalfont St. Giles); and Birch, *History of the Royal Society*, vol. 4 pp. 144–45, 147, 502.

57. Birch, *History of the Royal Society*, vol. 4 pp. 302–3 (11 June 1684), 431 (18 November 1685).

58. Birch, *History of the Royal Society*, vol. 4 pp. 20–21 (4 March 1680), 34 (22 April 1680), 44 (8 July 1680), 73–76 (16 March 1681), 110 (7 December 1681), 200 (18 April 1683), 522 (26 January 1687). One of Houghton's written contributions was Houghton (1699) "A discourse of coffee," *Philosophical Transactions*, 21, pp. 311–17.

59. Birch, *History of the Royal Society*, vol. 4 pp. 93 (29 June 1681), 99 (26 October 1681), 425 (28 October 1685). These connections included his brother in Virginia, who sent him answers to various queries; see p. 96 (27 July 1681).

60. Birch, *History of the Royal Society*, vol. 4 p. 533 (27 April 1687). For debate on the characteristics of the investigation of "matters of fact" in the late seventeenth-century scientific societies, see Lorraine Daston and Katherine Park (1998) "Strange facts," in *Wonders and the Order of Nature, 1150–1750* (Zone Books, New York) pp. 215–53; Steven Shapin and Simon Schaffer (1985) *Leviathan and the Air-Pump: Hobbes, Boyle, and the Experimental Life* (Princeton University Press, Princeton, NJ); and Barbara J. Shapiro (2000) *A Culture of Fact: England, 1550–1720* (Cornell University Press, Ithaca).

61. Walter E. Houghton Jr. (1941) "The history of trades: its relation to seventeenth-century thought," *Journal of the History of Ideas*, 2, pp. 33–60. A key example, produced at the same time

writing as "a very aged Vertuouso" to recommend books by Evelyn, Hartlib, and Mun; and an essay on malting by John Flamsteed.[62] Other material was taken directly from the *Philosophical Transactions*, including Charles Howard's discussion of saffron cultivation, Boyle's enquiries on agriculture, and extensive essays on earth, air, fire, and water published in the 1690s that drew on the work of Boyle, Hooke, John Wallis, and Edmond Halley. These also included Houghton's own experiments on snow, conducted at Gresham College with "my very good Friend, Dr. *Hans Sloan*, F.R.S."[63]

On its reappearance in 1692 Houghton's *Collection* was supported by a testimonial attested by twenty-eight gentlemen, including twenty-four identifiable as fellows of the Royal Society (see Figure 14).[64] Among them were such luminaries as Samuel Pepys, Robert Plot, Nehemiah Grew, Evelyn, Sloane, and Halley. This, it stated, was "to testifie our Knowledge of the approv'd Abilities and Industry of Mr. JOHN HOUGHTON, Citizen of London, and Fellow of the ROYAL SOCIETY there, in the Discovery and Collection of Matters worthy [of] Observation, and more particularly such as relate to the Improvement of *Husbandry* and *Trade*." It recommended him and his aim of "communicating the Effects thereof to the Publick" to the notice, assistance, and encouragement of gentlemen "desirous of promoting the Endeavours of a Person so qualify'd and dispos'd to the Service of his Country."[65] However, Houghton claimed more from his involvement with the Royal Society than just a testimony to his character and connections.[66] The Society was, he had stated in 1681, to provide him with a way of navigating knowledge of the difficult world of trade. While he followed Lewes Roberts's "Map of Commerce, *cap.*2" in arguing that trade "is nothing else but a Communication, Bargaining, Contracting, or Exchanging of one Man

as Houghton's *Collection* is Joseph Moxon (1683–84) *Mechanick Exercises on the Whole Art of Printing* (printed for J. Moxon, London).

62. Houghton, *Collection*, 16 January 1683, 15 June, and 20 July 1682. Flamsteed sent all the numbers of Houghton's *Collection* to the virtuoso Richard Towneley in Lancashire and noted his own contribution; see Eric G. Forbes and (for Maria Forbes) Lesley Murdin and Frances Willmoth (eds.) (1997) *The Correspondence of John Flamsteed, The First Astronomer Royal*, vol. 2: *1682–1703* (Institute of Physics Publishing, Bristol and Philadelphia) letters 459 (22 June 1682) and 480 (8 December 1682).

63. Houghton, *Collection*, 6 November and 11 December 1683; 4, 11, 14, 18 (for quotation), 21 and 28 May 1692, and 4 and 18 June 1692.

64. Hunter, *The Royal Society and Its Fellows*. It also included the naval administrator William Hewer, who had been unsuccessfully proposed as a fellow by Houghton in 1681.

65. Houghton, *Collection*, 30 March 1692; and Houghton, *Proposal*.

66. On the importance of such issues to the Royal Society's truth telling, see Steven Shapin (1994) *A Social History of Truth: Civility and Science in Seventeenth-Century England* (University of Chicago Press, Chicago).

with an other," and therefore a simple extension of specialized agricultural production, he also recognized the contentious nature of the discussion over trade and improvement in print.[67] To preempt accusations against him, he argued that in dealing with such matters he would be guided by something akin to the Royal Society's experimental program:

> I shall not, for fear of censure, swim down the Current of the Times, or swallow the vulgar Errors; no Authority shall prevail with me, though it comes from them that are esteem'd the most sagacious, unless I can apprehend it to agree with sense and reason; for you know our *Motto* is *Nullius in Verba*, and 'tis possible I may sometimes take upon me the Office of being an *Observator*.[68]

The Royal Society's production of knowledge, in the laboratory or in print, was of course by no means as straightforward as this suggests.[69] However, Houghton's rhetorical adherence to their creed that knowledge was a matter of sense and reason, and to be taken "On no man's word," offered him a position from which to write about commerce that could not simply be condemned as partisan and self-serving (see chapter 4). Yet, as James Jacob has shown of Houghton's reworking of the Society's ideology of natural philosophy in the service of Charles II in the 1680s, the *Collection* did have a political aim.[70] This was signaled by the "Office" Houghton suggested for himself. It was his commentary, as an "Observator," on "Mr Roger Coke's Reasons of the Decay of the English Trade," that produced the statements that so angered John Beale.[71] Moreover, the *Observator* was the title of the periodical published by Roger L'Estrange, the licenser of the press and arch Tory propagandist. It was L'Estrange's venture with which Houghton alligned "the design of these Papers," claiming that both publications sought "the Peace and Prosperity of your King."[72] The *Collection*, therefore, sought to deploy print and the Royal Society's knowledge-making practices in the interests of a Tory ideal of monarchy and national improvement.

---

67. Houghton, *Collection*, 8 September 1681, p. 3, is referring to Lewes Roberts (1638) *Merchants Mappe of Commerce* (London). See also [John Houghton] (1677) *England's Great Happiness. Or, a Dialogue Between Content and Complaint* (printed by J.M. for Edward Croft, London).

68. Houghton, *Collection*, 8 September 1681, p. 3.

69. Shapin and Schaffer, *Leviathan and the Air-Pump*; and Johns, *The Nature of the Book*.

70. Jacob, "Restoration ideologies and the Royal Society."

71. Houghton, *Collection*, 24 November 1681. The work discussed was Roger Coke (1670) *A Discourse of Trade* (London, printed by H. Brome and R. Home).

72. Houghton, *Collection*, 29 January 1684 (pp. 97–98).

Houghton's project also drew upon the practices of the Royal Society in promoting the widespread distribution of knowledge through printed communication.[73] His prefatory letter to John Beale, published when the *Collection* began in 1681, stated that "my design is often to publish Papers, as shall cause this Kingdom to be so well husbandry'd, as to exceed not only the United Provinces, but also what on another occasion, you were pleased to stile the Garden of the World, *Barbadoes*."[74] This was a recognition of Beale's "penchant for public communication" and an acknowledgment that, in pursuit of the goals of Bacon and Hartlib, "[h]is was a constant voice in the early Society for greater dissemination of information by correspondence and print."[75] Indeed, as early as 1662, Beale was suggesting to Henry Oldenburg, the Society's secretary, that they might use "the Mercuryes," or newssheets, for scientific communications.[76] Houghton learned from Oldenburg too. His *Collection* imitated the literary technologies of the first scientific periodicals, being made up of pages filled with a wide variety of matters of fact.[77] He argued that his model in "this my Miscellanious method hath been the Examples of Mr. *Oldenburgh*, Dr. *Grew*, and Mr. *Hook*, in their *Philosophical Transactions*, and *Collections*." He hoped that "this *Libertine* way of handling this Matter, may perhaps prove more useful, than had I bound up my self to the severest Rules." This was a response to both his audiences and the printing trades. For the former, Houghton published "in small parcels often" so "that they may do the greatest good in the least time, and that not only the Theorical Gentlemen, but also the Practical Rustic may enjoy their benefits." For the latter, he hoped that publishing this way would cause "a greater Consumption" and, as a result, "the *Bookseller* and I may have the better understanding."[78]

For Houghton it was not just in terms of the press that increased circulation meant an increase in prosperity. His aim was to promote the bringing of more agricultural goods to the market (particularly by encouraging enclosure of the commons), an expanded export trade, more consumption, an

---

73. Johns, *The Nature of the Book*; Marie Boas Hall (1965) "Oldenburg and the art of scientific communication," *British Journal for the History of Science*, 2:8, pp. 277–90; and Marie Boas Hall (1983) "Oldenburg, the *Philosophical Transactions*, and technology," in John G. Burke (ed.) *The Uses of Science in the Age of Newton* (University of California Press, Berkeley) pp. 21–47.

74. Houghton, *Collection*, 18 September 1681, p. 2.

75. Stubbs, "John Beale [part 2]," pp. 349, 326.

76. A. Rupert Hall and Marie Boas Hall (eds.) (1965) *The Correspondence of Henry Oldenburg*, vol. 1: *1641–1662* (University of Wisconsin Press, Madison) item 251, Beale to Oldenburg, 21 December 1662, p. 481.

77. Daston and Park, *Wonders and the Order of Nature*.

78. Houghton, *Collection*, 18 September 1681, pp. 3–4.

increase of population (including by immigration), and a greater number of traders. In 1677 he had published an imagined dialogue set in a coffeehouse (or "Complaining School") between the tea-sipping Mr. Content and the coffee-drinking Mr. Complaint. They debated each of these points in order to demonstrate the ideological position that despite the complaints about bullion exports, "over-high living," "too many Foreigners," enclosures, and excessive competition in trade, these were all in fact the causes of "England's Great Happiness" and that "[w]e have more Wealth now, than ever we had at any time before the Restauration of his Sacred Majestie."[79] Indeed, it was also these political and economic goals that were to be achieved through the medium of print in the *Collection*. For Houghton it was the widespread circulation of reliable information that was the key to improving the circulation of goods, capital, and people to the greater profit and strength of the nation and the monarch.

This improvement through circulation was to be achieved not only via the essays on improving agriculture and trade, but by all the material that Houghton published. For example, it was in pursuit of such goals, although rather overemphasizing his influence, that he began to print favorable reports on the progress of the war in 1692. Houghton argued that the eight thousand copies of the *Gazette* would not get these reports to ten thousand parishes and a population of eight million people, and "'twould be much for the Improvement of Trade to have them spread."[80] More importantly, increases in profitable trade were to be a direct effect of the numerical information that was published in the *Collection* from 1692. His table of commodity prices would "inform the *Corn-Merchant, Meal-man, Baker, Brewer, Feeders of Cattel,* &c. when 'tis best to buy; and the *Farmer* or *Maultster* when best to sell," and "inform, not only those that live in *Port*, but such who live at *distance*, the best time of laying in their *Coals*," as well as providing a similar service for buyers and sellers of hops, soap, tallow, and wool.[81] Houghton's publication of material abstracted from the Custom House bills of entry, which helped buyers and sellers of imported commodities to find each other, also intended to further "the Design of these my *Papers*," "that *Trade* may be better understood, and the whole *Kingdom* made as one *trading City*."[82]

79. [Houghton], *England's Great Happiness*, pp. 3, 20. Houghton presented a copy of his pamphlet to the Royal Society on 8 June 1681; Birch, *History of the Royal Society*, vol. 4 p. 90.

80. Houghton, *Collection*, 1 June 1692.

81. Houghton, *Proposal*, p. 1.

82. Houghton, *Collection*, 27 April 1692.

It was on these grounds, and to make the venture as profitable as possible, that Houghton aimed to achieve as wide a circulation as he could for his periodical. The *Collection* was sold from his house in Bartholomew Lane, just north of the Royal Exchange, and by booksellers in Cornhill, St. Paul's Churchyard, and Poultry, and at the New Exchange. Houghton also promised that "[a]ny one in *London* and *Westminster* may have these *Collections* in *Sets* or in *Sheets* as they come out, brought to their Houses, they paying only a penny the *Sheet*: and 'tis hop't that all who like them, will tell their Acquaintance."[83] When the *Collection* reappeared in 1692 he initially proposed to provide a weekly "Letter" by subscription for a guinea a quarter, or forty shillings a year, plus postage to be paid in advance. For those who thought that too much, he suggested that "several may *club*, making it as *easie* as they will, and have it sent to any publick place of their meeting."[84] Houghton used whatever channels of distribution he could. It was available from at least nine booksellers in both the City and Westminster, and from George Rose's premises in Norwich. The *Collection* was also sold by "Hawkers," and he promised that "any body in England may have them by the Post," either by personal subscription or collectively "from a Bookseller, Coffee-Man, or some other, who may afford to pay a Carrier, and sell them there for 2d. or at most 3d." These middlemen were even offered the papers on a sale-or-return basis.[85] Finally, he justified the adoption of more advertising as a way of increasing circulation since it would "particularly encourage the Advertizers to increase the vent of *my* Papers."[86] So, by 1693 he was claiming that his *Collection* was sold by "most of the Booksellers in *England*" and that while the larger print run for the *Gazettes* made them "the most universal Intellingencers," he could claim to be "their first Handmaid, because it goes (tho' not so thick, yet) to *most* parts: It's also *lasting*, to be put into Volumes with Indexes."[87]

The *Collection* also depended upon gathering intelligence in. To do so it relied upon what Houghton called his "Correspondence." This made the collecting of information into a collective enterprise with which the readers of the periodical were to be actively involved, contributing material for publication and responding to previous issues.[88] Once again, in undertaking

83. Houghton, *Collection*, 13 March 1683.
84. Houghton, *Proposal*, p. 2.
85. Houghton, *Collection*, 27 and 30 April and 7 May 1692.
86. Houghton, *Collection*, 21 July 1693.
87. Houghton, *Collection*, 30 June and 21 July 1693.
88. Glaisyer, "Readers, correspondents and communities."

his enterprise John Houghton drew upon the Royal Society. Indirectly, his
information network mirrored the correspondence with natural philosophers in Britain, Europe, and North America maintained, particularly by
Oldenburg, in the interests of the production, collation, and dissemination
of scientific knowledge.[89] More directly, Houghton informed his readers
that his enterprise was made possible by the Society, not only because he
participated in its "Discourse" and was able to use its library, but, as he
wrote, because "some of them (I thank 'em) have help'd me to such Correspondence, as will, by the help of my own Friends and Industry, furnish
me, I question not, sufficiently to make good my Title."[90] Houghton could
certainly have taken the short walk down Bartholomew Lane to Jonathan's
coffeehouse to gather stock prices himself (see figure 12), and he later reproduced them from other published sources as he had the bills of mortality
and the Custom House entries.[91] However, publishing prices for domestic
commodities from markets across the country required a real and extensive correspondence, as did drawing together the dispersed knowledge on
agricultural practice and the trades. In 1683 Houghton told his readers that
he had printed a letter to be sent to ten or twenty thousand "ingenious
Persons" from whom he expected to receive accounts "tending to the Increase of Husbandry or Trade" that he would publish. Indeed, he promised
minimal intervention between author and reader, suggesting that he would
"be meerly *Editor*" of these contributions.[92] In the 1691 *Proposal* Houghton
again argued that the nature of the project necessarily made it a collective endeavor, "and not the Work of one or few hands." He hoped, therefore, that he would "obtain, *every Week*, one Letter at least out of each
County of the Kingdom; to which purpose, all Lovers of this *Work* are invited."[93] In pursuit of this constant aim of increasing the number of places
from which he received information, he promised a copy of the *Collection*
to anyone willing to send him prices from their market. Correspondents
were also requested to "look over all the Prizes of their Town mentioned

---

89. Hall, "Oldenburg and the art of scientific communication"; and David S. Lux and Harold J.
Cook (1998) "Closed circles or open networks? communicating at a distance during the scientific
revolution," *History of Science*, 36, pp. 179–211.

90. Houghton, *Collection*, 18 September 1681.

91. Neal, "The rise of a financial press."

92. Houghton, *Collection*, vol. 2 no. 1, 1683 p. 3. On the relationships of credibility and
visibility governing editorial practice, see Robert Iliffe (1995) "Author-mongering: the 'editor'
between producer and consumer," in Anne Bermingham and John Brewer (eds.) *The Consumption
of Culture, 1600–1800: Image, Object, Text* (London, Routledge) pp. 166–92.

93. Houghton, *Proposal*, p. 1.

in my Paper, when they write to me, because I would be as exact as possible."[94]

This correspondence, and the production and circulation of the printed issues of the *Collection*, were necessarily part of the particular material geographies of information in late seventeenth-century England. The whole enterprise was impossible without printers, stationers, and booksellers. From 1692 the *Collection* appeared under the imprint of Randall Taylor, who was also the stationer responsible for one of Josiah Child's pamphlets defending the East India Company, the *Discourse of the Nature, Use and Advantages of Trade* of 1694. Houghton's papers would either have been printed at his premises near the Stationers Hall, or by another of a group of Whig stationers for whom he acted as "trade publisher." They were sold across London by a diverse range of booksellers from the "Whig grandee" Awnsham Churchill in Paternoster Row to the "crypto-Jacobite" Joseph Hindmarsh up against the Royal Exchange. Some of these, including Churchill (who later undertook the publication of Flamsteed's *Historia Cœlestis*) and Samuel Smith (who distributed Newton's *Principia*) had experience with the production and distribution of works of natural philosophy. Indeed, Smith had been the apprentice to the "Stationer-virtuoso" Moses Pitt, who, at the same time as producing the highly ambitious but ultimately unsuccessful *English Atlas* (1680–83) with the support of the Royal Society, had sold the first numbers of Houghton's *Collection*.[95]

Houghton had to negotiate personally with these stationers and booksellers, who kept a keen eye on their profits and had an often fraught relationship with authors and editors. He also had to work to maintain his network of correspondents, who, in addition to providing prices and other material, could further the sales of the *Collection*. For example, the geologist John Beaumont used Houghton as a go-between between himself and Randall Taylor, the publisher of both men's work, but also promoted Houghton's "papers," by showing them "to several Gentlemen who tell me they will subscribe for setts of them, wth myself, to ye Bookseller at Wells."[96] Houghton also had to be continually aware of the realities of his distribution mechanisms. As Richard Saunder informed him from Melton Mowbray,

---

94. Houghton, *Collection*, 6 April and 18 May 1692, and 20 October 1693. For readers' responses, see BL Stowe Mss. 747 f. 13, letter from Samuel Dale (of Braintree) to John Houghton, 8 March 1693; and f. 14, letter from Richard Saunder (of Leesthorp) to John Houghton, 20 March 1693.

95. Johns, *The Nature of the Book*, pp. 66, 68, 150, 573.

96. BL Stowe Mss. 747 f. 18, letter from John Beaumont jnr (of Wells) to John Houghton, 2 July 1693.

not only had Houghton made mistakes in publishing the prices Saunder had previously sent, but his copy of the *Collection* had been delivered to Worcestershire rather than Leicestershire.[97] More significantly, getting the *Collection* to its readers became entangled in the troubled relationships between the Penny Post (itself a joint-stock company, although probably an unprofitable venture) and the General Post Office, whose monopoly it threatened and which eventually took it over.[98] William Dockwra, the Penny Post's founder, wrote to Houghton to tell him that various places in Essex, including Chigwell, Ongar, and Epping, had been removed from the Penny Post "decreasing its Revenue, & abateing the people's conveniency in hopes of adding more profit to the Generall Post." Other places from which he returned undelivered copies of Houghton's papers were "such places & houses as are so sottish as not to take y^m in even gratis." These were poorer riverside parishes such as Stepney, Limehouse, Ratcliffe, and Poplar, and he advised Houghton to "cast no more of yor Pearles before such Swine."[99]

Although it was built from the partial, problematic, and uncertain relationships that governed the making and distribution of knowledge in print, Houghton's *Collection* imagined a quite different geography of information. What it presented to its readers was a single, internally differentiated national economic space. Again, it is worth contrasting this to Whiston and Woolley's provision of prices for traders in international commodities and Castaing and Freke's localized services for the stock market in Exchange Alley. In the *Collection* this imagined geography was understood in terms of the improvement of agriculture and trades in the national interest by sharing experience and techniques between places.[100] It was also understood through the tabulation of statistics that laid out a spatially differentiated, but comparable, set of figures on prices, trade, and demographics that, even when incomplete, suggested that it could begin to describe the nation.[101] This aim, and the uses to which it might be put, are well demonstrated by Houghton's publication in 1693, as a separate penny sheet, and later in the *Collection*, of "An Account of the Acres & Houses, with the Proportional

---

97. BL Stowe Mss. 747 f. 14.

98. Scott, *Constitution and Finance*, vol. 3 pp. 42–51.

99. BL Stowe Mss. 747 f. 93, letter from William Dockwra to John Houghton, 25 April 1698.

100. Stubbs, "John Beale [part 2]."

101. On publishing his second table of information, Houghton (*Collection*, 6 July 1692) argued that "Whoever compares this with the former, will see the *Rise* and *Fall* of *Markets*; it is easier than to make *Marks*" (p. v). He also addressed the problem of the diversity and comparability of weights and measures used in different places; Houghton, *Collection*, 9 and 23 June 1693. See Julian Hoppitt (1993) "Reforming Britain's weights and measures, 1660–1824," *English Historical Review*, 108, pp. 516–40.

Tax, &c. Of each County in England and Wales."[102] This table, a response to the new land tax, was in the same format as the *Collection* and also produced by Randall Taylor. It set out the acreage of each county (taken from calculations Edmond Halley had made ten years previously by weighing carefully dried cutouts of county maps), the proportion of the country that it made up in decimals, the number of houses that it contained, and the proportion of national taxation revenue that it provided. Houghton, as a Tory, had argued strongly in the 29 January 1684 edition of the *Collection* that it was "the Interest of the Subjects of England, and much for the Improvement of Husbandry and Trade, plentifully to supply their King." To provide ample taxation revenue for Charles II (and, implicitly, James II) was, he wrote, in the national interest and offered no threat to Parliament. Those Whigs who feared the monarch's "*Arbitrary Power*" were mistaken.[103] Ten years later, with a Whig ministry now holding the fiscal reins and funding King William's wars, Houghton was more circumspect. He submitted his table to both Houses of Parliament, along with calculations of how many members represented places he judged as over- or undercharged, noting that "[w]hether this *Table* may shew Reason for Alteration of the Method of *Taxing*, I submit to *proper Judges*. The *Matter of Fact* I here endeavour to Demonstrate."[104] What was clear was Houghton's argument that it was only with such carefully produced matters of fact, and with these new ways of bringing a single but differentiated nation into view, that reliable political decisions could be made.

Houghton's presentation of economic and political issues via the tabulation of numerical information was, of course, part of the broader development of late seventeenth-century political arithmetic. William Petty, Gregory King, Charles Davenant, and John Graunt all turned their attention to calculating economic, political, and demographic data.[105] What made

102. John Houghton (1693) *An Account of the Acres & Houses, with the Proportional Tax, &c. Of each County in England and Wales* (printed for Randall Taylor, London); and *Collection*, 20 and 27 January 1693.

103. Houghton, *Collection*, 29 January 1684 (pp. 97, 114); and Jacob, "Restoration ideologies and the Royal Society."

104. Houghton, *An Account of the Acres & Houses*. At least one reader wanted more. The botanist Samuel Dale told Houghton that "An account of what each county paid to the Quarterly Poll would not be unacceptable to the Publick if it were published in one of your papers." BL Stowe Mss. 747 f. 13ʳ.

105. Peter Buck (1977) "Seventeenth-century political arithmetic: civil strife and vital statistics," *Isis*, 68, pp. 67–84; Miles Ogborn (1998) "The capacities of the state: Charles Davenant and the management of the Excise, 1683–1698," *Journal of Historical Geography*, 24, pp. 289–312; and Mary Poovey (1998) *A History of the Modern Fact: Problems of Knowledge in the Sciences of Wealth and Society* (University of Chicago Press, Chicago).

Houghton's endeavors different was that his efforts were situated within the production of a printed periodical that aimed for widespread and regular distribution. His gathering and dissemination of this material depended upon print. A printed periodical was able to make visible what had not been readily available before. It made possible a reliable depiction, for subscribers anywhere in the country, of the standing of markets in other places and, over time, their rise or fall. Even when not every cell in the table of prices was filled, and even when the number of places from which there was information was limited, there remained the potential for comprehending the nation's husbandry and trade. In addition, Houghton was also distinctive in having what Nastasha Glaisyer describes as "a vision of the mutual dependence of husbandry, trade and the new financial order."[106] He therefore included in this tabulation of the nation the workings of the stock market. Moreover, he sought to explain this market to his readers in a way that was also shaped by the culture of information, knowledge, and fact that guided the *Collection* as a whole: the use of arguments from natural philosophy, an understanding of the benefits of circulation and the wide distribution of information, and the imagining of a national economic space.

Printing Prices, Explaining Markets

Houghton had included a discussion of investment in East India Company bonds, and the reasons for fluctuating interest rates on those investments, in the *Collection* for January and February 1683, and he had published a separate account of the Bank of Credit in the same year.[107] He had also been publishing stock prices for several years before he informed his readers in 1694 that "for a little Diversion from Natural History, and for the usefulness of the thing it self, I have thought fit to treat of *Joint-Stocks*, and of the various dealings therein, commonly called *Stock-Jobbing*."[108] There was no indication in the *Proposal* for the *Collection*'s relaunch that this would be among the subjects covered, and it seems most likely that Houghton was responding to the boom in joint-stock company foundations and investment in the early 1690s, which meant that around 85 percent of the 150 companies

106. Glaisyer, "Readers, correspondents and communities," p. 235.

107. Houghton, *Collection*, 16 January and 13 February 1683. [John Houghton] [1683?] *An Account of the Bank of Credit in the City of London* (printed by John Gain, London) considers a project that came before the Royal Society in 1682; see Birch, *History of the Royal Society*, vol. 4 p. 164.

108. Houghton, *Collection*, 8 June 1694.

in existence in 1695 had been formed since 1688.[109] Houghton recognized the problematic politics of information surrounding stock-jobbery and positioned himself as an interested outsider "not much concerned in Stocks, and therefore hav[ing] little occasion to Apologize for Trading therein." He explained his aim as being "to rectifie Men's Judgements, whom I find running into Errors and Mistakes." Bolstered by an extensive quotation in Latin from "the learned *des Cartes,*" which the Frenchman had used "to introduce his most excellent Philosophy into the World, he well knowing what difficulties he was to encounter... by reason that the generality of Men had imbibed certain wrong Notions, which they esteemed as Maxims not to be receded from," Houghton addressed himself to his readers. He asked them "to lay aside all Prejudice and Partiality, there being nothing that is so great hindrance to a Man's coming to the true Knowledge of any thing, as false Preconceptions and Notions of it." He wished them to act as natural philosophers should and to approach the world with an open and questioning mind. Without the adoption of a principle like that of *Nullius in Verba*, he noted, his discussion of stock trading would be to no purpose at all: "I am the larger on this, because I know many worthy Persons of great Honour and Probity, who deal in Stocks, that do abominate the least unjust Action, and would not for the World have an ill gotten Penny among the rest of their Estates; and it is a great hardship on such Gentlemen to undergo the Censures of Mankind, who inveigh against all Traders and Trading in Stock, tho' at the same time they know little or nothing of it."[110]

Houghton understood this new world of finance in terms of improvements in the circulation of capital. His account of the Bank of Credit in the City of London set it within a history of barter, money, bills of exchange, and banking that saw each develop as a mechanism to resolve the problems of the extension of trade and commerce over increasing spans of time and space.[111] Yet he was wary of the way in which this might be done without any solid basis or guarantees, leading to bank failures and creditors' losses. In contrast, the Bank of Credit simply allowed nonmonetary assets to become liquid. Deposits were made in goods (or, for the nobility and gentry, in mortgages on land, or plate and jewelry), which were valued and stored; a bill of credit was issued for something less than their value, and at an interest rate of 6 percent. These bills were as good as ready money with the

109. Scott, *Constitution and Finance*, vol. 1 p. 327.

110. Houghton, *Collection*, 8 June 1694.

111. [Houghton], *An Account of the Bank of Credit*; see also the discussion of "The Matter and use of Money" in Houghton, *Collection*, 27 October 1681.

traders connected to the bank. Houghton's publication of stock prices served a similar purpose. Stock trading, he noted, "however spoke against, makes a *quick Circulation of Money*."[112] His printed tables were to help bring new capital into the market. Thus they were "design'd, for Incouragement of those at distance to turn *Merchants*, and to inform them how their *Stock* goes, that they may thereby see when best to buy or sell."[113] This broadening of knowledge about the stock market through the medium of print would, he argued, be in the national interest:

> Altho they that live in *London*, may, every *Noon* and *Night* on Working-days, go to *Garraway*'s *Coffee-house*, and see what *Prices* the *Actions* bear of most *Companies* trading in *Joynt-Stocks*, yet for those whose Occasions permit not there to see, they may be satisfi'd *once a Week* how it is, and thereby the whole Kingdom may reap Advantage by those *Trades*: Also they may learn some of the *Cunning* of *Merchandizing*, and have this *Advantage*, by laying their Monies there, in one or two days time they may *sell*, and have their *Money* to supply their wants at any time. Without doubt, if those Trades were better *known*, 'twould be a *great Advantage* to the Kingdom; only I must *caution* Beginners to be very *wary*, for there are many *cunning Artists* among them.[114]

Realizing these benefits and avoiding the dangers was, therefore, a matter of spreading information. Houghton told his readers in 1693 that "[t]his buying and selling of *Actions* is one of the great Trades now on foot," but finding that "a great many understand not this affair," he sought to explain it to them.[115] In doing so, like many nonbrokers, he used the term "Stock-Jobbing" to mean simply the buying and selling of stocks.[116] Houghton listed the main joint-stock companies, and discussed the recent growth in smaller joint stocks as investors sought to use their money in ways that were "more easily shifted from hand to Hand" than land, houses, or commodities. He explained the mechanism as a "plain, honest Proceeding" for someone who "has thought of an Art or Invention, or discover'd some Mine, or knows, or thinks of some New (or New manner of) way of Trade" that cannot be funded by them alone. As he put it, "He then imparts it to some Friend or

---

112. Houghton, *Collection*, 24 February 1693.
113. Houghton, *Proposal*, p. 1.
114. Houghton, *Collection*, 6 April 1692.
115. Houghton, *Collection*, 17 March 1693.
116. Dickson, *The Financial Revolution*, p. 494.

Friends, who commonly consider or enquire of the Learned, Whether 'twill stand good in Law." If so, they agree on some principles of constitution, including the voting rights that shares bring, and on the form of remuneration for the "first Inventer" (in money or shares). Then "each brings in his Friend, 'till all the Shares be bought at such a Price as stated."[117] A committee and chair are then chosen to manage the affairs of the joint-stock for a year. Houghton's explanation of such companies as a circle of "Friends" effectively domesticated these new concerns and explained them to his readers in the very terms that he used to understand and extend the "community" formed through the *Collection*.[118] The familiarity and reliability of these ventures was also emphasized in Houghton's presentation of their orderly administration: committees "consult the best ways for carrying on the business"; bills are paid on time; all decisions and transactions are entered in the minute book by the clerk; and there is a careful subdivision of committees to manage particular parts of their affairs and to audit the accounts.[119]

The mechanics of stock trading were also shown as working just like any other commodity transaction:

> The manner of managing the Trade is this; The Monied Man goes among the *Brokers*, (which are chiefly upon the *Exchange*, and at *Jonathan*'s Coffee-House, sometimes at *Garaway*'s and at some other Coffee-Houses) and asks how *Stocks* go? and upon Information, bids the Broker buy or sell so many Shares of such and such Stocks if he can, at such and such Prizes: Then he tries what he can do among those that have Stocks, or power to sell them; and if he can, makes a Bargain.[120]

However, there was a recognition that the prices at which stocks were sold depended upon somewhat intangible conditions, since "if they [the sellers] have hopes of great Gain, they will not sell their *Share* for 10*l*. If they fear Loss, they'l sell for less; and so *Actions* rise or fall, according to hopes and fears."[121] These fluctuations in the market, on the basis of anticipations of the future, formed the conditions for a series of more complex deals in options known as the "refuse" and the "put." The refuse was where the option of buying the shares at a set price within a set time period was sold

117. Houghton, *Collection*, 15 June 1694.
118. Glaisyer, "Readers, correspondents and communities."
119. Houghton, *Collection*, 15 June 1694.
120. Houghton, *Collection*, 22 June 1694.
121. Houghton, *Collection*, 17 March 1693.

for a fixed fee per share. The potential buyer gambled on a rise in prices, the seller on their fall. The put was similar, although here the option sold was one to compel another trader to take the shares at an agreed-upon price within a particular time period. Houghton explained how these deals might be used by experienced traders "to sell to one, and buy of another different Shares of the same *Stock* for different prices, and so make Advantages." Most complex was "*the great* Mystery *of* Buying *more than* all." This involved wealthy men combining, by "Friendship," to secure as many of the shares in a stock as possible, and then securing promises from others that didn't hold any of the stock that they would sell it to them in the future (those sellers thinking that the market would not rise sufficiently to deprive them of a profit derived by buying the shares in time at a lower price). When the time comes due they are forced to buy "at such rates as they [the wealthy men] or their Friends will sell for; tho' Ten or Twenty times the former Price." However, unlike the critiques of stock jobbing, Houghton played down the manipulation of the markets by powerful, hidden forces, noting simply that "in small Stocks 'tis possible to have Shares rise or fall by the Contrivances of a few Men in Confederacy; but in great Stocks 'tis with more difficulty."[122]

Houghton's aim was to explain all of this "in plain *English*," stressing both the potential pitfalls and the possible profits for those thinking of entering the market.[123] He was also able to provide, in print, the wording of the transfer of shares in the companies' transfer books and the forms of contracts made for the "put" and "refuse" of shares. These seem to have been standard in Exchange Alley by this time, so those new to the stock market could use them to judge the legal security of such deals, even if they could not be guaranteed that they would make money on them. They might also check any contracts they did make against the printed versions in the *Collection*.[124] Houghton also published the names of "Exchange" brokers alongside those of brokers in corn, grocery, hemp, and silk and gave an indication of the rates of brokerage they charged per share. In providing this information he certainly had his own brokerage business in mind. Specific market information came at a price. So when he published an expanded list of over fifty stocks in May 1694 (Figure 15), he only gave the values for those that had been listed before, noting that "[a] great many desire a List

122. Houghton, *Collection*, 6 and 13 July 1694.
123. Houghton, *Collection*, 22 June 1694.
124. Houghton, *Collection*, 13 July 1694; and Carruthers, *City of Capital.*

of *Stocks*; and those that desire the Values of them to be Published may have them on very reasonable Terms."[125] Yet there was a broader goal here too. Houghton's report on the progress of the companies whose stock prices he made available aimed "to shew the Advantages to the Nation, tho' the Undertakers get or lose." Overall, having reviewed the fortunes of a wide range of joint-stock companies, he argued once more that investment in this new economy was in the national interest, "foreasmuch as I can see, if an Hundred Stocks more arise, we have no reason to be offended." As he had said before, "the whole Kingdom may reap Advantage by those *Trades*."[126]

Houghton has been defined as an economic optimist.[127] In terms of his publication of stock prices and his explanations of the stock market, this involved a version of "stock-jobbing" that presented it as part of, rather than separate from, the worlds of agricultural improvement and trade. Houghton's presentation of a stock market in which all could, perhaps even should, become involved and that might work in the national interest stood in strong contrast to contemporary pamphleteers' accusations of stock-jobbing and trickery and their identification of this as springing from and defining the working of the East India Company. In the early 1690s the stock market was a new phenomenon, and Houghton and others struggled to define what it meant and how to understand things like the fluctuations in share prices or the morality of options and futures deals.[128] As has been shown, different forms of print were used to offer different versions of the stock market. Satirical and overtly political pamphlets made charges of corruption and deception. Simple price listings sought to open its workings to commodity traders, and Houghton drew on the Royal Society's forms of knowledge making and ideals of printed communication to present reliable and trustworthy information to anyone in the country willing to pay a penny for it. Since the market was driven up and down by hopes, fears, anticipations, and speculations fueled by interpretations of various forms of knowledge, information, and fact, each of these forms of print was part of the workings of the stock market rather than simply a commentary upon it. Moreover, in a context where joint-stock companies were substantial holders of government debt

125. Houghton, *Collection*, 11 May and 6 July 1694.
126. Houghton, *Collection*, 30 March 1692 and 20 June 1694.
127. Appleby, *Economic Thought and Ideology*; and Hunt, *The Middling Sort*.
128. Dickson, *The Financial Revolution*, p. 516, notes that there were repeated attempts between the 1690s and the 1770s to control Exchange Alley, but they "were ineffectual, owing to their time-honoured but futile identification of economic pressures with human wickedness or folly."

and the sites of political conflict, it meant that these versions of the stock market were all involved in making and taking political positions. Some were obvious, others less so. In John Houghton's case, the overtly Tory position of support for Charles II through taxation uncontrolled by Parliament that he presented in the *Collection* in the early 1680s was less in evidence by the 1690s. However, in 1694 he supported the East India Company, associated with Tory interests and James II and under sustained attack from the Whigs, by arguing that "[t]he *India* has traded openly, besides private Trade, for a Million a Year; and I don't doubt if they were not hindred but they would make it Tenfold."[129] Unlike other Tories, he did not simply denigrate, via accusations of stock jobbery, the new financial order associated with the post-1688 regime. Instead, he recognized its importance, presented it in a national context, and sought to broaden the engagement with it far beyond the money men of Exchange Alley. In Houghton's publication of stock prices, therefore, it is possible to look beyond what is now simply a familiar listing of a commonplace commodity to see a distinctive project of fashioning a stock market for a Tory commercial nation and support for the East India Company against its enemies.

Conclusion

Ironically, it was a later attempt to work the stock market in favor of a Tory administration that began the first great stock market crash: the bursting of the South Sea bubble in 1720. This crisis demonstrated how the value of intangible commodities such as stocks could become detached from any material referent, with disastrous consequences. It was rumor, gossip, speculation, and reputation, in print and speech, that drove stock prices up and down. In the storm of engraved images and printed pamphlets and ballads that accompanied it, the bubble seemed to affirm with a vengeance the notion that the stock market was the site of lies, deception, and the immoral, even devilish, manipulation of appearances in the pursuit of greed and private interest. Morality was, it seemed, completely removed from economic transactions when it came to stock market dealing.[130]

Accusations of the immorality of the market were, of course, a regular feature of seventeenth- and eighteenth-century economic thought and

---

129. Houghton, *Collection*, 20 June 1694.

130. John Carswell (1960) *The South Sea Bubble* (Cresset Press, London); and, more popularly, Malcolm Balen (2002) *A Very English Deceit: The South Sea Bubble and the World's First Great Financial Scandal* (Fourth Estate, London).

writing.[131] As such, they were reflected upon by those who produced and con-sumed printed periodicals. In July 1712, the best-known eighteenth-century periodical, *The Spectator*, reported to its readers that it was going to become something like Houghton's *Collection*. Recognizing that "impertinent and unreasonable Fault in Conversation, for one Man to take up all the Dis-course," Addison and Steele's pages would be open to "all Persons of all Orders, and of each Sex" to send in material "for the profitable Information of the Publick." This was so that "the mechanick Arts should have their Place as well as the liberal," and because "The Ways of Gain, Husbandry, and Thrift, will serve a greater Number of People, than Discourses upon what was said or done by such a Philosopher, Heroe, General, or Poet." Do-ing this, Steele argued, would acknowledge a striking reconfiguration in the nature of "Greatness and Success" and how it might be achieved:

> Is it possible that a young Man at present could pass his Time better, than in reading the History of Stocks, and knowing by what secret Springs they have such sudden Ascents and Falls in the same Day? Could he be better conducted in his Way of Wealth, which is the great Article of Life, then in a Treatise dated from *Change-Alley* by an able Proficient there? Nothing certainly could be more useful, than to be well instructed in his Hopes and Fears; to be diffident when others exalt, and with a secret Joy buy when others think it their Interest to sell.

Such a means to greatness and success—replacing "Learning and Moral-ity" with utility and the pursuit of wealth, and replacing the "Wit [and] Humour" of the *Spectator* with the dry fare of something like Houghton's *Collection*—was not seriously proposed.[132] However, the suggestion served to draw attention once again, albeit in the language of an ironic urbane politeness rather than of rustic Christian virtue, to the same lack of overt consideration of questions of morality in publications like Houghton's that had so provoked John Beale in 1682.

In the early 1680s the problem was with Houghton's arguments. On Beale's reading, the "ugly business" was the *Collection*'s active promotion of prodigality, pride, luxury, and vanity as in the interests of the nation and the damage this would do to a Royal Society already under attack by atheists

131. Appleby, *Economic Thought and Ideology*; and Andrea Finkelstein (2000) *Harmony and Balance: An Intellectual History of Seventeenth-Century English Economic Thought* (University of Michigan Press, Ann Arbor).

132. Donald F. Bond (ed.) (1965) *The Spectator* (Clarendon Press, Oxford) vol. 4 no. 428 (11 July 1712) pp. 4–7.

and critics. From the 1690s, with the development of a stock market and the publication of stock prices, the problem was the immorality of the market written into "the History of Stocks" and played out in the deceptions of buying and selling them. Houghton had recognized this and had used the Royal Society's methods and the cultural resources of print in an attempt to make stock prices into matters of fact. These were to be available to all in black and white and, like the facts of the new natural philosophy, were presented as separable from their production, their interpretation, or the moral discourse about them. As Houghton said of trading in stocks, "it is not the Use but the Abuse of them that deserves Censure."[133]

Yet these facts were an artifact, a product of print. Houghton's separation of the truth-making practices of the Royal Society and the cunning tricks of Exchange Alley was really sustainable only in the pages and tables of his *Collection*. In the late seventeenth century, it was virtuosi as much as traders who frequented Exchange Alley coffeehouses. They rubbed shoulders in Jonathan's and Garraway's. These were regular places of resort after Royal Society meetings, a short distance away at Gresham College (Figure 12). Natural philosophers performed public experiments among the coffee cups and newspapers, and in December 1682 the Royal Society bought £250 worth of East India Company stock from John Evelyn.[134] Indeed, it was a rare day in the late 1670s and early 1680s when Robert Hooke did not pay a visit to Exchange Alley to discuss planetary motion and other matters with Sir Christopher Wren, John Wallis, John Flamsteed, Edmond Halley, and Nehemiah Grew; to write his lectures; to trade information and instruments with artisans and craftsmen; to find out the talk of the town; and to move against his enemies.[135] The business of making natural knowledge depended upon these coffeehouse transactions as well as on life in the laboratory.[136] And John Houghton, both broker and natural philosopher, was there too. Hooke, as well as nominating Houghton for Royal Society membership, knew him well enough to take care to record his moods, to invite him home, to walk the streets of London with him, and to discuss together the workings

133. Houghton, *Collection*, 8 June 1694. On the history of such facts see Lorraine Daston (1991) "Baconian facts, academic civility, and the prehistory of objectivity," *Annals of Scholarship*, 8:3–4, pp. 337–64; Poovey, *A History of the Modern Fact*; and Shapiro, *A Culture of Fact*.

134. Cowan, *The Social Life of Coffee*; Ellis, *The Coffee-House*; and John Evelyn (1955) *The Diary of John Evelyn*, E. S. de Beer (ed.) (Clarendon Press, Oxford) vol. 4: *Kalendarium, 1673–1689*, pp. 297–98.

135. Henry W. Robinson and Walter Adams (eds.) (1935) *The Diary of Robert Hooke, 1672–1680* (Taylor & Francis, London).

136. R. C. Iliffe (1992) "'In the warehouse': privacy, property, and propriety in the early Royal Society," *History of Science*, 30, pp. 29–68; and Johns, *The Nature of the Book*.

and future of the Society. Hooke filled his pipe with tobacco he bought from Houghton, and sweetened his cup with his chocolate.[137] Houghton's knowledge making was born out of the coffeehouse's combination of stock-jobbing and science, and he later reflected on coffeehouse knowledge. He certainly celebrated how they spread learning, as in his *Collection*: "*Coffee-Houses* makes all sorts of People sociable, the Rich and the Poor meet together, as also do the Learned and Unlearned: It improves Arts, Merchandize, and all other Knowledge, for here an inquisitive man, that aims at good Learning, may get more in an Evening than he shall by books in a Month." Yet there were always dangers lurking there for the ill prepared. As Houghton put it, signaling the ever-present difficulty of separating truth and trick, "but I must confess, that he who has been well Educated in the Schools, is the fittest Man to make good use of *Coffee-Houses*, and am fearful that too many make ill uses of them."[138] That went for those experimenting with East India stocks as well as with natural philosophy.

---

137. *Diary of Robert Hooke*, pp. 394, 443, 444, 454, 460. Hooke spells his name "Hauton" and "Haughton."

138. Houghton, *Collection*, 23 May 1701. The terms "Schools" refers to the universities.

CHAPTER SIX

The Work of Empire in the Age of Mechanical Reproduction

The previous two chapters detailed flourishing London print cultures, of pamphlets and politics and of periodicals and natural philosophy, within which the Company and its fortunes were implicated. In contrast, the uses of print by the Company in India were very restricted before the final quarter of the eighteenth century. When Streynsham Master was at Fort St. George, the only printed texts in use in the factory were the sheets of rules and regulations distributed by the Company directors to try to ensure uniformity of practice across their scattered settlements, and the small libraries of religious and legal books that were sent out from London or purchased from returning Company servants.[1] Everything else was in manuscript (see chapter 3). Individual employees, even in the lower ranks, did have their own book collections. For example, Nathaniell Whetham, a writer who died in September 1676, had among his possessions a bible, "Two printed books in quarto," and three other books. Yet ownership of print could not be taken for granted. William Calloway, another writer who died at the same time, was evidently more concerned with the world of script. Along with his store of trade goods—gold buttons, silver buckles, seventeen hats, and variety of textiles—he had a "China Escretore," an inkstand and fine paper, but no printed books.[2] For many years the only presses in India were those of the Jesuits. The Mughal emperors were interested in European books, but not in

---

1. *Rules, Orders, and Directions Appointed and Established by the Governour and Committees of the East-India Company, for the Well Regulating and Managing their Affairs in the Parts of* INDIA [London, 1680?]; Sir Richard Carnac Temple (1911) *The Diaries of Streynsham Master, 1675–1680*, 2 vols. (Indian Record Series, John Murray, London) vol. 1 p. 247; and Records of Fort St. George (1910) *Diary and Consultation Book, 1672–1678* (Government Press, Madras) p. 127 (19 December 1677) for the purchase of William Langhorn's library of twenty-eight law books.

2. Temple, *Diaries of Streynsham Master*, vol. 2 pp. 31–33, 37–39.

using printing themselves (see chapter 1). As for the English, Bhimji Parekh's attempts to establish a printing press at Bombay in 1674 came to nothing when the Company could not supply him with a typefounder.[3] By 1755 it remained the case that any "Publications"—circulars or notices—produced by the Company in India were composed and copied by hand.[4]

Company printing in India came with empire rather than trade. The deployment of British military strength on the subcontinent rapidly expanded as the Anglo-French global rivalries of the Seven Years War worked themselves out in the context of the decline of the Mughal empire and the intensified competition between its successor states.[5] As a result, the first press controlled by the English Company was captured from the French near Madras and run by a French printer overseen by two missionaries sponsored by the Society for the Promotion of Christian Knowledge.[6] However, it was in northern India that the fiscal and military demands of the Company's wars with Indian polities combined to lead Robert Clive to accept the position of the *dīwān* in 1765 and with it the rights to gather the revenues of the Mughal *subāh* of Bengal. Doing so involved constructing and legitimating the Company's new territorial power through existing forms of Mughal statehood and the idea of "the ancient Mogul constitution."[7] This also substantially altered what was required of the Company's administration by putting into its hands the taxation and judicial apparatus of Bengal, Bihar, and Orissa, along with dominion over the Northern Circars and the land around Madras (Figure 16). Consequently, it was Calcutta, the administrative, commercial, and social center of the Company's territorial conquests, that was to become the center of printing in British India from the late 1770s.[8]

This chapter investigates printing and empire in late eighteenth- and early nineteenth-century Bengal. To adapt Walter Benjamin, it sets out the work of empire in the age of mechanical reproduction in order to render problematic what has in an older historiography been seen as the straightforward diffusion from Europe of a portable and useful technology that promised the

3. Anant Kakba Priolkar (1958) *The Printing Press in India* (Marathi Samshodhana Mandala, Bombay).

4. IOR E/4/616: *Despatches to Bengal (Original Drafts), 28th November 1753–December 1759,* 31 January 1755.

5. Seema Alavi (ed.) (2002) *The Eighteenth Century in India* (Oxford University Press, New Delhi).

6. Graham Shaw (1981) *Printing in Calcutta to 1800* (Bibliographic Society, London).

7. Sudipta Sen (2002) *Distant Sovereignty: National Imperialism and the Origins of British India* (Routledge, London); and Robert Travers (2001) *Contested Notions of Sovereignty in Bengal under British Rule,* unpublished Ph.D. thesis, University of Cambridge.

8. Shaw, *Printing in Calcutta.*

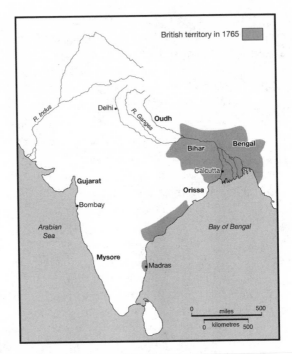

Figure 16 British territorial empire in India, 1765.

benefits of the uniform replication of large numbers of copies.[9] Printing was different in India. It was organized in particular ways, and the process and its products had specific meanings in that context. Understanding this also involves questioning the assumption that there was a simple European will to power built into the printing process and its products.[10] As Adrian Johns has argued, the outcomes assumed to be inherent to print—the fixity, standardization, and dissemination of texts—are instead the product of difficult and continuous, if often invisible, labor around the printing press, and in all the places where the products of the press are put into use. His call to make

9. Walter Benjamin (1992) "The work of art in the age of mechanical reproduction" in Walter Benjamin, *Illuminations* (Fontana, London) pp. 211–44; Katharine Smith Diehl (1990) *Printers and Printing in the East Indies to 1850*, vol. I, *Batavia 1600–1850* (Aristide D. Caratzas, New Rochelle, New York); and Dennis E. Rhodes (1969) *The Spread of Printing: Eastern Hemisphere—India, Pakistan, Ceylon, Burma and Thailand* (Vangendt & Co, Amsterdam).

10. Ngugi wa Thiong'o (1986) *Decolonising the Mind: The Politics of Language in African Literature* (James Currey, London); and Lucien Febvre and Henri-Jean Martin (1976) *The Coming of the Book: The Impact of Printing, 1450–1800* (Verso, London). See also Anindita Ghosh (2003) "An uncertain 'coming of the book': early print cultures in colonial India," *Book History*, 6, pp. 23–55.

visible the work that went into producing print's meanings and its "air of intrinsic reliability" turns attention to the social and cultural organization of the practices of printing and reading, and to the geographies of local print cultures.[11] In short, those who wished to make use of mechanical reproduction in the process of empire building in India had a lot of hard work to do.

It is, therefore, to the details of the relationship between printing and empire that this chapter turns. Making an empire in late eighteenth-century India meant building an East India Company-state in Bengal and constructing relationships between that state and the British state. Doing so involved difficult and contested issues of sovereignty, legitimacy, and administrative effectiveness. These were questions of both political power and cultural identity for which different people had different answers.[12] Any technology, such as printing, used in attempts to resolve these conflicts was inevitably part of this imperial cultural politics. As Michael Warner has demonstrated for the contemporaneous situation of anti-imperial Republican politics in the North American colonies, there can be no assumption of an automatic connection between printing and democratic politics, as embodied in the liberty of the press and the freedom of speech. He replaces these assumptions with a careful cultural history of the active construction of a relationship between print and politics in which both are transformed.[13] This was also true for printing and empire building in India. The relationships between print and the sovereignty, legitimacy, and effective functioning of a Company-state in Bengal had to be constructed and could not be assumed. Understanding Company printing in India involves situating it within the contested processes of doing imperial politics.

Overall, therefore, to set the task as one of understanding the relationship between printing and empire in British India is to pose the question in the wrong terms if it assumes that there are such different, separate, and already constituted entities as "printing" and "empire" and then asks how each affects the other. It is more productive to consider different forms of

---

11. Adrian Johns (1998) *The Nature of the Book: Print and Knowledge in the Making* (University of Chicago Press, Chicago) p. 3.

12. Peter J. Marshall (1987) *Bengal: The British Bridgehead, Eastern India 1740–1828* (Cambridge University Press, Cambridge); C. A. Bayly (1988) *Indian Society and the Making of the British Empire* (Cambridge University Press, Cambridge); Jon E. Wilson (2000) *Governing Property, Making Law: Land, Local Society, and Colonial Discourse in Agrarian Bengal, c. 1785–1830*, unpublished Ph.D. thesis, University of Oxford; Sen, *Distant Sovereighty*; and Robert Travers (2004) "'The real value of the lands': the *Nawabs*, the British, and the land tax in eighteenth-century Bengal," *Modern Asian Studies*, 38:3, pp. 517–58.

13. Michael Warner (1990) *The Letters of the Republic: Publication and the Public Sphere in Eighteenth-Century America* (Harvard University Press, Cambridge, Mass.).

imperial printing or different empires of print as specific configurations of cultural and material practices that attempted to produce particular organizations of power and knowledge. It is this that can, to twist Benjamin even further, reveal the work of mechanical reproduction in the age of empire. This begins, of course, with the recognition that the vast majority of imperial writing even in late eighteenth-century Bengal was done in script and not in print, and that these handwritten documents were subject to variable regimes of transformation and stability in relation to and separate from printing.[14] It is not, for example, the case that notions of fixity, standardization, and dissemination were inapplicable to the carefully organized procedures of manual copyists. That was exactly what Streynsham Master aimed to achieve (see chapter 3). Nor is it the case that ideas of publication, or of its use in regulating the relationships between a political power and its subjects, were incompatible with handwritten notices and circulars.[15] The use of print was not inevitable.

It will not, therefore, be argued that a ready-made "empire" used printing as a technology to further its ends, or that preformed printing techniques had their usual and predictably revolutionary effects on another part of the world newly annexed by Europeans. Instead, I will show how imperial politics were done through printing practices and how a print culture was constructed from imperial materials. In order to do this the chapter is primarily concerned with what became known as the Honourable Company's Press, and with its founding text: Nathaniel Brassey Halhed's *Grammar of the Bengal Language* (1778). The first part of the chapter details the powerful imperial promise that Halhed's *Grammar* set out for print and positions it within contemporary debates over the political philosophy of empire and the broader range of print cultures in British Bengal. The making of imperial print is then analyzed in terms of the work that was necessary if that promise was to be realized: the management of the politics of dissemination, of the social relations of the printing office, and of the technologies of the press itself in order to produce texts that could make claims to authority. This demonstrates the particularities and limitations of this empire of print and highlights the many forms of mental and manual labor, by both Europeans and Indians,

---

14. Bernard S. Cohn (1996) *Colonialism and Its Forms of Knowledge: The British in India* (Princeton University Press, Princeton, NJ); Huw V. Bowen (2005) "Methods: an empire in writing," in *The Business of Empire: The East India Company and Imperial Britain, 1756–1833* (Cambridge University Press, Cambridge) pp. 151–81; and C. A. Bayly (1996) *Empire and Information: Intelligence Gathering and Social Communication in India, 1780–1870* (Cambridge University Press, Cambridge).

15. IOR E/4/616, 31 January 1755.

involved in constructing it. Finally, the specific configuration of power and knowledge within this imperial print culture is demonstrated by detailing the difficulties involved in the publication in Calcutta in 1789 of a work critical of the British, Ghulam Husain Khan Tabatabai's *Seir Mutaqharin.*

Print's Imperial Promise

As late as the 1760s, printing was more a threat to the Company's authority in India than an instrument of its power. Understanding the construction of an empire of print in British Bengal means situating it within a range of print cultures that might undermine imperial power as well as promoting it. The potential implications of the press are well illustrated by the career of William Bolts. Described as a "Dutch Adventurer," Bolts came into conflict with the Bengal Council under Harry Verelst in the late 1760s as it tried to restrict the disruption private trade caused to the Company's commerce and its relationships with Indian polities.[16] Eventually, in a controversial extension of their powers, the council resolved to expel Bolts from Bengal, claiming to the directors in London that this was "absolutely and essentially necessary to the Peace, Order and Tranquility of your Settlement and throughout the Company's Possessions in Bengal."[17] Bolts suspected that the real reasons lay in their own private interests. The evidence that the council marshalled sought to ensure that the expulsion would be judged an act of state rather than one of interested individuals. Bolts, they said, had corrupted Company servants in order to gain access to confidential documents, and he had treated the government of Bengal with contempt by illegitimately assuming powers granted only to the president. Their final "Corroborating Proof" of Bolts's "turbulent Character" and his "factious Attempts to sow Seeds of Discontent in the Settlement" was that he had fixed the following handwritten notices to the door of the Council House and several other "Publick Places":

> TO THE PUBLIC
>
> Mr. Bolts takes this method of informing the public that the want of a printing press in this city being of great disadvantage in business, and making it extremely difficult to communicate such intelligence to the

16. N. L. Hallward (1920) *William Bolts: A Dutch Adventurer under John Company* (Cambridge University Press, Cambridge); and Sudipta Sen (1998) *Empire of Free Trade: The East India Company and the Making of the Colonial Marketplace* (University of Pennsylvania Press, Philadelphia).

17. IOR E/4/28 *Bengal Letters Received, 26th August 1767–April 1769*, 13 September 1768.

community as is of utmost importance to every British subject, he is
ready to give the best encouragement to any person or persons who are
versed in the business of printing to manage a press, the types and utensils
of which he can produce. In the mean time, he begs leave to inform the
public that having in manuscript many things to communicate, which
most intimately concern every individual, any person who may be induced by curiosity or other more laudable motives, will be permitted at
Mr. Bolts's house to read or take copies of the same. A person will give
due attendance at the hours of ten to twelve any morning.[18]

This threat was not realized in Calcutta. However, it traveled back to England with the deported Bolts. Once in London he devoted his energy to
attacking the Indian government and pursuing Verelst through the courts to
their eventual mutual ruin. Yet Bolts's most enduring line of attack did come
via the printing press. His *Considerations on Indian Affairs*, published in
London in 1772, presented itself as daring to speak out against the East India
Company by "exposing their secrets" in the public interest. It contained a
184-page appendix of "authentic papers," many in translation. Using these
documents Bolts accused the Company of being "an absolute government of
monopolists" that was destroying Bengal by restricting private trade. He also
argued that the Company's claims to be upholding "the old established laws
and actual form of government of the Mogul empire" was absurd "when
no such laws or empire exist."[19] His solution was that King George III, to
whom Bolts dedicated his book, should assume power over India, separating
sovereignty from commerce and opening the country up to private traders.

Bolts's book was an early intervention in an extensive public debate over
India, Britain, and empire from the late 1760s. This was a discussion that
continued in a variety of forms into the nineteenth century as ideas, and the
people, books, and papers that carried them, moved back and forth between
Asia and Europe.[20] The wealth of returning "nabobs," reports of the Bengal

18. IOR E/4/28 13 September 1768; and H. E. Busteed (1908) *Echoes from Old Calcutta: Being
Chiefly Reminiscences of the Days of Warren Hastings, Francis, and Impey*, 4th ed. (W. Thacker
& Co, London) p. 182.

19. William Bolts (1772) *Considerations on Indian Affairs; Particularly Respecting the
Present State of Bengal and its Dependencies* (printed for J. Almon in Piccadilly, P. Elmsley
in the Strand, and Brotherton and Sewell in Cornhill, London) pp. vi, xii, xiv, 49.

20. Eric Stokes (1959) *The English Utilitarians and India* (Clarendon Press, Oxford); Ranajit
Guha (1996) *A Rule of Property for Bengal: An Essay on the Idea of Permanent Settlement* (Duke
University Press, Durham); Wilson, *Governing Property, Making Law*; and Travers, "'The real
value of the lands.'"

famine, and concerns over the legitimacy and domestic implications of Britain extending a territorial empire in India were all canvassed as the nature of Company sovereignty and policy was debated. Bolts's book, Verelst's response—sponsored by the East India Company, and published as *A View of the Rise, Progress and Present State of the English Government in Bengal* (1772)—and Alexander Dow's *History of Hindostan* (1768–72), which made the case for India's inherent despotism, all fed into Parliament's formulation of the Regulating Act of 1773.[21] Warren Hastings, who was appointed governer general of Bengal under this act, was kept informed while in Calcutta of the reception of Bolts's and Dow's work in England. It was also the case that Calcutta's commercial printers were animated by the party political conflicts brought to the city by Parliament's appointment of the Bengal Council under the same legislation.[22]

There was, therefore, no imperial teleology to the origins of printing in British Bengal. The first press was established in 1777 by James Augustus Hicky, a man described as "the most objectionable rowdy that ever landed in Calcutta."[23] Indeed, Hicky set up as a printer to satisfy his creditors when imprisoned for debt after the failure of a trading voyage. His subsequent employment by the Company's Military Department to print its pay bills enabled him to proclaim on the cover of the *Bengal Gazette*, the newspaper that he started in January 1780, that he was "first and late printer to the Hon. Company." However, the *Gazette* itself became a thorn in the governor general's side. Hicky called it "A weekly Political and Commercial paper open to all parties but influenced by none," but used it to spread gossip and attack his political enemies. In particular, his barbs were directed at Warren Hastings, the supreme court judge Elijah Impey, and John Zachariah Kiernander, the missionary who helped publish Hastings's *India Gazette*. As a result, in November 1780 Hastings prohibited the circulation of the *Bengal Gazette* through the Post Office. Furthermore, in 1782 Hicky was prosecuted for libel, convicted, and imprisoned on Impey's orders. His newspaper was shut down.[24]

21. Harry Verelst (1772) *A View of the Rise, Progress and Present State of the English Government in Bengal* (printed for J. Nourse; Brotherton and Sewell; and T. Evans, London); and Alexander Dow (1768–72) *History of Hindostan*, 3 vols. (printed for T. Becket and P. A. de Hondt, in the Strand, London).

22. Travers, *Contested Notions of Sovereignty*.

23. Quoted in S. B. Chaudhuri (1968) "Early English printers and publishers in Calcutta," *Bengal Past and Present*, 87, p. 67.

24. Busteed, *Echoes from Old Calcutta*, pp. 182–222; Shaw, *Printing in Calcutta*; and Chaudhuri, "Early English printers," pp. 67–68.

Where Hicky had led, others followed. Calcutta's commercial printers multiplied to meet the needs of the European community. As Graham Shaw's invaluable research has shown, there were usually between three and five presses at work in the city in the 1780s and between seven and ten in the 1790s. They were generally organized around the production of a newspaper. Twenty-four papers were started before 1800 with the *Calcutta Gazette* (1784), *Bengal Journal* (1785), *Oriental Magazine* (1785), *Calcutta Chronicle* (1786), *Harkaru* (1793), and *Indian World* (1794) all having some success. These printers also produced calendars and almanacs; lists of Company servants; works on law, medicine, revenue, and horses; and a variety of ephemeral invitations, notices, and forms that have rarely survived. They operated in difficult conditions. There was no copyright protection. Their production costs, particularly for paper, ink, and engraving, were high, and would-be authors often found it cheaper to send their work to England for printing.[25] Finally, the government was suspicious of their activities. Hastings's actions against Hicky are a case in point, and a number of other newspaper editors were also expelled from India.[26] More systematically, Lord Wellesley imposed severe wartime restrictions on the press between 1799 and 1818. There was to be no printing outside Calcutta. Printers within the city had to put their name on all items produced, and all newspapers required prior government approval.[27] As we shall see, the Company's printing in Bengal was always part of this more heterodox milieu of inky-fingered men seeking to make a profit, and perhaps some friends and enemies, from working the press.

The other significant culture of print in British Bengal was that constructed by the Baptist Missionary Society (BMS) after 1798. This was built upon an almost unshakeable belief in the fundamental necessity of creating Christian converts by putting into heathen hands the Scriptures printed in their own languages and teaching them the word of the Lord.[28] At its heart were three self-made men of the book: William Carey, Joshua Marshman, and William Ward. Carey was a shoemaker who "manifested an avidity

---

25. Shaw, *Printing in Calcutta*; Chaudhuri, "Early English printers"; and Mofakhkhar Hussain Khan (1976) *History of Printing in Bengali Characters up to 1866*, unpublished Ph.D. thesis, School of Oriental and African Studies, University of London.

26. Most notably William Duane, the Irish-American editor of the *Indian World*, and James Silk Buckingham of the *Calcutta Journal*; see Chaudhuri, "Early English printers."

27. M. Siddiq Khan (1962) "The early history of Bengali printing," *Library Quarterly*, 32, pp. 51–61.

28. Catherine Hall (2002) *Civilising Subjects: Metropole and Colony in the English Imagination, 1830–1867* (Polity Press, Cambridge).

for books of science, history, and travels" and whose own conversion, the "vital change of heart," came by reading Robert Hall's *Help to Zion's Travellers*.[29] Despite his poverty Carey built up a small library when working as a schoolmaster. It was while reading Captain Cook's *Voyages* and teaching his students geography "that Mr Carey was led to contemplate the moral and spiritual degradation of the heathen, and to form the noble design of communicating the Gospel to them."[30] However, it took several years before the BMS was formed. Carey arrived in India only in 1793 and did not acquire a printing press until 1798.

In 1799 Carey was joined by a group of missionaries including Marshman and Ward. Joshua Marshman was a weaver's son with little formal schooling but a "thirst for reading," which meant that "[h]e thought little of walking a dozen miles for the loan of a book."[31] This prodigy, who had read a hundred volumes by the time he was twelve, was apprenticed to a Holborn bookseller. He read the books in the street as he delivered them, and he read by candlelight at night. Baptized in 1794, he became interested in missionary work by reading the periodical accounts of the BMS. William Ward was the son of a carpenter and builder, and a studious child. It was reported that "[t]he time given by his playmates to recreation he devoted to reading, and he always appeared to have some intellectual pursuit in hand." Apprenticed to a Derby printer, he advanced in both his trade and his learning to become the editor of the *Derby Mercury*. In his youth, attracted by the democratic ideals of the French Revolution, he joined the Derby Constitutional Society. The rules he drew up for them, and a political address "adapted to the revolutionary sympathies of the day," attracted the attention of the authorities. However, he escaped prosecution, and on being baptized in 1796 "renounced all interest in politics and journalism, and gave his heart and soul to the nobler vocation of communicating Divine truth to his fellow-countrymen."[32] He later thought to do this as a missionary.

In India, Carey would oversee the translations; Ward would print them; and Marshman would run a school. These were men whom Robert Southey called "low-born and low-bred mechanics" in the *Quarterly Review* of 1809.[33] They had, however, felt themselves utterly transformed and elevated by what they did with words, books, and print. So when the first sheet

29. John Clark Marshman (1859) *The Life and Times of Carey, Marshman, and Ward*, 2 vols. (Longman & Co, London) vol. 1 pp. 1, 4; and Robert Hall (1781) *Help to Zion's Travellers* (Bristol).

30. Marshman, *Carey, Marshman, and Ward*, vol. 1 p. 9.

31. Marshman, *Carey, Marshman, and Ward*, vol. 1 pp. 100–101.

32. Marshman, *Carey, Marshman, and Ward*, vol. 1 pp. 93, 94, 96.

33. Quoted in Marshman, *Carey, Marshman, and Ward*, vol. 1 p. 378.

of their Bengali translation of the Bible came off the press on 18 March 1800, it was a magical moment that kindled a "feeling of exultation" and "bright visions of future success."[34] Within this missionary culture of print, therefore, printing and printed objects could mean something quite different than for the Calcutta commercial press. The missionaries' response to their first complete Bengali New Testament is indicative. It was bound and placed on a communion table in the chapel, "and a meeting was held of the whole mission family and the newly baptized heathen, to acknowledge their gratitude to God for the completion of this important work."[35] For the missionaries printing was a sacramental act.

What the missionaries had in common with Calcutta's newspapermen was an ambivalent relationship with the East India Company. Before 1813, rules prohibiting missionary work forced them to locate themselves and their press at the Danish settlement of Serampore, sixteen miles upriver from Calcutta. They had established their operation at the time of Wellesley's restrictions on the press and, depending on the opinions of the current governor general and the strength of antimissionary feeling, they were periodically under threat of expulsion, of removing their press to Calcutta, and of censorship by the authorities. The Company was concerned about the effect that the missionaries' distribution of tens of thousands of native-language bibles and tracts calling for conversion to Christianity might have on the stability of the empire. However, the mission was also informally assured that it would not be interfered with if it remained at Serampore and agreed to "print nothing on the subject of politics."[36] Moreover, in 1801 William Carey was engaged to teach Bengali at the Company's new College of Fort William, and the press began printing dictionaries, grammars, and editions of Asian-language texts for the students.[37]

These varied and competing cultures of print show that any enduring connection between the practices of printing and those of imperial governance could not be assumed and had to be constructed. What this process involved can be shown by considering the first book published in Bengal (and, indeed, the first book produced using Bengali types), Nathaniel Brassey Halhed's A Grammar of the Bengal Language, printed by Charles Wilkins on Mr. Andrews's press at Hugli in 1778 (Figure 17). This volume was also

34. Marshman, Carey, Marshman, and Ward, vol. 1 p. 129.

35. Marshman, Carey, Marshman, and Ward, p. 141.

36. BMS IN/17(A) William Ward's Journal, 1799–1811, 4 vols., 29 June 1800.

37. Kapil Raj (2000) "Colonial encounters and the forging of new knowledge and national identities: Great Britain and India, 1760–1850," Osiris, 11, pp. 119–34.

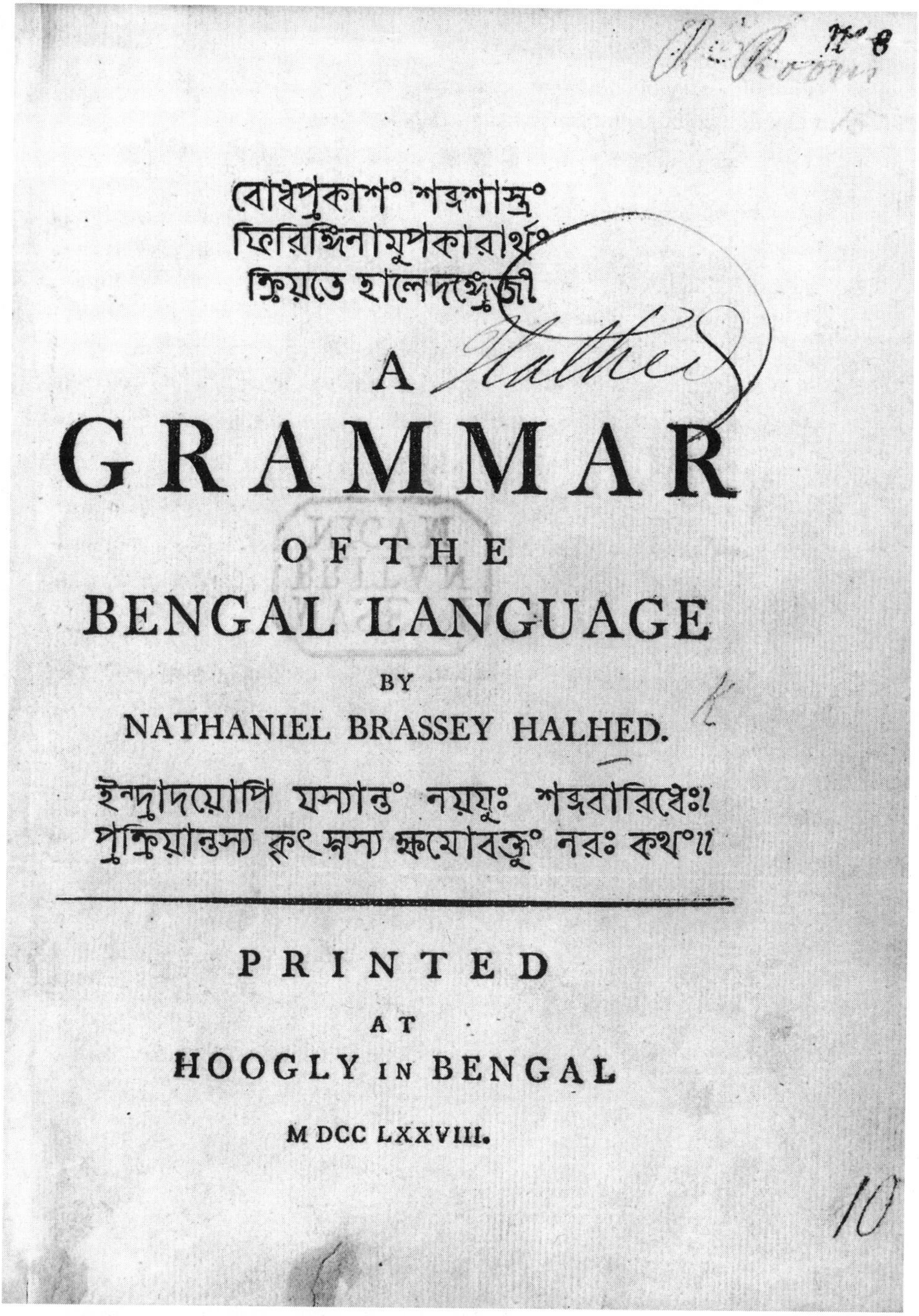

বোধপ্রকাশং শব্দশাস্ত্রং
ফিরিঙ্গিনামুপকারার্থং
ক্রিয়তে হালেদঙ্গ্রেজী

A
GRAMMAR
OF THE
BENGAL LANGUAGE
BY
NATHANIEL BRASSEY HALHED.

ইন্দ্রাদয়োপি যস্যান্তং নযযুঃ শব্দবারিধেঃ।
প্রক্রিয়ান্তস্য কৃৎস্নস্য ক্ষমোবক্তুং নরঃ কথং।।

PRINTED
AT
HOOGLY IN BENGAL
MDCC LXXVIII.

Figure 17 Title page of *A Grammar of the Bengal Language* (1778). (By permission of the British Library, 1560/1944.)

the founding text of the Honourable Company's Press, which produced a third of all the works printed in Calcutta before 1800.[38]

Halhed, the author (Figure 18), was the son of a London merchant whose own father "had been a broker in Exchange Alley, where he acquired a considerable estate."[39] Educated at Harrow and Oxford, Halhed excelled in Latin and Greek and began to learn Persian. Nominated as an East India Company writer by Verelst in 1771, he polished his language skills learning the silk trade in Kassimbazar before returning to Calcutta, where he "struck out the plan of the Bengal translatorship to the Board of trade."[40] He was subsequently selected by Warren Hastings to translate, from the Persian, the code of laws commissioned from a team of learned Brahmin pandits and intended to form the basis of the Company's legal apparatus. The *Code of Gentoo Laws* was published in London in 1776, receiving a great deal of attention.[41] Wilkins, the printer (Figure 19), was born in Somerset and became a Company writer through the patronage of his great uncle, a London banker. Arriving in India in June 1770, he became superintendent of the factory at Maulda in 1774. Through the encouragement of his friend Halhed, Wilkins began to learn Sanskrit, in addition to Bengali and Persian. These two men with deep and shared linguistic interests worked closely together to produce the *Grammar*.[42]

Halhed and Wilkins were also men of the book, although their identities were formed in relation to the printing press and its products in quite different ways from Calcutta's commercial printers and the missionaries of Serampore. For Halhed and Wilkins, composing and printing the *Grammar* was an intellectual exercise that found its rationale in language learning and imperial service. The production of the *Grammar* was pursued outside the formal requirements of their official positions—although they clearly sought to profit from it through the patronage of Warren Hastings—but Halhed's preface presented their endeavor in stridently imperial terms. The

38. Shaw, *Printing in Calcutta*.

39. J. Grant (1856) "Warren Hastings in slippers," *Calcutta Review*, 26, p. 62.

40. IOR E/1/73 *Home Correspondence, 1783*, p. 258 (18 November 1783).

41. Nathaniel Brassey Halhed (1776) *A Code of Gentoo Laws; Or, Ordination of the Pundits* (London); J. Duncan M. Derrett (1968) *Religion, Law and the State in India* (Faber & Faber, London); and Rosane Rocher (1983) *Orientalism, Poetry, and the Millennium: The Checkered Life of Nathaniel Brassey Halhed, 1751–1830* (Motilal Banarsidass, Delhi).

42. Mary Lloyd (1979) "Sir Charles Wilkins, 1749–1836," *India Office Library and Records: Report for the Year 1978*, pp. 9–39; and Fiona G. E. Ross (1999) *The Printed Bengali Character and Its Evolution* (Curzon, London).

Nathaniel Brassey Halhed.
Born 1751 – Died 1830.
From a painting made in 1771.

Figure 18 Nathaniel Brassey Halhed in 1771. (By permission of the British Library, 10815.e.20.)

Figure 19 Charles Wilkins c. 1830. (National Portrait Gallery, London.)

Bengali epigraph on the title page presented the text "for the benefit of the foreigner,"[43] and the preface elaborated an imperial politics of language:

The wisdom of the British Parliament has within these few years taken a decisive part in the internal policy and civil administration of its Asiatic territories; and more particularly in the Kingdom of Bengal, which, by the most formal act of authority in the establishment of a Supreme Court of Justice, it has professedly incorporated with the British Empire. Much however still remains for the completion of this grand work; and we may reasonably presume, that one of its most important desiderata is the cultivation of a right understanding and of a general medium of intercourse between the Government and its Subjects; between the Natives of Europe who are to rule, and the Inhabitants of India who are to obey. The Romans, a people of little learning and less taste, had no sooner conquered Greece than they applied themselves to the study of Greek: They adopted its Laws even before they could read them, and civilized themselves in subduing their enemies. The English, who have made so capital a progress in the Polite Arts, and who are masters of Bengal, may, with more ease and greater propriety, add its Language to their acquisitions: that they may explain the benevolent principles of that legislation whose decrees they inforce; that they may convince while they command; and be at once the dispensers of Laws and of Science to an extensive nation.[44]

This was an active intervention into both the ideology and practice of imperial governance in Bengal in the late 1770s. Under Warren Hastings, and in response to the charges of corruption and the plundering of Bengal under Robert Clive, the Company had first farmed out the collection of the revenue and then taken over its direct management. There were also alterations in the judicial system. Hastings established civil and criminal courts in which English judges would apply Muslim and Hindu law, and the Regulating Act instituted a supreme court to guarantee the rule of law in India.[45] In Halhed's short opening paragraph, the terminology that was used, the ideas that were drawn upon, and the policies that were suggested all established political positions within the ongoing debate over how the Company should rule India. All of these positions could be challenged.

43. Ross, *The Printed Bengali Character*, p. 33.

44. Nathaniel Brassey Halhed (1778) *A Grammar of the Bengal Language* (printed at Hoogly [sic] in Bengal) pp. i–ii.

45. Huw V. Bowen (1991) *Revenue and Reform: The Indian Problem in British Politics, 1757–1773* (Cambridge University Press, Cambridge).

Halhed's presentation of "the Kingdom of Bengal" being incorporated into the British empire by the parliamentary wisdom evident in the Regulating Act of 1773 offered an idealized mode of imperial sovereignty rather than a straightforward description of Indian political geography. As a Mughal province, Bengal had been ruled by a *nizām* and a *dīwān*, a characteristically pluralistic office-holding structure.[46] In the early eighteenth century there had been a process of "regional centralisation" under Murshid Quli Khan. He had been sent to Bengal as *dīwān* by the Emperor Aurangzeb in 1700 in order to tighten administrative control and increase the revenues. The actual outcome was the creation of "a semi-autonomous state" that gained power as the Mughal empire declined.[47] The Company, intent on legitimating its own rule according to "the ancient Mogul constitution" and encouraged in that by Mughal political elites, had initially maintained the separation of powers and acted only as *dīwān*. This became contentious after 1772 in the face of Hastings's attempts to build a more centralized polity, particularly through his judicial reforms.[48] In 1775, Hastings's opponents on the Bengal Council appointed Muhammad Reza Khan, who had also held high office under Clive, as *nāib subāh* (deputy governor) and put him in control of the *nizāmat*. Yet, by the time of the publication of the *Grammar* the tide was turning. Muhammad Reza Khan was sacked in 1778 as Hastings reasserted his control, and the working of the supreme court under Elijah Impey increasingly meant that retention of the separation of powers between *nizām* and *dīwān* was seen as unworkable.[49] Halhed's invocation of this singular "Kingdom" supported Hastings's imperial vision, as did the notion that the establishment of the supreme court in 1774 meant that Bengal had been incorporated into the British empire. There had been no such official declaration of dominion. The Parliament and the Crown refused to assume sovereignty and sheltered under the fiction of continued Mughal rule. As Philip Francis, Hastings's greatest enemy on the Bengal Council, put it, "[T]he People at present have either two Sovereigns or None." The nature of imperial sovereignty in Bengal remained far more unsettled and unsettling in 1778 than it appeared in Halhed's *Grammar*.[50]

46. John F. Richards (1995) *The Mughal Empire* (Cambridge University Press, Cambridge).

47. Muzaffar Alam and Sanjay Subrahmanyam (eds.) (1998) *The Mughal State, 1526–1750* (Oxford University Press, New Delhi) pp. 46, 53.

48. Bayly, *Empire and Information*.

49. Travers, *Contested Notions of Sovereignty*.

50. Guha, *A Rule of Property for Bengal*, p. 152; B. B. Misra (1959) *The Central Administration of the East India Company* (Manchester University Press, Manchester); and Sen, *Distant Sovereignty*.

These idealizations of the nature of political power also lay behind Halhed's separation of "the Natives of Europe who are to rule, and the Inhabitants of India who are to obey." Indeed, this was a simplification not only because there continued to be Indian "rulers," but due to the considerable debate over how these forms of rule should work in practice.[51] There was intense disagreement between Warren Hastings and Philip Francis. Francis argued that English and Indian forms of government were fundamentally incompatible. This meant that while overall sovereignty should be in the hands of the English king, India should be administered by Indians. In particular, local political, judicial, and revenue-gathering powers should be put into the hands of the *zamīndar*s, who he understood to be the owners of hereditary landed property. Hastings agreed with Francis on the necessity of overall English sovereignty and the need for security of property. He also agreed that India should be ruled, according to the "ancient Mogul constitution," through the indigenous legal systems and the customary forms of revenue collection that were appropriate to the country. However, his mode of governance was an adaptation of Indian "despotism" that distrusted the *zamīndar*s and relied upon an active role for English officials, as judges and revenue officers, in the administration of Indian law and the revenue apparatus.[52]

It was this vision that underpinned Halhed's emphasis on language learning in the *Grammar*, just as it had underpinned the intended uses of the *Code of Gentoo Laws* by English judges in India. First, the language that Halhed emphasized was Bengali, rather than Persian, Sanskrit, Hindustani, or English.[53] This was, he argued, the everyday language of Bengal, used for accounting and for all personal and epistolary communication. This meant that "Bengal Interpreters" were required in the Company's commercial, judicial, and revenue departments. Many of the documents that governed the powers and obligations of revenue payers and collectors were, Halhed stressed, in Bengali, and the revenue farmers and the "Jentoo Zemindars" often did not understand the Persian documents they were also faced with.[54] The "confidential Bengal Translator" and the *Grammar* itself would remove the need to trust native officials. They would also, through the courts and

---

51. Marshall, *Bengal: The British Bridgehead.*

52. Sophia Weitzman (1929) *Warren Hastings and Philip Francis* (Manchester University Press, Manchester); Guha, *A Rule of Property for Bengal;* and Travers, *Contested Notions of Sovereignty.*

53. Hastings's mode of government was by English administrators speaking and writing Bengali and Persian; Francis's would be via Indian officials speaking and writing English; see Guha, *A Rule of Property for Bengal.*

54. Halhed, *Grammar,* p. xv.

"the appointment of gentlemen of mature experience in the manners and customs of the natives to the several divisions and districts of Bengal, to act as justiciary arbitrators between the head farmer and his under tenants," secure the property of "the indigent villager" against "the exactions of an imperious Landlord or grasping Collector."[55] For Halhed, as for Hastings, English officials proficient in Indian languages were the foundation of legitimate and effective imperial rule in Bengal.

Arguing for these forms of governance meant fending off the critics who proclaimed that it was inevitable that Indian despotism, and, more dangerously, English rule through Indian despotism, would eventually threaten political liberties in England. Halhed's appropriation of Roman imperial history countered this by suggesting that the English could become more rather than less civilized by learning from India's ancient civilization as they conquered it. For both Halhed and Wilkins that began with language learning, although in this case that meant Sanskrit, India's "classical" language, not Bengali.[56] Yet this was not the only way to read Rome's conquest of Greece for late eighteenth-century imperialists. In 1770 the archbishop of Mexico, Francisco Antonio Lorenzana y Buitrón, complained of the continued failure to teach Castilian to the Amerindians by pointing out that "[t]here has never been a Cultured Nation in the World, that when it extended its Conquests, did not attempt the same with its Language: the Greeks took for barbarians those other Nations that remained ignorant of its speech; the Romans, after conquering the Greeks, required them to accept their Latin Language...with such rigor that they did not allow anyone who spoke another, foreign language to conduct business in the Senate."[57] Lorenzana's version of classical history, and his unfavorable comparison of Nahuatl to Latin, Greek, and Hebrew, did not stop him from expending considerable resources on collecting, studying, and printing Amerindian sources. Indeed, Jorge Cañizares-Esguerra argues that the archbishop should be understood as one of the founders of what he calls "patriotic epistemology" in the Americas. This held that it was only learned Amerindians and Creoles, rather than Europeans, who could properly understand the history of the New World.[58]

55. IOR E/1/73 p. 258b; and Halhed, *Grammar*, pp. xvi.

56. Muhammad Abdul Qayyum (1982) *A Critical Study of the Early Bengali Grammars: Halhed to Haughton* (Asiatic Society of Bangladesh, Dhaka).

57. Quoted in Walter D. Mignolo (2003) *The Darker Side of the Renaissance: Literacy, Territoriality, and Colonization*, 2nd ed. (University of Michigan Press, Ann Arbor) pp. 59–61.

58. Jorge Cañizares-Esguerra (2001) *How to Write the History of the New World: Histories, Epistemologies, and Identities in the Eighteenth-Century Atlantic World* (Stanford University Press, Stanford).

Thus, for quite different imperial situations, and deploying diametrically opposed versions of the same history, Lorenzana and Halhed both constructed positions on language that suggested that they had the moral legitimacy to rule overseas empires.

Matters of legitimacy were also evident in Halhed's justification of language learning in order to explain the "benevolent principles" of the law to those who had to obey it. It was not, however, clear where these laudable principles were to spring from. In Halhed's preface the source seems to be the English Parliament, whose laws were guaranteed by its own "ancient constitution."[59] Yet the use of English law was limited in India. Hastings himself was strongly opposed to any extension of English law to Bengal. The purpose of the *Code of Gentoo Laws* had been to set out India's own laws and, in the process, to demonstrate the strength of Indian law-making traditions. In practice, however, as Robert Travers has shown, any attempt to invoke ancient constitutional principles, either Indian or English, was always an interpretation and reworking of versions of legal and political history and theory in the context of present concerns. Doing so, for Hastings, always seems to have involved attempts to improve Indian law by the application of universal principles.[60] The same notion of "improvement" pervades Halhed's preface. English imperial sovereignty, pursued through law and bolstered by language learning, was intended to be both legitimate rule in India on ancient Indian principles and at the same time a force to civilize and improve Indians and their conquerors. As such, the *Grammar* was a political document aimed as much at audiences in England as at Company servants in India keen to learn Bengali.[61]

This version of Britain's Indian empire was to be made through the printing press. Halhed's preface, printed by Wilkins, outlined the ways in which printing's promise of standardized reproduction and wide dissemination would work. First, printing in Bengali with Wilkins's types promised the centralization, consolidation, and security of crucial parts of the imperial information order:

The gentlemen at the head of Indian affairs do not want to be told of the various impositions and forgeries with which Bengal at present abounds, in Pottahs, (or Leases) in Bonds and other written securities of property; in

59. J. G. A. Pocock (1957) *The Ancient Constitution and the Feudal Law: A Study of English Historical Thought in the Seventeenth Century* (Cambridge University Press, Cambridge).

60. Travers, *Contested Notions of Sovereignty.*

61. Indeed, Qayyum, *A Critical Study of the Early Bengali Grammars,* questions its usefulness for the practical business of language learning.

Rowanahs and Dustucks, in Orders and Notices of government issued in the country languages; as well as in all the transactions of commerce: and also in the Processes, Warrants and Decrees of the supreme and inferior Courts of Judicature; all of which afford ample scope for the exertion of Mr. Wilkins's ingenuity.[62]

Printing was, therefore, understood as a technology for securing the power of the imperial state through existing forms of paperwork. This was, quite literally, to reproduce Mughal forms of rule. The *pattās* (leases), *rawānas* (permits), and *dastak*s (passes) were part of an earlier administration of revenue and trade in an increasingly "literacy aware" society (see chapter 1).[63] By the eighteenth century, the middling sort in north India "had greater recourse to fixed texts and to rational rules for the conduct of religion, business and government."[64] They were part of a world of commerce and politics in which proficiency in reading and writing reached well beyond the higher castes and the upper levels of the administrative hierarchy.[65] As well as long-standing Mughal investment in the arts of calligraphy and bookbinding, it has also been argued that it was during Aurangzeb's reign (1658–1707) that a "paper empire" emerged. This was based on an increasing scribalization of power that sought to formalize the gathering of information and the routines of administration, and to make imperial epistles and orders (*parwānas*) into more regularized statements of policy.[66] At the same time, however, the process of regional centralization in early eighteenth-century Bengal also benefited those in command of the written word as "[c]ommercial men, scribal families and local gentry consolidated their power at the expense of the centre."[67] As a result, the idealization of a singular sovereignty in the "Kingdom of Bengal," with relatively uniform modes of governance, could be attributed to the Mughal empire and its paperwork, and reproduced in print by the British.

However, this work of reproduction was also part of the process of improvement. The printing press would, Halhed suggested, restore what had

62. Halhed, *Grammar*, p. xxiv.

63. H. H. Wilson (1855) *A Glossary of Judicial and Revenue Terms* (William H. Allen and Co., London).

64. Bayly, *Empire and Information*, pp. 36, 43.

65. Anindita Ghosh (2006) *Power in Print: Popular Publishing and the Politics of Language and Culture in a Colonial Society* (Oxford University Press, New Delhi).

66. Alam and Subrahmanyam, *The Mughal State*, p. 31. For processes of documentary formalization, see S. M. Azizuddin Husain (ed.) (1990) *Raqaim-i-Karaim (Epistles of Aurangzeb)* (Idarah-i Adabiyat-i Delli, Delhi); and John F. Richards (1986) *Document Forms for Official Orders of Appointment in the Mughal Empire* (E. J. W. Gibb Memorial Trust, Cambridge).

67. Bayly, *Indian Society and the Making of the British Empire*, p. 4.

become debased forms of writing. It would take these Indian forgeries and impositions and make them into the true and trustworthy instruments of a singular, centralized, and rule-governed polity. In turn this would be part of a broader imperial improvement brought about via Wilkins's types. Halhed, ignoring the dynamism of Bengali commercial society, used the stadial theories of Scottish moral philosophy to argue for the benefits that printing would bring.[68] It would, he said, "enable Great Britain to introduce all the more solid advantages of European literature among a people whom she has already rescued from Asiatic slavery: to promote the circulation of wealth, by giving new vigour and dispatch to business, and to forward the progress of civil society by facilitating the means of intercourse."[69] Thus, the promotion of increased trade and circulation through print was strongly linked to ideas of improvement, even if, in practice, the East India Company sought to retain its commercial monopolies.[70]

This combination of faithful reproduction and necessary improvement, and of the centralization and certainty of imperial rule, with the promotion of new forms of trade was most clearly articulated by Francis Balfour when praising Wilkins's new Persian types:

> The only printed Persian character that has hitherto been in use, except in exhibiting fair copies of Dictionaries and Grammars, has been subservient to no public purpose; and is but ill calculated for becoming the Channel of authority, or the Medium of business, over an extensive empire, where it is almost unknown, and scarcely understood; whereas the Types which Mr. Wilkins has invented, being a perfect imitation of the *Taleek*, the character in which all Persian books are written and consequently familiar and universally read, are not only well calculated for promulgating the Edicts of Government; but for every Transaction in business, where the Persian character is required.[71]

For Persian, the language of the Mughal empire itself, the arguments for the continuities of imperial rule were even stronger. They also chimed with

---

68. Rajat Datta (2000) *Society, Economy and the Market: Commercialisation in Rural Bengal, c. 1760–1800* (Manohar, New Delhi).

69. Halhed, *Grammar*, pp. xxiv–xxv.

70. Sen, *Empire of Free Trade*; and Sudipta Sen (2004) "Liberal empire and illiberal trade: the political economy of 'responsible government' in early British India," in Kathleen Wilson (ed.) *A New Imperial History: Culture, Identity and Modernity in Britain and the Empire, 1660–1840* (Cambridge University Press, Cambridge) pp. 136–54.

71. Francis Balfour (1781) *The Forms of Herkern* (printed at Calcutta) pp. 6–7.

the political ideas of the Persianate and scribalized ranks of the old Mughal bureaucratic elite.[72] Balfour's *The Forms of Herkern* (1781), dedicated to Warren Hastings, was a translated and printed version of a book that, as Balfour argued, "is put into the hands of almost every beginner; [since] it contains the common forms of business and correspondence," including how to write *parwānas* and *dastaks*. Printing it with Wilkins's types was a response to "the defects of Moonshies and Manuscripts": the inadequate knowledge of native language teachers (munshis) and the multiple versions of manuscripts with their misshapen characters, indistinct word spacings, lack of punctuation, and inconsistent usages.[73] Balfour believed, as many others have done, that these problems in oral and scribal cultures could be removed to produce a fixed and standardized version of the text that could be widely disseminated. It should be evident by now that the desirability of doing so in late eighteenth-century Bengal was because such notions of fixity, standardization, and dissemination were understood through an imperial political philosophy that sought to achieve a single, effective, and rule-bound political entity.[74] This would be based on a faithful reproduction of Mughal forms of governance, but would also create a civilized and ordered commercial society in the "Kingdom of Bengal." With its promise of combining precise replication and improved communication, printing seemed to provide a technical solution to these political problems.

However, putting this technology into place, and achieving these effects, was by no means automatic. Realizing any part of the promise of printing that was set out within this imperial political philosophy required careful work to manage the dissemination of texts, the organization of the press, and the operation of the printing technologies themselves. As I will show, where this work was left undone the claims to authority made by imperial texts began to falter. Moreover, even where it was done the effects were both limited and unpredictable. First, the requirements of mechanical reproduction meant that the authority of print was confined to certain sorts of imperial texts and relationships. Second, something novel was always being created in what appeared to be the process of uniform replication.

---

72. Kumkum Chatterjee (1998) "History as self-representation: the recasting of a political tradition in late eighteenth-century eastern India," *Modern Asian Studies*, 32:4, pp. 913–48.

73. Balfour, *The Forms of Herkern*, pp. 4, 5.

74. Jon E. Wilson (2005) "'A thousand countries to go to': peasants and rulers in late eighteenth-century Bengal," *Past and Present*, 189, p. 104, notes the commitment of the British to "a more absolutist approach to revenue collection than their predecessors" in the 1770s and 1780s.

An Empire of Print?

The Politics of Dissemination

The work of Orientalist scholars in British Bengal such as Halhed, Wilkins, Balfour, Francis Gladwin, and William Jones is well known, although its relationship to imperial power is hotly debated.[75] These men were all engaged in learning languages and in obtaining and translating Persian, Sanskrit, and Bengali texts via very close, if unequal, collaborations with Hindu pandits and Persian-speaking munshis. They sought to produce versions of Persian and Sanskrit manuscripts, often in translation; legal codes and digests; learned discussions of Indian religion, science, and law; and dictionaries and grammars. Just as importantly, all of them sought to have their work printed, whether in Calcutta or London. Halhed's *Grammar* was part of this broader production of printed texts, and many of these works were produced with Charles Wilkins's types and on his printing presses. This raises two questions. What needed to be done for these works to be disseminated? And what part did the dissemination of printed materials play within debates over British empire building in Bengal?

The dissemination of printed works was not simply a product of the press's ability to produce multiple standard copies. The sorts of texts produced by these scholars were not viable as commercial products of the Calcutta presses. They required political patronage in order to be brought into existence. Prior to Halhed's *Grammar of the Bengal Language* the government of Bengal had extended such aid to scholars but not to the printing process in India. Hastings encouraged Halhed and Wilkins to undertake the *Grammar* and paid the salaries of a pandit and scribe to assist in its compilation.[76] The governor general then presented the page proofs to the Bengal Council in January 1778 and recommended it "as a Work highly meriting their Countenance & Patronage." He praised, in equal measure, its composition and its printing, suggesting that the government should understand it as "part of its Duties to incourage the Effects of Genius, or to facilitate the Introduction of new arts, by which the Dispatch of Business may be quickened or even the general Intercourse of Society rendered more practicable." His suggestion was that the government undertake to purchase the entire

75. Carol A. Breckenridge and Peter van der Veer (eds.) (1993) *Orientalism and the Postcolonial Predicament: Perspectives on South Asia* (University of Pennsylvania Press, Philadelphia).

76. IOR P/50/8 *Bengal Revenue Consultations, 3rd February to 11th March 1778*, 20 February 1778; and J. L. Brockington (1989) "Warren Hastings and Orientalism," in Geoffrey Carnall and Colin Nicholson (eds.) *The Impeachment of Warren Hastings: Papers from a Bicentenary Commemoration* (Edinburgh University Press, Edinburgh) pp. 91–108.

print run of one thousand copies at thirty rupees a copy for resale to its employees.[77] To demonstrate his confidence in this novel proposition Hastings even offered to enter his own bond against the thirty thousand rupees required. This was, however, rendered unnecessary by Philip Francis's willingness to support the work, albeit limiting the council to the purchase of only five hundred copies, the Board of Directors in London being encouraged to buy the rest.[78]

The vital importance of the personal patronage of Warren Hastings as governor general is witnessed by numerous dedications and prefatory letters and minutes in Orientalist works.[79] Such projects required political or institutional intervention to get them into print. Legal texts such as Halhed's *Code* and William Jones and Henry Colebrooke's *Digest of Hindu Law* (1798) were heavily supported by Company funds, under Hastings and Cornwallis respectively, on the basis that they would transform the administration of justice in Bengal.[80] Wilkins's translation of the Bhagavadgītā, the first scholarly translation of a Sanskrit text into any European language, was printed in London in 1785 under the authority of the directors of the East India Company.[81] Other authors found other means. The first volume of the *Asiatick Researches* (1789) was taken on by the Company's printer only following the promise that every member of the Asiatick Society of Bengal would buy a copy for twenty rupees each.[82]

These forms of patronage and finance shaped the patterns of dissemination and distribution. Copies of legal codes were sent to the provincial courts as well as those in Calcutta, and the intention was that Halhed's

77. IOR P/50/7 *Bengal Revenue Consultations, 2nd January to 27th January 1778*, p. 115 (9 January 1778), and P/50/8, pp. 949–50 (20 February 1778).

78. IOR E/4/37 *Letters Received from Bengal, 4th February 1778 to 8th January 1779* (20 March 1778); P/50/9 *Bengal Revenue Consultations, 28th April to 22nd May 1778* (28 April 1778); E/1/70 *Home Correspondence, 1782* (28 March 1782); and P/50/57 *Bengal Revenue Consultations, 7th January to 10th March 1785* (21 January 1785).

79. Peter J. Marshall (1973) "Warren Hastings as scholar and patron," in Ann Whiterman, J. S. Bromley, and P. G. M. Dickson (eds.) *Statesmen, Scholars and Merchants* (Oxford University Press, Oxford) pp. 242–62. For example, see Balfour, *Forms of Herkern*; Francis Gladwin (1783) *Ayeen Akbery: Or, The Institutes of the Emperor Akber*, 3 vols. (Calcutta); and Charles Wilkins (1781) *A Translation of a Royal Grant of Land by One of the Ancient Rajaas of Hindostan* (printed at Calcutta by the translator).

80. Henry T. Colebrooke (1798) *A Digest of Hindu Law On Contracts And Successions*, 4 vols. (printed at the Honourable Company's Press, Calcutta); and Derrett, *Religion, Law and the State*.

81. *The Bhagvat-Geeta, or Dialogues of Kreeshna and Arjoon*, translated by Charles Wilkins (printed for C. Nourse, London, 1785).

82. *Asiatick Researches*, vol. 1 (printed and sold by Manuel Cantopher, at the Honourable Company's Printing-Office, Calcutta); and Khan, *History of Printing in Bengali Characters*.

Grammar would be sold to Company servants across Bengal. Printing was part of the creation of a new imperial geography of authors, texts, and readers. In particular, it meant that the geography of these printed objects was one that worked between Calcutta and London, as well as within Bengal.[83] Printing allowed readers, writers, books, and pamphlets to become active participants in the political debates over the British empire. Halhed's *Code* was printed in London before it appeared in Calcutta, Warren Hastings arguing that it "wou'd not only be serviceable in itself in enabling the Board to decide with confidence and without reference to others in such Cases as turn on points of the Gentoo Law in their Capacity of Judges in the Adawlat but that it may prove also worthy the attention of the public & remove the false prejudice which seems to have prevail'd in England respecting the Laws of this Country."[84] The success of Hastings's attempts to govern India as a theocratic state with a civilized and ancient religious and political culture depended upon arguing the case in public, and that meant producing printed texts.[85] Thus, Francis Gladwin's (1783) *Ayeen Akbery* announced that "it comprehends the original constitution of the Mogul Empire," enabling it to "show where the measures of their [the Company's] administration approach to the first principles, which perhaps will be found superior to any that have been built on their ruins."[86] Finally, the publication in London of Wilkins's *Bhagvat-Geeta* (1785) was intended to convince readers that India possessed "a theology accurately corresponding with that of Christian dispensation" and to effect a reconciliation between the Indians and the British. As Hastings put it, "its public reception will be the test of its real merit."[87] Indeed, such was the significance of Wilkins's text that William Jones argued in 1789 that "Europeans" who "wish[ed] to form a correct idea of *Indian* religion and literature" should "begin with forgetting all that has been written on the subject, by ancients and moderns, before the publication of the *Gítà*."[88] Printing, therefore, reconfigured the power relations of dissemination. Within texts such as the *Asiatick Miscellany*, which published in Calcutta "fugitive and miscellaneous pieces" on "Oriental subjects," this

83. Giles Barber (1982) "Book imports and exports in the eighteenth century," in Robin Myers and Michael Harris (eds.) *Sale and Distribution of Books from 1700* (Oxford Polytechnic Press, Oxford) pp. 77–105.

84. IOR P/49/42 *Bengal Revenue Consultations, 2nd November to 31st December 1773*, pp. 3937–38 (10 December 1773).

85. Cohn, *Colonialism and Its Forms of Knowledge.*

86. Gladwin, *Ayeen Akbery*, pp. ix–x.

87. *Bhagvat-Geeta*, pp. 10, 13; Marshall, "Warren Hastings as scholar and patron."

88. *Asiatick Researches*, vol. 1 p. 355. Notably, 150 copies of the printed version were sent to India; Lloyd, "Sir Charles Wilkins," p. 18.

opened the possibility for a sort of Anglo-Indian epistemology whereby "an English reader in India" might compare historical travel accounts "with what he sees on the spot" and "from these premises, may draw conclusions more just, respecting their present state, then the most sagacious politicians have been able to do at home."[89] If the circulation of printed texts enabled an imperial sphere of "social communication" to be constructed that included readers and writers in India and in England, then this could not necessarily be assumed to be a consensual interpretative community.[90]

The rationale for Orientalist scholarship was that it recovered the original texts of ancient civilizations, whether in Persian or in Sanskrit. It can, of course, be shown that the inevitable processes of selection, translation, and interpretation could not but produce something new.[91] Indeed, in the spheres of what the British understood as law and custom, their belief in the possibility of constructing authoritative texts with which to govern was based on profound misunderstandings of the dynamic, negotiated, and contested nature of Indian society and politics. It was certainly the case that printing was used to try to make texts authoritative and to fix the grounds of interpretation. It is also evident that this was an imperial project that was destined to fail as long as it rested on the notion that those texts could be generated from Indian manuscript or ethnographic sources.[92] However, it is easy to overestimate the novelty of British attempts to disseminate standardized texts and the singularity of their role in creating them. As C. A. Bayly has shown, by the late eighteenth century large numbers of cheap scriptural texts were available in manuscript in Bengali. A few rupees would buy versions of parts of the great epics, the Mahābhārata and Rāmāyaṇa, and dispersed sectarian groups were active in attempts to "disseminate standard texts and precepts for people spread across the country."[93] Moreover, it was the collaboration between the British and the Brahmin pandits working on Indian texts that emphasized the importance of Sanskrit and the Sanskritic roots of Bengali language and literature. Together they attempted to create a set of texts in a purified language that was returned to the classical roots from which both British administrators and high-caste Hindus drew their

89. The Asiatick Miscellany (printed by Daniel Stuart, Calcutta, 1785) pp. iv–vi.

90. In contrast, the term "social communication" was used by Warren Hastings to signal the building of knowledge and affection between the Britons and Indians, "which lessens the weight of the chain by which the natives are held in subjection; and ... imprints on the hearts of our own countrymen the sense and obligation of benevolence"; see Bhagvat-Geeta, p. 13.

91. For legal texts, see Derrett, Religion, Law and the State.

92. Wilson, Governing Property, Making Law.

93. Bayly, Empire and Information, p. 42.

power and legitimacy. It was this collaboration that succeeded in creating something new.[94]

What was certainly novel was that printing changed the material form of texts, and with it their audiences, their geographies, and their meanings. Indian manuscript forms and European book forms were quite different. Even though it was produced in Bengal, Halhed's *Grammar* adopted European conventions rather than those of, for example, the Bengali *punthī*.[95] That such differences in form meant differences in meaning was particularly the case with sacred texts. Unlike the Europeans, and contrary to the assumed importance of the *śāstras*, many Brahmins saw writing as an inferior mode of preserving sacred knowledge, valuing instead particular forms of spoken repetition and memory work. When Brahmins read they read differently from Orientalist scholars. This was also true of Muslim relationships to script and print, where the spoken word was the repository of trust.[96] For others, texts were important. Some Bengali manuscripts were sacred objects that were kept for worship and could be touched only by particular people. There are also reports that printing was understood as threatening the sacred. Mofakhkhar Hussain Khan quotes an early nineteenth-century report that when printing was first introduced Hindus would refuse to look at any printed page because they thought it would destroy their religion. There is other evidence that orthodox Hindus believed that printing holy works using ink containing animal fat or employing low-caste artisans in the printing process would defile those texts.[97] There were also differences in the meanings of the book as a work of art. For Persian manuscripts, different sorts of paper and different forms of calligraphy were appropriate to different purposes. These differences were not entirely lost on Europeans. William Jones noted the beauty of those works written with the blackest ink "upon fine silky paper, the ground of which is often powdered with gold or silver dust," and bound with illuminated leaves into books perfumed with roses or sandalwood. "It is not strange," he told the readers of his *Grammar*, "that they prefer their manuscripts to our best printed books."[98]

94. Ghosh, *Power in Print*; and Qayyum, *A Critical Study of the Early Bengali Grammars*.

95. Khan, *History of Printing in Bengali Characters*.

96. Jonathan Parry (1985) "The Brahmanical tradition and the technology of the intellect," in Joanna Overing (ed.) *Reason and Morality* (Tavistock, London) pp. 200–225; and Francis Robinson (1993) "Technology and religious change: Islam and the impact of print," *Modern Asian Studies*, 27:1, pp. 229–51.

97. Khan, *History of Printing in Bengali Characters*, pp. 409, 411; Priolkar, *The Printing Press in India*, p. 128; and Ghosh, *Power in Print*.

98. William Jones (1771) *A Grammar of the Persian Language* (printed by W. and J. Richardson, London) pp. 149–50.

Printing was, therefore, a matter of changing the material form of texts, and with it their geographies and audiences. Printed compilations of Hindu law, collections of Indian letterforms, or religious texts could then circulate between Calcutta and London, where they might become part of public debates, even being reviewed in the periodical press.[99] The implications were different for different sorts of texts. Some, such as Halhed's *Grammar* and his *Code*, had been created to take part in exactly these exchanges. Others, such as Wilkins's *Bhagvat-Geeta* and Gladwin's *Ayeen Akbery*, were older texts finding new audiences and new purposes in their translated and printed form. Printing placed these texts within an imperial circuit of production, dissemination, and reception.[100] However, since the making and distribution of these texts was a matter of patronage and politics, their fortunes were subject to change. Hastings's attempts to secure Company funds for these works were not always as successful as he would have wished. There was a long delay before Halhed and Wilkins were paid for their *Grammar*.[101] The Company also refused to support the publication of Gladwin's *Ayeen Akbery*, and Hastings was forced to subscribe for fourteen sets himself.[102] Dissemination was, therefore, always a matter of the shifting politics of empire. Changes in imperial administration, and in the ways it was understood in Britain, came about through the making and the consumption of print between Calcutta and London. In turn, these changes produced the political and institutional conditions for the dissemination of new sorts of texts with new implications for imperial administration and politics.[103]

Composing the Printing Office

Before these works could be disseminated they had to be printed. As Adrian Johns has argued in relation to Restoration natural philosophy, the production of authoritative texts depended upon control of the printing press, and

99. Michael J. Franklin (2002) "'The Hastings Circle': writers and writing in Calcutta in the last quarter of the eighteenth century," in E. J. Clery, Caroline Franklin, and Peter Garside (eds.) *Authorship, Commerce and the Public: Scenes of Writing, 1750–1850* (Palgrave, London) pp. 186–202. See also the comparative use of Halhed's *Code* detailed in Alex Drace-Francis (2006) "A provincial imperialist and a *Curious Account of Wallachia*: Ignaz von Born," *European History Quarterly*, 36:1, pp. 61–89.

100. Michael J. Franklin (ed.) (2000) *Representing India: Indian Culture and Imperial Control in Eighteenth-Century British Orientalist Discourse*, 9 vols. (Routledge, London) vol. 1 p. x.

101. IOR P/50/57 21 January 1785.

102. Gladwin, *Ayeen Akbery*, pp. ix–xi.

103. Wilson, *Governing Property, Making Law*; and Travers, *Contested Notions of Sovereignty*.

that in turn depended upon the management of the social relations of the printing office.[104] In British Bengal the crucial site was what became the Honourable Company's Press under Charles Wilkins himself. Indeed, rather than suggesting that Halhed's *Grammar* was the first text produced by the Honourable Company's Press, it is in fact the case that the press was produced by Halhed's *Grammar*. Once again, Warren Hastings's patronage was crucial. At his instigation Wilkins applied to the Bengal Council on 13 November 1778 to establish a Company press for printing in Bengali and Persian characters. Hastings strongly supported the application, stating that it would prevent printing in those languages from being lost following the publication of the *Grammar* and would "apply it to Public Use." He proposed a one-year trial to ascertain the public utility of the press and its commercial potential. It was Wilkins, the only person with the adequate experience, who was to be appointed superintendent of the Honourable Company's Press, with a salary of 350 rupees a month and generous allowances for house rent, servants, and assistants. He would also be paid "a reasonable Price" for every paper that he printed.[105] The press would be a private enterprise, but one run by a salaried appointee undertaking government printing.

By 1778 control of the Bengal Council was back in Warren Hastings's hands. George Monson's death in 1776, and the continued support of Richard Barwell, had given Hastings the casting vote on a council of four, despite the continued opposition of Philip Francis and Edward Wheler. Hastings's opponents did indeed oppose the establishment of the press. Wheler stated that he would support it as a purely commercial venture, but that it represented an illegitimate increase in public expenditure "at a Period when Oeconomy and Frugality appear to me particularly necessary." Francis agreed, arguing that he was "at all times very unwilling to increase Establishments, or to engage the Company in new Expences."[106] This was not, however, simple penny pinching. Francis, well aware of the political power of the press, saw it as an instrument of civil society rather than of the state.[107] His objection was that if "the State of the Country or the general uses of Society be not such as to demand the Introduction of the Press and find sufficient Employment for it," then it should not be propped up by the government.[108] It

104. Johns, *The Nature of the Book*.

105. IOR H/207, pp. 464, 468 (13 November 1778).

106. IOR H/207, pp. 471, 473 (13 November 1778).

107. Travers, *Contested Notions of Sovereignty*, argues that in the late 1760s Francis (under the pseudonym "Junius") published a series of "true Whig" tracts critical of King George III's new government.

108. IOR H/207, pp. 473–74 (13 November 1778).

was Francis's deployment of the "country ideology" of Bolingbroke and the Rockinghamite Whigs against Hastings's expansion of the state apparatus in India that provided an intellectual rationale for supporting a commercial press, but not a government-funded one.[109] In the absence of Barwell, Hastings was forced to play for time. He agreed with Wheler that the issue should be considered further. He then ensured that when it came before the council again on 22 December 1778, the continued opposition of Wheler and Francis was overruled by Barwell's support and his own casting vote.[110]

Tensions between Hastings and Francis were high. They had clashed many times since 1774 and in August 1780 they would fight a duel prompted by comments in the council chamber. However, there was also a degree of common ground between them. Despite their differing political philosophies, the two men shared an understanding of the need for a single overall sovereignty, the benefits of "a uniform and permanent system" of government, and the improvements that would be brought by increased trade.[111] This was the basis of the promise outlined by Halhed and Balfour for Wilkins's types. It was also what was emphasized in the particular, and somewhat restricted, forms of print that Hastings used to justify the council's decision to bring the art of printing in Bengali and Persian under the "Patronage of Government." He reported to London that the Company's press would be "for the purpose of printing such papers as are Confined to a settled Form, whether in the Persian, Bengal or English Character."[112] While there might be disagreement over the need for Orientalist scholarship, and over its uses for imperial governance, there was greater consensus on these more mundane administrative texts.[113] In his initial application Charles Wilkins's primary argument had been that "[t]he beneficial Consequences that would arise from the introduction of printed papers in the different Departments of this Government, for Pottahs[,] Coboolyets, Amulnamahs, Perwannahs, Rowannahs[,] Dustucks, Choar Chitties &c. and all such others as are confined to settled form must be so self evident that it

---

109. Travers, *Contested Notions of Sovereignty*.

110. IOR H/207 (22 December 1778). This was also the case when the press was continued for another year in 1779, see IOR P/50/20 *Bengal Revenue Consultations, 19th October to 17th December 1779* (10 December 1779).

111. Travers, *Contested Notions of Sovereignty*, p. 289.

112. IOR H/207 p. 464 (13 November 1778), and E/4/38 *Bengal Letters Received, 1st February 1779 to 14th March 1780*, p. 40 (9 February 1779).

113. Travers, *Contested Notions of Sovereignty*, p. 222 n. 157, cites an instance in June 1778 where Hastings attempted to use Halhed's *Code* to resolve a case of *zamīndarī* inheritance before the Revenue Board, and Francis questioned his interpretation by noting the ambiguity in the printed book itself.

would be unnecessary to point them out."[114] This was a striking assumption of agreement in a context where every aspect of governance was subject to scrutiny and dispute. Hastings and Francis fundamentally disagreed on how to understand land-holding *zamīndars*, the *raīyats* who undertook cultivation, and their relationships to each other and to the Company-state.[115] They could, however, agree that they would be better regulated by uniform, centrally produced, and widely disseminated printed forms.[116]

This was an extension of what was also attempted in manuscript. For example, *pattās*, documents that authorized the cultivation of a plot of land, were an important instrument of the Company's revenue reforms. Robert Travers shows Henry Vansittart, supervisor of the Dijnapur district of Bengal, issuing fixed *pattās* in the hope that they "would make revenue extraction less oppressive and more 'regular.'"[117] Later they were a feature of Cornwallis's Permanent Settlement in 1793, which laid down that the "rents" the *raīyats* should pay to the *zamīndars* "should be fixed and formally recorded in documents called *pottas*."[118] In both cases a standardized documentary form underpinned attempts to regulate relationships between landholders and tenants, and between governors and the governed. Print, as Halhed's preface indicated, would act to guarantee this process. It offered the centralized "European supervision" that all shades of Western opinion on the Bengal revenue sought.[119] Wilkins was therefore to receive orders from "Provincial Councils and Collectors, and...all other heads of Offices" for the papers that they wanted printed, in Persian, Bengali, or Roman characters, "leaving Blanks for Names, Dates, and other Occurrences as are liable to alter, and Specifying the Number of each form they usually issue in the Course of a year." Responding to the difficulties involved in "Composing and preparing the Work for the press," he would then "Strike off as many Copies as may be Required for three or four years."[120] Printing did not simply

114. IOR H/207, pp. 466–67 (13 November 1778). *Kabūliyat*s were written agreements to pay the assessed revenue; *āmil-nāma*s were warrants for revenue collectors; *chhoṛ chiṭṭhī*s were deeds of release. See Wilson, *Judicial and Revenue Terms*. Wilkins also added "Chelanes" (*chalān*s), invoices, and "Tullub Chitties" (*talab chiṭṭhī*s), summonses for revenue defaulters, along with Orders for Confinement and a variety of commissions, warrants, balance sheets, receipts, and certificates; see H/207 p. 470.

115. See, for example, "Mr. Lind's Abstract of the Several Plans transmitted relative to the future Government of Bengal," in IOR H/339 *Home Miscellaneous Series*.

116. Guha, *A Rule of Property for Bengal*, pp. 131–32, discusses Francis's attitude to *pattās*.

117. Travers, *Contested Notions of Sovereignty*, p. 139.

118. Wilson, *Governing Property, Making Law*, p. 124.

119. Marshall, *Bengal: The British Bridgehead*, p. 117.

120. IOR P/50/15 *Bengal Revenue Consultations, 5th January to 5th February 1779*, pp. 202–3 (8 January 1779).

produce and guarantee standardization, it required a prior standardization in order to proceed. This offered a somewhat limited sphere of operation for imperial print. It could function authoritatively only where the way already appeared to be well-trodden.

After a year of operation Wilkins listed twenty different "forms" that he had printed in Persian and Bengali, and several more in English, for the use of the Company. There were the "Amulnaamas," "Cabooleats," "Rowannas," "Dustucks," "Pottahs," and "Perwannahs of Reference" that Wilkins had anticipated printing. The revenue department had also requested that he produce "Raazee Naamas" (*razi-namas*, or deeds of assent) and "Baaze Zemeen Choor Chitties" (*bāzi zamīn chhor chiṭṭhī*s, or notifications of land exempt from revenue). These were among a range of documentary mechanisms of enforcement for taxation and law, many of which retained their English names despite being printed in other languages. They including orders for measuring lands, summonses, demands of revenue, warrants for imprisonment, bail bonds, and bills of sale. The fewer forms that were printed in English were evidently intended more for use within the Company's administration itself, including commissions for the army, warrants to surgeons and pilots, and passes for ships.[121] The other crucial aspect of government printing in Indian languages was the publication of volumes of regulations for the courts. Wilkins printed the *Regulations for the Administration of Justice in the Courts of Mofussil Dewannee Adaulut, and in the Sudder Dewannee Adaulut* in English in 1781 and in Persian in 1782. Francis Gladwin, who succeeded Wilkins as the Company printer, produced Jonathan Duncan's translation of these regulations into Bengali in 1785 along with Bengali and Persian translations of Pitt's India Act, or at least "such parts of the late Act of Parliament as were necessary to be notified to the Natives." Other regulations were regularly published and, from 1793, following Cornwallis's reforms, the *Regulations passed by the Governor-General in Council* were printed in English, Persian, and Bengali and bound into annual indexed volumes by the superintendent of the Company's press.[122]

121. IOR H/207, pp. 481–83 (10 December 1779); and Wilson, *Judicial and Revenue Terms*.

122. *Regulations for the Administration of Justice in the Courts of Mofussil Dewannee Adaulut, and in the Sudder Dewannee Adaulut* (at the Hon'ble Company's Press, Calcutta, 1781); *Translation of the Persian Abridgement of the Regulations...for the Sudder and Mofussul Dewanny Adaulets* (at the Hon'ble Company's Press, Calcutta, 1783); and *Regulations for the Administration of Justice in the Courts of Dewannee Adaulut...With a Bengal Translation* (at the Honorable Company's Press, Calcutta, 1785). For the 1784 India Act, see IOR P/50/60 *Bengal Revenue Consultation, 24th August to 9th September 1785*, p. 36 (9 September 1785). The English Short Title Catalogue lists thirty-one items printed by order of the Governor General in

The volume of this work was substantial. The uneven survival of this sort of printing and the attention given to more didactic forms of Orientalist scholarship should not undermine the importance of these forms and regulations. The Revenue Department spent over 147,000 rupees on printing between 1778 and 1786. The office of the buxey (the military paymaster) also had printing valued at 83,000 rupees undertaken in the same period, contributing to a total expenditure on government printing since the establishment of the press of 236,287 rupees (about £23,000). This included a huge variety of forms, proclamations, advertisements, and regulations in English, Persian, and Bengali that were often produced in large numbers. Over eight thousand copies of the regulations for the courts, and their supplements and abridgements, were printed in all languages, amounting to over 159,000 printed sheets between 1782 and 1785. Print runs of the more routine bureaucratic documents were also regularly in the thousands. For example, between May 1780 and July 1781 Charles Wilkins produced over 33,000 printed copies of ten different papers for the Revenue Department, including 6,285 copies of "the Additional Clauses to the Caboleats Amulnamahs in Persian and Bengali" and 2,024 *parwānas* relating to the *qānūngos* (Muslim judicial officers and keepers of land records).[123]

Despite its volume, this work was done with care. Fiona Ross notes that an early example of a blank form printed in Bengali using Charles Wilkins's first set of types displayed a higher standard of typography than Halhed's *Grammar*.[124] The political impact of this routine printing was also significant. In 1783, Halhed stated of the *Grammar* that "it served to introduce & establish the use of printing in the Bengal Type all official papers applied in the Routine of the public business of Government, to the very great promotion of regularity Œconomy & dispatch in every department of the Company's civil and public affairs." Again, in 1785, he and Wilkins claimed that "[i]nnumerable examples prove the utility of the [Bengali] Types and they are at once a saving of great expence to the Government, and infinitely facilitate all the Channels of intercourse."[125] Although in each instance these pronouncements were part of arguments for positions or requests for

Council from 1781 to 1792. See also *Regulations Passed by the Governor General in Council of Bengal*... vol. 1: *1793, 1794, and 1795* (J. L. Cox, London, 1828) p. 346.

    123. IOR P/3/28 *Bengal Public Consultations, 5th September to 28th September 1787*, pp. 499–541 (28 September 1787).

    124. Ross, *The Printed Bengali Character*. The form is to be found in BL OIOC Add. Mss. 5660F.

    125. IOR E/1/73 p. 258c (18 November 1783), and P/50/57, p. 155 (21 January 1785).

payment, their authors must have been confident that they would not fall upon deaf ears.

Printing was, therefore, a political technology. The desire for fixity, uniformity, and wide dissemination with centralized supervision was widely shared within the administration of the Bengal revenue, and had been sought in script before the promise of print became a local possibility. Print's promise to separate the technical problems of administration from political debate underpinned a degree of consensus on imperial governance, even though it effectively restricted printing's utility to the sorts of forms and regulations that could be standardized before their printing rather than by it. Yet, the assumption that these documents had what could be taken as a settled form, and the claim that the legitimate documentation could be sifted from the "forgeries" and "impositions," was also to take a political position.[126] To do so was to assume that the nature of law, custom, and practice, and the relationships between them, could be known to those who governed and fixed in ways that would remain stable. This was far easier in rhetorical statements such as Halhed's *Grammar* than it was in practice.[127] The difficulties involved meant that there was no simple printing revolution in the government of India. For example, the Company's accounts show that, despite their rhetorical prominence, little if any printing of *pattās* was undertaken between 1778 and 1786 for the Revenue Department.[128] It is likely that the same resistance, from either *raīyats* or *zamīndars*, depending on the geographical balance between land and labor, which restricted the imposition of fixed agreements in the early 1770s and after 1793, also severely restricted the utility of printed *pattās* in the early 1780s. In each instance, these forms of documentation were rejected by those for whom they imposed restrictive conditions and unhelpfully solidified what were better kept as more fluid and negotiable political and economic relationships.[129]

Where fixity, standardization, and wide dissemination were required by the Company-state, and either could be imposed or were accepted, they were not automatic products of the press. The printing office had to be organized to produce them. The potential pitfalls are well illustrated by those who

---

126. Marshall, *Bengal: The British Bridgehead*, p. 61.

127. Wilson, *Governing Property, Making Law.*

128. They do not feature in the account of all printing given in IOR P/3/28, pp. 499–541 (28 September 1787). They are mentioned in the list of forms produced in the first year of operation, but without figures on how many were printed; see IOR H/207, pp. 481–83 (10 December 1779).

129. Wilson, "'A thousand countries to go to'"; and Travers, *Contested Notions of Sovereignty.*

sought to use Calcutta's commercial press. Having successfully produced a series of standard pay and allowance forms for the Company's army, albeit with some difficulty over agreeing on a price for what was a novel enterprise, James Augustus Hicky was asked in 1779 to print Sir Eyre Coote's digest of the military regulations. To do so he employed several men to help with the printing, as well as a blacksmith, a carpenter, and brassmen to service the press and provide the brass rules such a publication required. This, and buying the appropriate paper, had meant borrowing four thousand rupees from his friend Captain Price. Hicky had proofs of the first nine sheets approved by Coote, and proceeded to print 2,135 copies. However, Coote had to leave Calcutta for Madras to try to break the siege of Fort St. George set by Haider Ali, the ruler of Mysore. Hicky was informed that he should submit his bill to Governor General Warren Hastings. This put him in a difficult position. Hicky later noted that since he "unfortunately laboured under the displeasure of the Honble Governor General and Council, he thought it would not be prudent to trouble them with an application of the kind, at that time."[130] He would wait for Coote to return.

Unfortunately, Coote was a problem too. He had presented the job to Hicky as a simple matter of producing "a digest... of the Existing Orders and Establishments free from the confused and voluminous State in which they were comprized."[131] However, the process of rationalizing a decade's annulments, amendments, and revisions in the regulations was seen by other high-ranking Company officers as an assault on some of their cherished privileges by a "Kings Officer." Coote received anonymous death threats and, in his absence, Hicky reported that "every possible obstruction was thrown in the way of the Printer in order to impede the Publication of the said Regulations." Coote's opponents, including his own aide de camp Colonel Owen, kept Hicky's printing office idle by regularly taking five or six days to approve a single sheet of proofs. They also blocked Hicky's attempts to employ soldiers as printers. A Private Keely, who had been brought up a printer, was denied to Hicky on Colonel Morgan's report "that he did not understand much about Printing and that he was fond of liquor." Another soldier was withheld by Colonel Pearse who "positively declared that the most worthless or useless Man under his Command should not lend the least assistance to Genl Coote's Regulations unless it was to make them into Skie

---

130. IOR P/3/12 *Bengal Public Consultations, 9th May to 26th May 1785*, p. 379 (3 June 1785).

131. IOR P/3/37 *Bengal Public Consultations, 2nd July to 29th August 1788*, p. 952 (22 August 1788).

Rocketts."[132] Eventually the new regulations were withdrawn. Hicky was paid for his labor only in 1793, fourteen years after he had set his press to work and after spending some years back in prison for his debt to Captain Price. Moreover, it remained unclear exactly what had been printed and on what basis, the Bengal Council having concluded that "Sir Eyre Coote was not authorized by the Board to give the Orders he did to Mr Hicky."[133]

The ability to fix a reasonable rate for the job and the establishment of clear lines of authority over the printing process and accountability for its products, as well as the capacity to carry the work out effectively and on time, were all problems that Hastings's and Wilkins's plans for the Honourable Company's printing office sought to resolve. It would mean that the Company-state would have the authority to print, and could actually get its printing done. Having its own press would ensure that the Company was not reliant on men so embedded in the fractious world of Calcutta party politics as James Augustus Hicky. This was important in achieving the production of official regulations and administrative forms in English, Bengali, and Persian. It was also necessary for more overtly political texts. In 1782 Hastings was able to require his appointee Charles Wilkins and the Honourable Company's Press to print, for an audience in both Calcutta and London, his justification of the "reduction" by force of Rajah Cheit Singh, the *zamīndar* of Benares, for not paying his military subsidy. Hastings was heavily criticized at the time, and this was one of the actions central to his subsequent impeachment. The 53-page *Narrative*, with over 150 pages of appended official documentation and affidavits, made Hastings's case that Cheit Singh's rebellion was illegal and his own use of violence justifiable, since the "Rajah" was merely a *zamīndar* who now owed allegiance to the Company's sovereign authority, rather than being a "tributary Prince." This was despite the fact that the translation of Cheit Singh's *sanad* (deed or decree) referred to his obligatory payments as "tribute" rather than "revenue." Hastings argued, and Wilkins dutifully printed, that "it is not on the translation, nor on the will of a translator, that the rights of the Company depend."[134] It is difficult to imagine Hastings paying Hicky, his enemy and a supporter of Philip Francis, or any other of Calcutta's commercial printers to produce a text that so clearly denied the political rights of *zamīndars*

132. IOR P/4/18 *Bengal Public Consultations, 4th January to 4th March 1793* (1 February 1793).

133. IOR P/3/37 p. 957 (22 August 1788) and P/4/21: *Bengal Public Consultations, 17th June to 26th July 1793* (26 July 1793).

134. *A Narrative of the Insurrection Which Happened in the Zemeedary of Banaris* (printed by order of the Governor General, Calcutta, 1782) pp. 9, 10.

and affirmed the overwhelming power of the Company-state to intervene in Indian affairs.[135]

The establishment of a Company printing office also sought to resolve a range of other problems that were endemic to Calcutta's commercial press. William Jones complained that "the compositors in this country are shamefully inaccurate," and his letters detail the laborious process of printing the first volume of the *Asiatick Researches*.[136] John Gilchrist, in the preface to his *A Dictionary, English and Hindoostanee* (1787–90), described himself as "a wretch whose head is fairly wedged in the press." His plan to produce his dictionary and grammar by subscription and in monthly numbers had gone badly wrong, trapping him between demanding (or slow-paying) readers, unscrupulous agents, and grasping printers, munshis, and pandits. Only money would move things along: "One day the press would bound with the agility of an antelope, and for weeks afterwards assume almost the retrograde gait of a crab, just as an influx of cash, and spirits, roused or benumbed its conductors." His financial resources being "swallowed up by the all devouring press," he spent ten years in "literary slavery" attempting to correct the work of the compositors, "every one more ignorant than another of the subject they were engaged on." Every emendation produced worse errors. Every four-page proof sheet took two or three weeks. The "slavish drudgery of correcting the press" reduced him to "living like a hermit in his cell," where he complained he had "confined myself so closely to my desk, as to have not only emaciated my whole frame, and ruined my eyesight, but nearly subjected myself to lameness for life." When his work was finally complete he was forced to apologize for "the badness of the paper, and all defects in the execution."[137]

For Gilchrist, these problems in producing his dictionary were filtered through the cultural politics of race in imperial Calcutta. This shaped the work of book production.[138] His difficulties with the compositors were

---

135. Reviewers in England paid equal attention to the *Narrative*'s political arguments and to its typographical merits and documentary veracity; see *Monthly Review* (1786) 74, p. 387, and *Critical Review* (1786) 61, p. 313.

136. Quoted in Shaw, *Printing in Calcutta*, p. 23.

137. John Gilchrist (1787–90) *A Dictionary, English and Hindoostanee* (Stuart and Cooper, Calcutta) pp. ix, xii–xv.

138. Betty Joseph (2004) *Reading the East India Company, 1720–1840: Colonial Currencies of Gender* (University of Chicago Press, Chicago) pp. 96–100 notes both the complexities of the politics of race and moments when the racialized order hardened. For example, between 1786 and 1791, just as Gilchrist was at work on his dictionary, Lord Cornwallis enacted statutes that called for the removal of Indians and people of mixed race from military and civil posts.

probably a result of the use by the vast majority of Calcutta printing office proprietors (including his printers Daniel Stuart and Joseph Cooper) of Indian labor unfamiliar with the range of languages in Gilchrist's dictionary; this allowed them to cut their printing rates by half.[139] In addition, Gilchrist's own financial situation—the lack of funds (and of political clout) that so severely slowed the work of his linguists and of the press—was also more than just a matter of economics. He understood it and its implications in terms of the social hierarchies and expectations of race and empire. As he put it, "To the natives, a leper is not half so loathsome, as the wight whose narrow means force him to be an economist, and whose want of influence with the great, is gorgon enough to petrify the humanity and literary spirit of all India." Overall, the author's lack of control over the publishing process, and the pain and anguish that he had personally gone through because he could not delegate that work to trusted others, were deeply felt and were expressed in the language of racial animosity. Gilchrist wrote, as a warning to those who might follow, that "my people could not crawl one inch in the business without my own personal superintendence, and their eternal treacherous behaviour when my work put me both in their and the printers power, generally disqualified me from all serious continued study, while tormented as I was with a posse of black unprincipled knaves."[140]

Securing the control over the press that could not be guaranteed when engaging commercial printers was also a crucial issue for the Baptist missionaries. Their spectacular achievements—the production of 212,000 volumes in forty languages between 1801 and 1832—were based on a distinctive combination of religious zeal and a form of social organization modeled on the Moravian missions.[141] As William Ward put it in 1800:

This week we have adopted a set of rules for the government of the family. All are equal, all preach & pray in turn, one superintends the family for a month, & then another; Bro. C[arey] is treasurer & has the regulation of the Medicine Chest; Brother Fountain is his librarian. The Saturday evening is devoted to the adjusting of differences, & pledging ourselves to love each other.[142]

139. Khan, *History of Printing in Bengali Characters*, p. 408.
140. Gilchrist, *Dictionary*, pp. xii, xv.
141. M. Siddiq Khan (1966) "Early Bengali printed books," *Gutenberg Jahrbuch*, pp. 200–208; and Marshman, *Carey, Marshman, and Ward*.
142. BMS IN/17(A) *William Ward's Journal*, p. 66 (18 January 1800).

This "family situation, in which all personal interests were swallowed up in the interest of the whole," was difficult to sustain.[143] The combination of strict equality of property and labor with a strongly hierarchical vesting of authority in Carey, Marshman, and Ward led to a revolt among the younger brethren and fears that the mission would not last beyond its founders. However, in the early years in India it did channel scarce resources into printing and preaching, despite the privations suffered by the missionaries and their dependants. Indeed, it made the printing press so obviously central to their lives that the Indians, "hearing Mr Carey's description of its wonderful power, pronounced it to be a European idol."[144]

Eventually the missionaries constructed a purpose-built printing office at Serampore with separate composing and binding rooms. By 1805 the office employed thirty workmen, who produced their own paper, types, and ink. Indeed, the missionaries sought to achieve control over all aspects of the production of their texts. It was their own command of the languages into which they translated the Scriptures, their intense scrutiny of every proof sheet, and their strict control of the press that gave them confidence that they were finally spreading the true word to the heathen.[145] Yet this could never be fully achieved. The mission press always relied upon the mental and manual labor of others. The printing office was used to give employment to potential converts, but it also depended upon workers who would never convert and were willing to openly object that while they worked "they were compelled . . . to hear so many blasphemies against their gods."[146] Ward, after a faltering start, spoke fluent Bengali. It was said of him that "[h]e surpassed his colleagues in a knowledge of the character and habits of the natives, and few Europeans have ever been so successful in managing them."[147] Yet this belies the complex social relations of the printing office. Ward later had composed and printed, by these very workmen, his assessment that "everyone who has been obliged to employ the Hindoos, has had the most mortifying proofs that if the vices of lying, deceit, dishonesty, and impurity, can degrade a people, the Hindoos have sunk to the lowest depths of human

143. BMS IN/17(A) *William Ward's Journal*, p. 231 (10 May 1802).

144. Marshman, *Carey, Marshman, and Ward*, vol. 1 p. 80.

145. BMS IN/17(A) *William Ward's Journal*, p. 424 (17 June 1805). Marshman, *Carey, Marshman, and Ward*, vol. 1 pp. 397, 180, notes early concerns that "The construction of the sentences of the first edition [of the New Testament], which the flattery of the pundits had pronounced to be perfect, was so entirely at variance with the idiom of the language that the work was barely intelligible" (p. 180).

146. BMS IN/17(A) *William Ward's Journal*, pp. 229 (5 May 1802), 369 (2 April 1804); and Ross, *The Printed Bengali Character*.

147. Marshman, *Carey, Marshman, and Ward*, vol. 2 p. 279.

depravity."[148] His own Indian workforce had clearly caused him a great deal of trouble as those he had to rely upon to produce his sacred texts. They were also problematic as readers. The missionaries could have no confidence that their translations would translate into conversions. As Ward put it, "The natives, we believe, read our things; but what the effects are we cannot tell."[149]

The Company's production of authoritative texts attempted to resolve these difficulties through a distinctive form of organization of the economy and society of the printing office. The press was organized as both a private enterprise and a branch of government. As a private enterprise it could take on other business and would not expect to survive on government printing alone. Its rates were kept comparable with those of other Calcutta printers and, unlike missionary printing, it was expected to be cost-effective. Yet it was also an arm of the state, and supported by it, guaranteeing that the government was not simply at the mercy of the commercial printers. Until 1785 the Company paid a regular salary to the superintendent, who was expected to run the press as part of his household, just as the printers of early modern London had done.[150] Again, until 1785 the Company paid the rent and equipped the household with servants. The initial printing office had two compositors for Bengali and Persian, and one for English, a sorter, a pandit and a munshi, a head pressman, eight other pressmen (presumably working in shifts on the two wooden presses), four footmen, a sweeper, and a bookbinder.[151] Regardless of who was actually doing the work of composition, printing, and collating, it was the superintendent who was directly accountable for its quality, efficiency, economy, and security.[152] Even after the salary and rent were no longer paid, on the grounds that the work itself was being paid for, the position of Company printer, and the monopoly that it carried with it, could be handed on from printer to printer only with the approval of the council.[153] The Honourable Company's Press, and the social organization of its printing office, had to be judged capable of producing texts that could make claims to authority, and it required government

148. William Ward (1811) *Account of the Writings, Religion, and Manners, of The Hindoos*, 4 vols. (printed at the Mission Press, Serampore) vol. 1 p. xx.

149. BMS IN/17(A) *William Ward's Journal*, p. 92 (3 August 1800).

150. Johns, *The Nature of the Book*.

151. IOR P/3/28, pp. 499–541 (28 September 1787), and H/207 (13 November 1778).

152. IOR P/2/37 *Bengal Public Consultations, 29th May to 3rd August 1780*, 29 May 1780.

153. *Fort William-India House Correspondence. Vol. IX Public Series, 1782–1785* (National Archives of India, Delhi, 1959) p. 216; and IOR P/3/24 *Bengal Public Consultations, 3rd January to 23rd February 1787*, 3 January 1787.

support to do so. It is, therefore, telling that when Hicky was imprisoned in 1782 for a libel against Warren Hastings, his printing equipment was sold to the Company printer at a knock-down rate.[154]

The Management of Characters

The production of texts that could claim authority required control not only over the social organization of the printing office, but also over the materials and technologies of the printing process itself. By the late eighteenth century the printing press had remained relatively unchanged for several centuries. Printers in India could import wooden flatbed hand presses or have them constructed, as Hicky did, by local craftsmen. Types for Roman characters were also brought from Britain.[155] Those required for printing in Indian languages were, as Nathaniel Halhed recognized, much more problematic:

> That the Bengal letter is very difficult to be imitated in steel will readily be allowed by every person who shall examine the intricacies of the strokes, the unequal length and size of the characters, and the variety of their positions and combinations.[156]

Once again it was William Bolts who demonstrated the difficulties of imperial printing. Bolts's desire to print either for the Company or against it led him to commission Joseph Jackson, a skillful London typecutter and founder, to produce a font of Bengali types. This font, which appeared in an inventory of Jackson's premises in 1773 as "BENGAL or modern SHANSCRIT," was, however, unusable.[157] This was the fault of Bolts's designs rather than Jackson's technique. The problem was that the separate characters that Bolts had isolated for typecutting were insufficiently abstracted from the particular handwritten examples that he had used to identify them. He had not recognized the different styles of Bengali script, failing to distinguish between cursive and formal hands. He had also failed to disaggregate characters from their contexts, mistaking different versions of the same character for entirely different characters. As a result, the types themselves "incorporated

---

154. Khan, *History of Printing in Bengali Characters.*
155. Khan, *History of Printing in Bengali Characters.*
156. Halhed, *Grammar,* p. xxiii.
157. Edward Rowe Mores (1778) *A Dissertation Upon English Typographical Founders and Foundaries* (privately printed) p. 83, suggests this was done in the 1720s. Khan, "The early history of Bengali printing," argues that Bolts had been commissioned by the English East India Company to prepare a grammar of the Bengal language.

[the] personal idiosyncracies, unorthodoxies, and inaccuracies" of the scribes on whose writing they were based.[158]

Therefore, the initial problem those keen to construct Bengali types faced was that notions of uniformity and fixity at the level of the character itself had to be imposed upon or drawn out of the heterogeneous multiplicity of Bengali scribal culture. The lineaments of standardized characters had to be conceived and ways found to bring them into view before they could actually exist. In the process of doing this it was impossible to separate the technical from the social. As Halhed noted, Bolts, "who is supposed to be well versed in this language," had "egregiously failed in executing even the easiest part, or primary alphabet" because of his failure "to procure a writer accurate enough to prepare an alphabet of a similar and proportionate body throughout, and with that symmetrical exactness which is necessary to the regularity and neatness of a fount."[159] Bolts's failure was that he was unable to gather the right people around him to make the technology work.

How then did Charles Wilkins succeed? One significant difference is that for Wilkins understanding the nature and shape of the characters of the Bengali alphabet was an important intellectual project. As part of the discovery of the classical roots of vernacular languages English Orientalist scholars had argued that the characters used for Bengali were the same as those used for writing Sanskrit.[160] As Wilkins himself later put it, "the Alphabet in common use throughout the provinces of *Bengal, Assam*, and, perhaps, *Orissa*, is essentially the same as the *Dēva-nāgari*, in which the *Sanskrit* language is usually written in the upper parts of *Hindustān*, it being formed upon the same principles; and how muchsoever the two characters differ at present, there can scarcely be a doubt of their having been originally identically the same."[161] Identifying Sanskrit as the ancient language of the Hindus, and therefore seeing it as the medium of the legal and religious prescriptions that they wished to use to govern India, Englishmen like Wilkins and Halhed were keen to learn it. In the early 1770s, however, they found themselves frustrated by what they thought were Brahminical prohibitions that meant that "to sully its purity by imparting the slightest knowledge of it to strangers was ever cautiously avoided as the most inexpiable crime."[162]

158. Khan, "The early history of Bengali printing," p. 56; and Ross, *The Printed Bengali Character*.

159. Halhed, *Grammar*, p. xxiii.

160. For example, the comparison of the alphabets in Halhed, *Code of Gentoo Laws*.

161. OIOC Eur Mss. K430 *Memoir of Sir Charles Wilkins*, p. 1.

162. Halhed, *Grammar*, pp. iii, x.

The initial reluctance of these Brahmin scholars turned out to be more practical and political than strictly religious. This indigenous intelligentsia was ensuring that wherever possible they were not too dependent upon their new British rulers when they collaborated with them.[163] Initially, therefore, the Orientalists had to explore other avenues.

One way forward was through the idea that a language's characters were not simply a medium of expression but important carriers of knowledge in themselves. In his preface, Halhed traced the history and geography of Bengali lettering and, in particular, its relationship to Sanskrit, which he called "the Parent of almost every dialect from the Persian Gulph to the China Seas." This history of descent was, he argued, traceable in words, monosyllables, numbers, and the names of "such things as would be first discriminated on the immediate dawn of civilization." It was also visible in the "characters upon the medals and signets [and coins and seals] of various districts of Asia."[164] It is, of course, in the later work of William Jones that these ideas were most fully expressed. Jones's famous Third Anniversary Discourse, when he argued that Sanskrit, Greek, and Latin were all derived from a common source, used evidence from "the roots of verbs" and "the forms of grammar." Perhaps learning from Halhed, his arguments were also based on the shapes of letters, which, for Sanskrit, "are believed [by Brahmins] to have been taught by the divinity himself . . . in a voice from heaven." Jones's own notions of the historical geography of alphabets were somewhat more prosaic, but no less important in securing identities and histories. He argued that India itself could be defined as "that whole extent or country, in which the primitive religion and languages of the *Hindus* prevail at this day with more or less of their ancient purity, and in which the *Nágarì* letters are still used with more or less deviation from their original form."[165] He also reasoned that rather than relying on unreliable manuscripts, the history of Indians, Arabs, and the other Asian peoples could be accurately reconstructed only through "their *language, letters,* and *religion,* their ancient *monuments,* and the certain remains of their *arts.*"[166] This meant, among other things, a close scrutiny of characters.

---

163. Rosane Rocher (1989) "The career of Rādhākānta Tarkavāgīśa: an eighteenth-century pandit in British employ," *Journal of the American Oriental Society,* 109:4, pp. 627–33

164. Halhed, *Grammar,* pp. iii–iv.

165. *Asiatick Researches,* vol. 1 pp. 419, 423; and Brockington, "Warren Hastings and Orientalism."

166. *Asiatick Researches,* vol. 2 (printed and sold by Manuel Cantopher at the Honourable Company's Printing Office, Calcutta, 1790) p. 4.

It is these ideas that underpinned Charles Wilkins's work with type. He was experimenting with making Bengali types prior to the production of the *Grammar*, and he was subsequently the first Englishman to learn Sanskrit from the Brahmin pandits. These ideas also help to explain his continuing experimentation. He designed several fonts for the Honourable Company's Press, including those used to print the first works in Sanskrit. However, because he had not perfected a set of Devanāgarī types, he continued to work on them on his return to England in 1786. His attempts to do so ended in a disastrous fire at the small country home where he had set up his printing press. Most of the types were lost or destroyed, "having been thrown out and scattered over the lawn," but fortunately Wilkins was able to save his books and manuscripts, and also most of the punches and matrices from which the types could be recast.[167]

It was because the study of the shape of Indian characters was an important intellectual project that Charles Wilkins was willing to engage so intimately and for so long with the artisanal process of making and using types. However, how far this actually meant getting his own hands dirty is an important question. In England in 1795 he had set up his domestic printing press "with the assistance of such mechanics as a country village could afford."[168] In India, by contrast, Halhed reported that the successful creation of the Bengali font was a matter of Wilkins's "solitary experiment," in which "he has been obliged to charge himself with all the various occupations of the Metallurgist, the Engraver, the Founder and the Printer."[169] This, as many historians have pointed out, is a myth.[170] Charles Wilkins could not have done it on his own. He succeeded because of the work of others.

First, Wilkins and Halhed needed, as has been seen, to find a writer or writers who could prepare an appropriate alphabet. This was, in Halhed's opinion, "no easy task" because for "the bulk of the modern Bengalese," "Their forms of letters, their modes of spelling, and their choice of words are all equally erroneous and absurd."[171] Something is known of the pandits, munshis, and scribes who worked with Halhed and Wilkins and taught

167. Ross, *The Printed Bengali Character*; and Charles Wilkins (1808) *A Grammar of the Sanskrita Language* (printed for the author, London) p. xii.

168. Wilkins, *Grammar*, p. xii.

169. Halhed, *Grammar*, p. xxiv.

170. Khan, *History of Printing in Bengali Characters*; Shaw, *Printing in Calcutta*; and Ross, *The Printed Bengali Character*.

171. Halhed, *Grammar*, pp. xxiii, 179.

them the Bengali alphabet. From what the forms of transliteration of Bengali words in the *Grammar*, and in the manuscripts in Halhed's collection, reveal about the pronunciation of his Indian language teachers, Muhammad Abdul Qayyum has argued for the presence of an East Bengali Muslim munshi, who taught Halhed Bengali via Persian in the early 1770s, and a Bengali Brahmin pandit, perhaps from the court of the raja of Krishnanagare, who took over the Englishman's instruction in Bengali sometime in 1774. This may have been the pandit paid for by Hastings to work on the *Grammar*. Qayyum argues that this man, whom Halhed referred to as "an intelligent Brahmin," was instrumental in Sanskritizing the version of Bengali presented in the Englishman's *Grammar* and drawing it away from the vibrant, heterogeneous, and syncretic everyday vernacular of Bengal. He provided Halhed with a list of "the most Ancient and Authentic Bengal Books," which included the Mahābhārata and the Rāmāyaṇa, upon which the *Grammar* extensively drew.[172] It is perhaps not too fanciful to see their language lessons as blending different traditions of reading and study. On the one hand, this was an English gentleman assisted by a paid employee in acquiring a difficult and valued skill with languages.[173] On the other, it was a rather more private form of the Bengali *kathakatā*, in which a Brahmin read from and subsequently interpreted written materials, particularly the accessible and exciting epics, for illiterate and semiliterate audiences (see chapter 1). What could be more appropriate for the instruction of those, like Halhed and Wilkins, who, at least in the mid-1770s, were taking their first steps in Bengali?[174] These sessions were based upon manuscript materials, gathered by Halhed via his Indian collaborators, that also contributed directly to the construction of the *Grammar*. We know that these were copied for Halhed's use by various copyists including Tārāchandra Dās Ghosh of the village of Malangā, Sheikh Jamal Muhammad of Kalinga, and Ātmārām Dās Ghosh, a *Kāyastha* from Calcutta. However, the precise identity of the scribes who produced the alphabet from which the types were to be produced remains unknown.[175]

Wilkins also needed help in cutting the punches to impress the matrices from which the types were cast. In this he was assisted by a former Birmingham die-sinker and gem and seal engraver, Joseph Shepherd, whom

172. Quoted in Qayyum, *A Critical Study of the Early Bengali Grammars*, pp. 56, 59.

173. For example, for William Jones's work as a language tutor, see Garland Cannon (1990) *The Life and Mind of Oriental Jones* (Cambridge University Press, Cambridge).

174. Ghosh, *Power in Print*; and Qayyum, *A Critical Study of the Early Bengali Grammars*.

175. Qayyum, *A Critical Study of the Early Bengali Grammars*; and Ross, *The Printed Bengali Character*.

John Gilchrist described as an "ingenious artist." Shepherd worked in the commercial world of decorative engraving in Calcutta, and there is evidence of his skillful production of "exotic" letterforms.[176] Finally, Wilkins needed a typefounder. This role was undertaken by Panchānan Karmakār, a blacksmith from Tribeni in the Hugli district of Bengal. Panchānan later worked with Henry Colebrooke to produce the smaller and finer font that was used to print the Cornwallis Code in 1793. He was also crucial to the initial establishment of the Serampore Mission Press. His legacy there was continued after his death in 1803 or 1804 by his son-in-law Manohar, who produced many new fonts for the missionary printers.[177]

It is impossible to recover the relationships between these men from different caste, class, and religious backgrounds, and also the other workers who were undoubtedly involved around the press as pressmen, peons, and sweepers. It is clear, however, that solving the intellectual and practical problems of printing in Bengali in British Bengal required the bringing together of complementary skills in the combination of which there was no clear division between mental and manual labor. Just as with the contemporaneous practices of surveying and mapping, printing in Bengali types was a coproduction dependent upon various skills rather than the simple playing out of the imposition of a European technology through the rigid grids of the colonizer and the colonized.[178] It was also profoundly local. In order to constitute an effective printing press these different characters had to be brought together in a particular place and the relationships between them worked out in ways that would be productive. Once again, the solutions to the problems of print were an amalgam of the social and the technical. As Kapil Raj argues, this always occurred in situations of asymmetrical reciprocity, where exchanges required cooperation but were structured by imperial relationships of power.[179] This is, of course, most evident in Halhed's presentation of the production of the Bengali types as Wilkins's "solitary experiment." Besides being an obvious imperial gesture, this also works as an attempt to resolve the problems of credibility and trust that hung around such a novel venture. Working between Calcutta and London, William Bolts

---

176. Gilchrist, *Dictionary*, p. xlii; Shaw, *Printing in Calcutta*; and Ross, *The Printed Bengali Character*, p. 11.

177. Shaw, *Printing in Calcutta*.

178. Kapil Raj (2003) "Circulation and the emergence of modern mapping: Great Britain and early colonial India, 1764–1820," in Claude Markovits, Jacques Pouchepadass, and Sanjay Subrahmanyam (eds.) *Society and Circulation: Mobile People and Itinerant Cultures in South Asia, 1750–1950* (Permanent Black, Delhi) pp. 23–54.

179. Raj, "Circulation and the emergence of modern mapping."

had been unable to resolve the differences between linguists, scribes, and typecutters. His assistants remained differentiated and individualized and his types remained unable to perform the work of replication. Wilkins, in Hugli, was part of an artful combination and alignment of linguists, scribes, engravers, founders, and printers that made the negotiations between them more or less invisible. This allowed Halhed to take Wilkins himself—the master printer—to stand for them all and through him to offer up the promise of perfect replication.[180]

Yet, once again, this could never simply be a matter of mechanical re-production. The management of the press may have been effective enough to claim the authority that came with the accurate replication of Bengali language and script in print, but it also bore the signs of the work that had been done. For example, Muhammad Abdul Qayyum identifies the many errors of spelling, transcription, and translation committed by Halhed (such as rendering *kheyuḍā* as "sacred odes" rather than "vulgar ditties"), as a result of his "wild guesses" and "false derivations." Halhed also included rules from three quite different schools of Sanskrit grammar without realiz-ing that he was doing so.[181] The composition of characters around the press at Hugli created a new social and cultural fragment in British Bengal, and the process of making types and composing them into texts also created some-thing novel. In part this was a matter of purification. Figure 20 shows pages from their *Grammar* where Halhed and Wilkins sought to demonstrate the current condition of the Bengali language. Unlike the rest of the book, where Halhed had drawn examples from poetry on the grounds that the demands it placed on both author and scribe made it "most conformable to the true genius of the language," these pages showed a petition on a revenue matter. On the left was Wilkins's copperplate engraving of a manuscript in the hand of Halhed's East Bengali Muslim munshi or his scribe. On the right was the same document printed with the new types.[182] The text had been cho-sen by Halhed because it was "replete with foreign expressions; . . . which serves to shew how far the modern Bengalese have been forced to debase the purity of their native dialect by the necessity of addressing themselves to

180. There are parallels here with the resolution of successful early modern "authors" of natural philosophy into "bipartisan writer-Stationer alliances" in Johns, *The Nature of the Book*, p. 158.

181. Qayyum, *A Critical Study of the Early Bengali Grammars*, pp. 70–71.

182. Halhed, *Grammar*, p. 36. Ross, *The Printed Bengali Character*, identifies the hand as that of a scribe who copied a collection of administrative documents for Halhed and the type as Wilkins's second Bengali font. For Qayyum, *A Critical Study of the Early Bengali Grammars* it is the hand of Halhed's munshi.

their Mahommedan Rulers." It was also, he noted, badly written, displaying "most of the vitiated forms of letters used in expeditious writing; and will introduce [the reader] to the irregular hands, which are constantly found in matters of business."[183] He proceeded to detail which words were Persian, Bengali, Arabic, Hindustani, and Sanskrit, and to point out examples of bad orthography and poor calligraphy. There was an evident politics here. Halhed's identification of Bengali with a "pure Sanskritized form" was both an attack on the Islamic languages of Bengal and on previous forms of rule.[184] It was a "purification" of Perso-Arabic elements of Bengali encouraged by the Brahmin pandit, who, as Qayyum argues, had eventually ousted from his place as Halhed's teacher the Muslim munshi whose form of writing was being derided. The new alliance of Brahmin and British through a Sanskritized Bengali is evident in the *Grammar* and in Figure 20. The identification of linguistic hybridization as a problem in speech was, therefore, also present in the promotion of print.[185] Jones's Persian grammar had decried the illegibility of "the common broken hand used in India," and the printed version of the manuscript petition in Halhed's *Grammar* presented it as "in the proper character."[186] The separation of different languages and scripts, something that could be more effectively pursued through print, was a response to the problematic mixtures of Mughal Bengal. Yet, once again, this was not simply an imperial imposition. It was part of a process of Sanskritizing and standardizing the Bengali language that was also being undertaken by Bengali political and economic elites. Ironically, this "purification" produced a more complex linguistic structure.[187]

These linguistic transformations were, therefore, another set of "improvements" carried out in the name of faithful reproduction. Fiona Ross's close scrutiny of the Bengali types produced under Wilkins has demonstrated what was new at the level of the written word itself. Most evident in Figure 20 is the introduction of interword spacing, which is not a feature of Bengali manuscripts. She also notes how the structure of the characters, while recognizable, did not match those of contemporary penned forms. The printed

183. Halhed, *Grammar*, pp. 207–8.

184. Khan, "The early history of Bengali printing," p. 53.

185. Qayyum, *A Critical Study of the Early Bengali Grammars*. Gilchrist, *Dictionary*, p. v, referred to the "barbarous gabble" of Hindustani "jargon." See also Javed Majeed (1995) "'The Jargon of Indostan': an exploration of jargon in Urdu and East India Company English," in Peter Burke and Roy Porter (eds.) *Languages and Jargons: Contributions to a Social History of Language* (Polity Press, Cambridge) pp. 182–205.

186. Jones, *Grammar*, pp. xx; and Halhed, *Grammar*, p. 208.

187. Marshall, *Bengal: The British Bridgehead*; Bayly, *Indian Society and the Making of the British Empire*; and Ghosh, *Power in Print*.

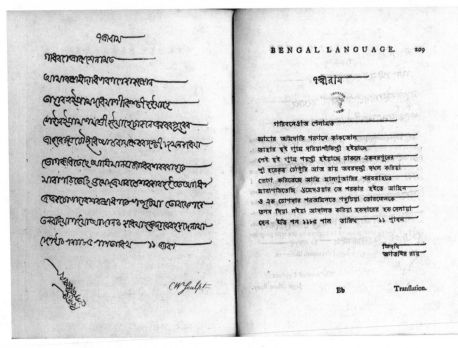

Figure 20 Script and print in *A Grammar of the Bengal Language* (1778). (By permission of the British Library, 1560/1944.)

characters are differently shaped to varying degrees and are not joined by the same configuration of strokes. Moreover, the *Grammar* is itself internally inconsistent since characters were modified during the printing process. Inevitably, as Ross demonstrates, the translation from script to print involved a whole series of choices and decisions. The choice of the scribe on whose hand the characters were based. The choice to base the types on the formal style of Bengali handwriting with its headline (evident in Figure 20) rather than the informal cursive. The choices between particular regional preferences for character shapes. Printing both required and produced characters that could be understood as separate elements in writing that could then be combined and recombined. This helped dictate the decisions made. Informal scripts, as Bolts had discovered, "retain[ed] the personal characteristics of the writer, and thereby generat[ed] almost infinite varieties with no accepted standard."[188] Moreover, the characters in the informal hands were themselves contextual and relational. It was often possible to identify them only in situ and in relation to the other characters to which they were

188. Ross, *The Printed Bengali Character*, p. 82.

joined. Once isolated they became illegible and unfixable. In contrast, formal hands were already becoming more standardized, partly as a result of
their use as the medium of instruction for Bengali language learners themselves.[189] Imperial printing became possible, therefore, on the basis of these
prior indigenous standardizations of Bengali script.

Printing also imposed other demands. Unlike English, Bengali is pho-
netic. It is spelled the way it sounds. The combinations of characters (ak-
sharas) that represent these sounds are not simply juxtapositions of the
fifty-two basic alphabetic characters, as they are in Roman script. Instead,
these are combined in ways that change the shapes of the characters when
they are brought together, producing a range of contextual variants for which
it can be difficult to tell which characters have been joined.[190] Effectively,
for the printer if not for the scribe, this created numerous new characters,
which potentially required a font of vast size to be cast.[191] Wilkins and
his collaborators solved the problem by using Halhed's understanding of
the structure of the Bengali language. For example, Halhed emphasized the
role of adjunct letters, or phalās, of which he identified twelve. These were
then printed using reduced versions of particular characters that joined with
a range of other characters to produce the variant combinations (this is visible in the upper and lower parts of printed characters in Figure 21). This
meant that Wilkins, Shepherd, and Panchānan did not have to design, cut,
and cast characters for every combination. Instead they could design reduced letterforms that, side-by-side with other characters on the composing
stick, would produce "the appearance of being one integral character."[192] As
a result these printers were able to cut only 170 sorts, including 48 reduced
letterforms and 32 special sorts and ligatures. However, this translation of
Bengali from script to type produced, through these more constrained forms
of combination, a range of new and previously unknown shapes on the
printed page. Since subsequent printers worked from these characters rather
than from manuscript hands, "Wilkins's" types had a substantial impact
on the production of subsequent Bengali fonts, and therefore on the shape
of the Bengali printed character.[193] It is also significant that the phalā was
one of only two (out of forty, the rest being Sanskrit) Bengali grammatical

---

189. Ross, The Printed Bengali Character.

190. S. P. Murdur, L. S. Wakankar, and P. M. Ghosh (1986) "Text composition in Devanagari,"
Sesame Bulletin, 1:1, pp. 18–27.

191. Khan, History of Printing in Bengali Characters, estimates that Bengali has 455 sorts,
compared to about 150 for Roman.

192. Ross, The Printed Bengali Character, p. 16.

193. Ross, The Printed Bengali Character.

terms that Halhed used in his *Grammar*. This was neither a complete San-
skritization nor a faithful replication of the existing vernacular.[194] It was
something new and unfamiliar. These experiments were at the beginning of
a process continued by the missionaries who reported that their translations
so grated on local ears that they were termed "Christian or Englishman's
Bengali." Moreover, as the *Native Schools Reports* of 1817 noted, "[I]t is not
easy to find a man who can read printed character with any readiness, not
withstanding its superior clearness and regularity."[195]

Achieving the publication of the texts of imperial government in British
Bengal was not simply a matter of applying a ready-formed printing technology to a new set of texts and a new political and cultural context. As this
detailed discussion of printing practices has shown, the composition, production, and dissemination of these texts were complex social, cultural, and
political processes that intended to recruit fixity, uniformity, and dissemination into the service of empire. The production of authoritative imperial
texts in print required the careful management of printing technologies
within which the composition of characters—of classically educated linguists and scribes, commercially trained artisans, and skilled blacksmiths—
around the printing press shaped the composition of alphabetic characters,
and even the very shape of those letters, on the printing press itself. Wilkins
was able to succeed in printing in Indian languages where Bolts and Bhimji
Parekh had failed before him because the Company was now willing and
able to order the social relations of the printing office. This combined Indian and European mental and manual labor within an enterprise that was
both commercially run and a branch of government. The impetus to do so
came from an understanding of the role of the routine texts of governance in
constructing an appropriate imperial polity. Under the control of the superintendent of the Honourable Company's Press, the printing office had the
capacity to produce these texts—forms, regulations, notices, commissions—
in vast numbers across the range of government activity. It had this capacity
because it was able to adequately resolve the problems of authority over the
printing process that Hicky had been unable to negotiate and which had subsequently disabled the production of Coote's military regulations. This was,
in many ways, an institutionalization of the forms of patronage that shaped
the dissemination of the works of Orientalist scholarship produced under
Warren Hastings. For all these works, however, this was no neutral process
of diffusion. Instead, it was the making of a new geography of knowledge

194. Qayyum, *A Critical Study of the Early Bengali Grammars.*
195. Quoted in Ghosh, "An uncertain 'coming of the book,'" pp. 38, 45.

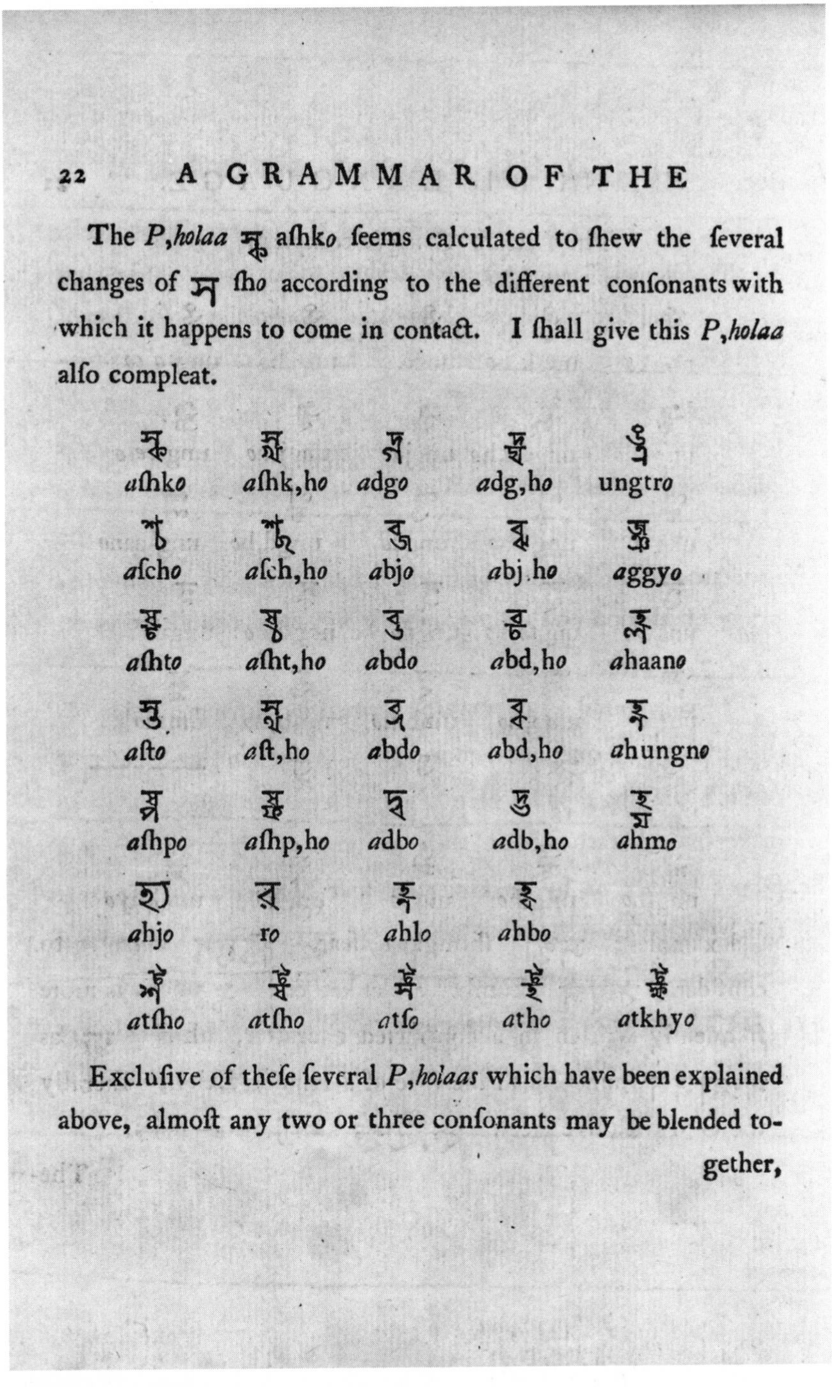

The *P,holaa* 쭉 aſhko ſeems calculated to ſhew the ſeveral changes of 지 ſho according to the different conſonants with which it happens to come in contaᴄt. I ſhall give this *P,holaa* alſo compleat.

aſhko	aſhk,ho	adgo	adg,ho	ungtro
afcho	afch,ho	abjo	abj.ho	aggyo
aſhto	aſht,ho	abdo	abd,ho	ahaano
aſto	aſt,ho	abdo	abd,ho	ahungno
aſhpo	aſhp,ho	adbo	adb,ho	ahmo
ahjo	ro	ahlo	ahbo	
atſho	atſho	atſo	atho	atkhyo

Excluſive of theſe ſeveral *P,holaas* which have been explained above, almoſt any two or three conſonants may be blended to-

gether,

Figure 21 The printing of a *phalā* in *A Grammar of the Bengal Language* (1778). (By permission of the British Library, 1560/1944.)

and practice through the transformation in the material forms of texts. In each case, those texts both traveled and were used differently in the building of an empire when they were in print. However, in this process of production, dissemination, and use, the supposedly precise replication of texts in print always created something new—in terms of artifacts, languages, and imperial politics. Recognizing this means acknowledging the contingency of successful imperial printing, its confinement to certain sorts of texts, and the ways that print both produced and was produced by asymmetrical relations of power in British Bengal. A final publication story can establish this case.

Amusements of Posterity: Imperial Critique and the Problems of Print

The tale is of the 1789 publication in Calcutta of the *Seir Mutaqharin* or "View of Modern Times" composed by Ghulam Husain Khan Tabatabai in 1781–82.[196] This chronicle of Indian history was written in Persian by a high-born Bihari official whose Persian father had served the Mughal emperor and whose mother was related to Alivardi Khan, the *nawāb* of Bengal.[197] Ghulam Husain had grown up at Alivardi's court, and held official bureaucratic positions in Delhi and Patna before entering the service of the *nawāb* of Awadh. He was part of a class of elite Mughal administrators whose world was profoundly disrupted by Company rule.[198] He was forced to seek the patronage of both the English and their Indian allies and opponents.[199] His history, which may have been written at the instigation of British officials allied with Warren Hastings, offered a stern critique of Company rule in Bengal and (with the exception of Hastings) its turning away from Mughal traditions of legitimate governance.[200]

---

196. Seid-Gholam-Hossein-Khan (1789) *A Translation of the Seir Mutaqharin; or, View of Modern Times*, 3 vols. (Printed for the Translator, Calcutta).

197. A manuscript copy is preserved in the Bodleian Library as Persian MS Ouseley Add. 156–58.

198. C. E. Buckland (1999) *Dictionary of Indian Biography*, 2 vols. (Cosmo Publications, New Delhi), s.v. "Ghulam Hussein Khan Tabatabā, Syad," vol. 1 p. 164; and Chatterjee, "History as self-representation."

199. Gulfishan Khan (1998) *Indian Muslim Perceptions of the West during the Eighteenth Century* (Oxford University Press, Karachi).

200. J. S. Grewal (1970) *Muslim Rule in India: The Assessments of British Historians* (Oxford University Press, Calcutta) p. 33; and Rajat Kanta Ray (1998) "Indian society and the establishment of British supremacy, 1765–1818," in Peter J. Marshall (ed.) *The Oxford History of the British Empire*, vol. 2: *The Eighteenth Century* (Oxford University Press, Oxford) pp. 508–29.

Ghulam Husain's criticisms of the "hat-wearers" highlighted the dangers of according power to the *zamīndars*, the disabling effects of party politics, the lack of appropriate justice offered by a supreme court that trespassed on local traditions, and, most importantly, the distance that had grown between the governors and the governed. Constant changes in personnel, from the governor general downwards, meant that "this country seems to have no master at all."[201] The hatred English gentlemen showed for appearing in public audiences, and their "extreme uneasiness, impatience and anger" when they did so, meant that they were unable to govern either effectively or legitimately since the people "never see any thing of that benignity and that munificence which might be expected from people that now sit on the throne of kings." He criticized the English for listening only to the self-interested advice of the "newmen" whom they had put at "the summit of power," particularly Mohammed Reza Khan.[202] The English rulers' distance from their subjects was, Ghulam Husain insisted, effectively produced by their reliance upon writing:

[T]he English commenced acquiring a knowledge of the usages and customs of the country: for it was a standing rule with them, that whatever remarcable they heard from any man versed in business, or even from any other individual, was immediately set in writing in a kind of book composed of a few blanc leaves, which most of them carry about, and which they put together afterwards, and bind like a book for their future use.[203]

Englishmen were, he observed, interested in becoming acquainted with "noblemen and other persons of distinction" only in order to pump them for knowledge of laws or revenue matters, and they "would immediately set it down in writing, and lay it up in store for the use of another Englishman." As a result, "the Books and Memorandums composed by the English upon such interested reports...have come to be trusted as so many vouchers: whereas they are only some faint idea of the exterior and bark, but not of the pith or real reason, of those institutions."[204]

201. Seid-Gholam-Hossein-Khan, *Seir Mutaqharin*, vol. 2 pp. 509, 580.
202. Seid-Gholam-Hossein-Khan, *Seir Mutaqharin*, vol. 2 pp. 408, 597.
203. Seid-Gholam-Hossein-Khan, *Seir Mutaqharin*, vol. 2 p. 402.
204. Seid-Gholam-Hossein-Khan, *Seir Mutaqharin*, vol. 2 pp. 404, 406. Wilson, "'A thousand countries to go to,'" argues that "[c]olonial categories were forged in the exchange of texts between officials, not in moments of encounter with Bengali everyday life" (p. 107).

Like Nathaniel Halhed, Ghulam Husain offered a critique of empire based on the failures of written communication. However, these failings were to be rectified not by print but by conversation. It was the spoken word that was to be trusted as authoritative.[205] Under the current government, which had solidified negotiated practices as laws, and had closed the Mughal channels of information and intelligence, "the gates of communication and intercourse are shut up betwixt the men of this land and those strangers, who are become their masters; and these constantly express an *aversion to the Society of Indians, and a disdain against conversing with them*; hence both parties remain ignorant of each other's state and circumstance."[206] It was only "conversation" that would allow the British "to learn what aches these poor natives." Without such social intercourse "a Company of Hindians having business with their English rulers, looks very much like a number of pictures set up against the wall." The English needed to be aware, Ghulam Husain suggested, that without "communication of ideas," "no benefit is reaped by either description of men from such an intercourse," and "no love, and no coalition . . . can take root between the conquerors and the conquered."[207] He was well aware of the American Revolution and warned the Company of the dire consequences of failing to open a conversation with the old Mughal elite about how to govern India properly.

Ghulam Husain's manuscript was, of course, part of this conversation. It demonstrated knowledge of European political philosophy and addressed itself to, among others, a British readership. The author was clearly aware of the power of writing. This related both to his own ability to compose a "valuable book, by making use of so inconsiderable an instrument as the slit-tongued reed," and to the uses that might be made of it by others.[208] In the course of his history he noted how, in the case of the politically controversial hanging of Hastings's opponent the Brahmin Nandakumar for forgery in 1775, the crucial testimony was "written down in the English Language and character; and the whole being bound together in the form of a book, was sent to England, from which such a vast number of copies were drawn out, that this subject became famous, and an object of much curiosity in that nation."[209] He also noted that Philip Francis had done the same with his accusations against Hastings, drawing "the whole into a methodical

205. Robinson, "Technology and religious change."

206. Seid-Gholam-Hossein-Khan, *Seir Mutaqharin*, vol. 2 p. 545

207. Seid-Gholam-Hossein-Khan, *Seir Mutaqharin*, vol. 2 pp. 587, 588, 553–54.

208. Seid-Gholam-Hossein-Khan, *Seir Mutaqharin*, vol. 2 p. 611.

209. Seid-Gholam-Hossein-Khan, *Seir Mutaqharin*, vol. 2 p. 465; and J. Duncan M. Derrett (1960) "Nandakumar's forgery," *English Historical Review*, 75, pp. 223–38.

writing which he got bound like a book; and after having obtained it's being registered, he thought himself strong enough with such a piece, and he departed this country."[210] It was not, however, to be so easy to turn his own manuscript into a mobile text with geographically extensive and reliable political effects.

Somehow, and it is not clear exactly how, the task of translating, printing, and disseminating Ghulam Husain's work fell to a Constantinople-born, French-educated, and English-speaking Muslim. Known both as Haji Mustapha and Monsieur Raymond, and referring to himself at times as a Turk and at others as a "Semi-Englishman," he had been employed and imprisoned as a spy by both the British and the French. A man of many parts, Haji Mustapha was a book collector, seraglio proprietor, and commercial traveler who picked up languages as quickly as he adopted disguises.[211] According to his own account he had, "with a mediocre dictionary and a bad grammar" and through conversing with the ship's captain, "learned enough of English in a journey of 19 days from Bombay to Balassor, as I might delight in Bolingbroke's Philosophical works." It was certainly enough to secure himself a position translating for Robert Clive, until he fell out of favor and got himself arrested trying to make contact with the French. He subsequently made another attempt to travel cross-country to Pondicherry armed only with a twelve-inch Dutch kitchen knife and disguised as a "dismounted trooper...incommoded with a hernia." He was arrested again, and spent sixteen months in prison.[212]

After his release, and after the loss of his small fortune, his cabinet of curiosities, and his library at Mecca in 1770—generating in him a strong aversion to Oriental books—Mustapha was driven to the edge of madness by the brutal murder of a young woman from his *zenāna* by the man he had forced her to marry. Finding himself talking to a portrait of Warren Hastings as if he were conversing with the man himself, Mustapha was saved, on his own account, by the fortuitous discovery of Ghulam Husain's manuscript:

> On my going into one of the Navvab's seats, an old woman, among other articles of sale, offered me some broken leaves of a decayed book, in which the author talked with encomiums of the English parliament in

210. Seid-Gholam-Hossein-Khan, *Seir Mutaqharin*, vol. 2 p. 524.

211. Buckland, *Dictionary of Indian Biography*, s.v. "Raymond," vol. 2 p. 353; and Seid-Gholam-Hossein-Khan, *Seir Mutaqharin*, vol. 2 appendix: letter from Mustapha to William Armstrong, 15 May 1790, p. 17.

212. OIOC Eur Mss. Orme OV 6 item 1: letter from Mustapha to William McGuire, Calcutta, 16 March 1761, footnote f p. 3 and p. 17.

Europe, and with some asperity of the English Government in Bengal. A Persian discourse upon English Politicks! strange indeed! I took the broken leaves, and perused some of them in the Garden; and the style, as well as the matter, having awakened my curiosity, I seized on this opportunity to afford some relief to my wearied mind.[213]

Having initially thought to translate a little to soothe his soul, then having decided to publish the translation so that he could fund an English education for his children, Mustapha finally resolved to undertake the task "as a matter of information, which it is incumbent upon me to impart to my adopted Countrymen (the English); and as a warning which I owe to their prosperity." Mustapha, therefore, presented himself as a valuable conduit of information to the British: a correspondent from the old Nawabi capital of Murshidabad, who spent his evenings in the streets and bazaars in Indian dress, listening to the talk of the town and picking up "a variety of information, which is often out of the power, and always out of the way, of any other European."[214]

Haji Mustapha was also well aware of the power of script and print. He was, he said, "won't [sic] ever to be a Scribbling upon a journey," and had tried to use his "travelling papers" to extricate himself from prison. In 1768 he wrote from India to Luke Scrafton in London to defend himself against the charge of spying printed in Scrafton's *Observations on Mr. Vansittart's Narrative* (1766).[215] His intervention certainly recognized the ephemerality of such works. As Mustapha said, "I have seen the World, and Know what fate soon or Late overtakes pamphlets." But he acknowledged their influence too. Considering himself "a man of honor . . . intitled at least to common justice," and wishing to arrive untainted when he realized his ambition of traveling to England, he trusted to the independent judgment of English readers: "Men born thinkers, or made so by the very nature of the Government: Men not to be imposed upon for any course of time; amused for a while with a flying Report, but upon any one Standing against it, ready

---

213. "Translator's preface," in Seid-Gholam-Hossein-Khan, *Seir Mutaqharin*, vol. 1 pp. 21–22. The editors of the 1902 Indian edition of *Seir Mutaqharin* (T. D. Chatterjee, Calcutta) argued that Hastings gave it to Mustapha to translate. For manuscript collecting in late eighteenth-century India, see Maya Jasanoff (2005) *Edge of Empire: Lives, Culture, and Conquest in the East, 1750–1850* (Alfred A. Knopf, New York).

214. Seid-Gholam-Hossein-Khan, *Seir Mutaqharin*, vol. 1 pp. 22–23.

215. OIOC Eur Mss. Orme OV 6 item 3: letter from Mustapha to Luke Scrafton, Calcutta, 2 January 1768, pp. 35, 37.

to listen, and equally able and willing to think for themselves."[216] Like Ghulam Husain, Mustapha understood the power of writing as a tool of political change which the English, at least in England, would heed. Indeed, by the time *Seir Mutaqharin* was printed he had already published in Calcutta a lengthy critique of the unjust workings of the Company's civil and criminal courts at Murshidabad.[217] While differing from Ghulam Husain in his suggestions for judicial change, Mustapha was able to use the manuscript's "preaching of sedition and anarchy" to reinforce his contempt for Muhammad Reza Khan and to issue thinly veiled threats that bolstered his own pleas for reform. In doing so Mustapha allied himself, Ghulam Husain, and William Bolts as those who dared to speak out against the "glaring injustices" of defective imperial government and as the only visible signs of "a subterraneous vein of national resentment" that the English should heed.[218]

However, making the manuscript into a book that would be heeded was no simple matter. It required that the transformations necessary for its translation, printing, and dissemination be effectively undertaken and the work involved in doing so rendered as invisible as possible. In particular, it was important that Haji Mustapha's own role be carefully managed to maximize the autonomy and authority of the text. He first sent a manuscript copy of the translation to England in a "rough state" to have it prepared for the press in order that it might be used as part of the defense in Warren Hastings's impeachment. Unfortunately, the "eminent historian" charged with the task was dying, and the merchant Mustapha had entrusted it to for the journey could find no time, nor any other qualified person, to correct it. Mustapha then resolved to try a different geographical audience, "supplicating the British public in Bengal, instead of addressing the British public in London."[219] He was, however, unable to afford the costs of publication himself. While the book was to be dedicated to Warren Hastings and was intended to be of some assistance to him, Hastings was no longer in any position to help the book come into being. Mustapha therefore resolved to publish

216. OIOC Eur Mss. Orme OV 6 item 3, pp. 35, 38, 39.

217. *Some Idea of the Civil and Criminal Courts of Justice at Moorshoodabad, in a Letter to Capt. John Hawkshaw, At Behrampore, of the 30th May 1789* (printed for the author, Calcutta, 1789). Mustapha claimed his other works included books and pamphlets printed in London, including a piece of futurology entitled *State of Europe in 1800*. Unfortunately, no trace of this has been found.

218. Seid-Gholam-Hossein-Khan, *Seir Mutaqharin*, vol. 2, appendix pp. 21–22.

219. Seid-Gholam-Hossein-Khan, *Seir Mutaqharin*, vol. 1, publication proposal p. 5.

it by subscription. Doing so successfully, however, required resources that he simply did not possess. He decried "the scanty reception which my own work has met with from the public" and compared his failures to the success of Francis Gladwin's *Vocabulary*:

> If then so very small a work has produced so large a sum, it was because the author's reputation, as an author, and a man of letters, is formed, known, and established; whereas no one knows any thing of me. Secondly because he is an Englishman, a man high in station; and of course has many friends: whereas I am next to nobody, and my station is immediately after nothing.[220]

Even more gallingly, another of Gladwin's volumes had made money only because the Company's directors had taken two hundred copies at 120 rupees each. In contrast, Mustapha found himself faced with the prospect of giving refunds to angry subscribers.

The reason the subscribers were up in arms was that they judged the book defective. Mustapha was unhappy with it too. His attempts to render the pronunciation of "oriental names" via "the Italian or rather Scotish Alphabet," using 8s and double dots to represent long "o" sounds were undone by the printing process: "the work having been printed by several Printers, this has been disregarded by some, who have adhered to the usual spellings." Indeed, it was "teeming with faults," since, as the distraught translator put it, the "Printers have been guilty of an infinity of alterations both through chance and wilfulness."[221] Farming the text out to multiple printers had multiplied errors so that, for example, the page numbering of volume two was hopelessly confused, with numbers missing, misplaced, and inverted, and with some pages numbered twice. Mustapha's readers questioned his translation and objected to his orthography and grammar (see Figure 22, part of an unknown reader's "corrections" of the first ninety pages of volume 2). Anonymizing himself as "Nota-Manus," Mustapha sought to reassure an audience "displeased with that wretched performance of mine" that they were hearing the original author's true voice and that his translation had been assisted by William Jones himself.[222] However, he also admitted that he had animated some of the text by elaborating on motivations and

---

220. Seid-Gholam-Hossein-Khan, *Seir Mutaqharin*, vol. 2, appendix pp. 10, 11; and Francis Gladwin (1780) *A Compendious Vocabulary English and Persian* (Malda).

221. Seid-Gholam-Hossein-Khan, *Seir Mutaqharin*, vol. 1, publication proposal pp. 1, 2, 4.

222. Seid-Gholam-Hossein-Khan, *Seir Mutaqharin*, vol. 2, appendix p. 30.

emotions in ways that "shall possibly surprise those Gentlemen employed at the public offices of Calcutta in translating Persian Letters," but which should appeal to the European readers excluded by dry Orientalist scholarship, including "women; and all those that read for amusement."[223]

Doubts over the quality of the translation and the printing clearly raised questions about the reliability of the book and its contents. There was a danger that it would not be taken seriously, with "no less than twenty persons at table, at a Mr Brown's at Chouringhi" having "declared that there was no reading two pages of it."[224] Against this Mustapha argued that any changes to the text that might make it more fluent "would detract from the genuineness of the translation, and of course from the veracity and integrity of the evidence."[225] In all of this Mustapha suspected that the reading of his book was overdetermined by the social relations of empire. A newspaper article had revealed that he was "Nota-Manus." This, he argued, had had disastrous consequences, because "it is only since I have wrote history, that is since, I have thrown away the mask, and given myself for what I was, that I have been taken for a foreigner, and have been thought so greatly defective in language." These readers were too willing to assume that all the errors of syntax and any "solecism"—a term that tellingly refers to both grammatical mistakes and bad manners—were his responsibility alone.[226] Haji Mustapha's editorial disguise was being stripped away and with it went the authority of his text.

Mustapha's problem was a simple one. Like John Gilchrist, and unlike Halhed and Wilkins, he did not have sufficient control over the presses on which his work was being produced:

No work within my knowledge in Calcutta, has been tolerably printed, but where the author himself was the owner of the printing office or a partner; or where the Printer had purchased the propriety of the work; or at least... made a sharer in the fate of the book being printed, that is, by being promised for his trouble one half of the author's profits.[227]

Commercial printers would "reserve the two or three good hands that may be in a Printing Office" for the most lucrative work. All but one Calcutta

223. Seid-Gholam-Hossein-Khan, *Seir Mutaqharin*, vol. 2, appendix p. 11.
224. Seid-Gholam-Hossein-Khan, *Seir Mutaqharin*, vol. 2, appendix p. 16.
225. Seid-Gholam-Hossein-Khan, *Seir Mutaqharin*, vol. 2, appendix p. 4.
226. Seid-Gholam-Hossein-Khan, *Seir Mutaqharin*, vol. 2, appendix pp. 11, 18.
227. Seid-Gholam-Hossein-Khan, *Seir Mutaqharin*, vol. 2, appendix pp. 4-5.

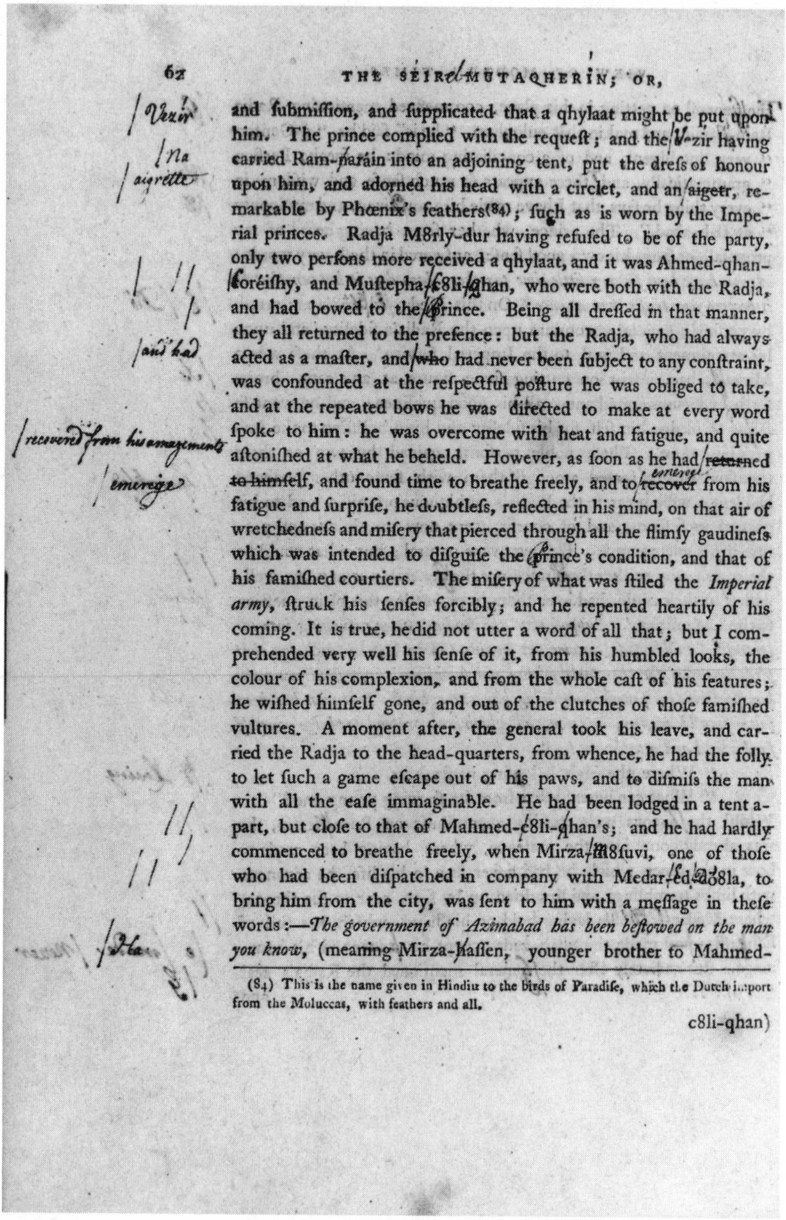

/ Vezir

/ na
/ aigrette

/ and had

/ received from his amazement
/ emerge

and fubmiffion, and fupplicated that a qhylaat might be put upon
him. The prince complied with the requeft; and the Vezir having
carried Ram-parain into an adjoining tent, put the drefs of honour
upon him, and adorned his head with a circlet, and an aigetr, remarkable by Phœnix's feathers(84); fuch as is worn by the Imperial princes. Radja M8rly-dur having refufed to be of the party,
only two perfons more received a qhylaat, and it was Ahmed-qhan-
Coréifhy, and Muftepha-c8li-qhan, who were both with the Radja,
and had bowed to the prince. Being all dreffed in that manner,
they all returned to the prefence: but the Radja, who had always
acted as a mafter, and who had never been fubject to any conftraint,
was confounded at the refpectful pofture he was obliged to take,
and at the repeated bows he was directed to make at every word
fpoke to him: he was overcome with heat and fatigue, and quite
aftonifhed at what he beheld. However, as foon as he had returned
to himfelf, and found time to breathe freely, and to recover from his
fatigue and furprife, he doubtlefs, reflected in his mind, on that air of
wretchednefs and mifery that pierced through all the flimfy gaudinefs
which was intended to difguife the prince's condition, and that of
his famifhed courtiers. The mifery of what was ftiled the *Imperial
army*, ftruck his fenfes forcibly; and he repented heartily of his
coming. It is true, he did not utter a word of all that; but I com-
prehended very well his fenfe of it, from his humbled looks, the
colour of his complexion, and from the whole caft of his features;
he wifhed himfelf gone, and out of the clutches of thofe famifhed
vultures. A moment after, the general took his leave, and carried the Radja to the head-quarters, from whence, he had the folly
to let fuch a game efcape out of his paws, and to difmifs the man
with all the eafe imaginable. He had been lodged in a tent a-
part, but clofe to that of Mahmed-c8li-qhan's; and he had hardly
commenced to breathe freely, when Mirza-M8fuvi, one of thofe
who had been difpatched in company with Medar-Ed-208la, to
bring him from the city, was fent to him with a meffage in thefe
words:—*The government of Azìmabad has been beftowed on the man
you know*, (meaning Mirza-Haffen, younger brother to Mahmed-

(84) This is the name given in Hindiu to the birds of Paradife, which the Dutch import
from the Moluccas, with feathers and all.

c8li-qhan)

Figure 22 Reader response: marginal annotations in *Seir Mutaqharin* (1789) volume 2. (By
permission of the British Library, 153.h.2.)

(Ali-*k*han)· *and yourself have been appointed divan to both the govern-*
ments of Azimabad and Ilah-abad; for which I wish you joy. This
particular has been affirmed by perfons of great credit; who add,
that Mirza Us-hac, who happened to come in at that particular moment, and to be joined to Mirza Musuvi in the fame commiffion,
could not help expreffing his furprife at the numerous improprieties
of the meffage. Both the envoys excufed themfelves, and refpect-
fully reprefented, " that fuch a meffage was not yet of feafon;
" and that they were certain it would never produce any good at
" all." But thefe objections having anfwered no other purpofe than
that of putting the general in a paffion, he fent one of his guards
to bring the Radja; and on his coming in, he informed him of the
difpofitions that had been made. The man very wifely inclined his
head in token of confent, congratulated the new governor on his
promotion, afked his protection, and then took his leave and went
to his tent: in the evening he fent a trufty man (it was Mahmed-
Afac, the cutwal of Azimabad) to inform the general, *he was going*
to town to refresh himself. The general anfwered: " You may; I
" wish you a good journey." At thefe words, fome perfons pre-
fent repeated to him the advice I had given; but he heard it with
deteftation, and faid, that *it was downright treafon: nor was any*
fuch matter stipulated in the treaty. They anfwered, " that they pro-
" pofed no harm at all; but that it was not ftipulated neither, that
" he should return to his caftle." At thefe words, the general's
indignation rofe to a height; he refufed to liften to any thing more
on that fubject, and added this remark: *Let him return to his caftle,*
as much as he pleafes: Where is the man daring enough to ftand the
blunt of my unfheathed fabre?
" A blind man having by chance ftumbled on an opportunity,—
" Made nothing of it himfelf, nor would let others put it to fome
" profit."
On the Radja's coming out of the tent, a perfon, upon an ele-
phant, received him into the chair, by the general's directions; and
he offered to carry him to the city. The Radja confented; but after having gone a few hundred paces with him, he excufed himfelf
on his being, as a Gentoo, forbidden drinking water in the fame

place

printing office was, he noted, "worked by natives, who print in a Printing-Office, just as they copy in a Counting-House, without understanding the language." Paying by the sheet or, even worse, paying in advance, only purchased unskillful apprentices and "supercilious inattention." Poor composing produced error-strewn proofs. Inaccurate correcting saw "a dozen of faults corrected, seven or eight preserved, and a novel crop of half a dozen new one's."[228] All of this delayed the printing by months, angering the subscribers.

Once again Mustapha understood that the social relations of the printing office were imperial ones. His identity mattered, and it damaged his text:

> And what if the Printer despising your quality of a Foreigner, thinks an Englishman *whatever*, has a right to know more of the language than you, and takes the liberty of correcting you, of sneering at your emendations, of substituting his own, and of throwing away without answer some angry notes in which you *inform him that he is paid for printing what is before him, and not for correcting* it?[229]

Haji Mustapha reported that he was constantly told that what he was writing "may be English, but not English of Europe, only English of Bengal." To him this was a matter of the printers' own ignorance. What they didn't understand, however grammatically correct, became "Bengalee English" that was despised and expunged.[230] Mustapha's problem was that the printers had the power to make the text speak the way they wanted it to no matter what he intended it to say. In the end they even changed the title. Mustapha reported that the printer he shared with the unfortunate Gilchrist, Joseph Cooper, showed it to "a gentleman skilled in oriental languages" who "dashed the *Amusements of Posterity* and substituted *Review of Modern times.*"[231] If what had been produced was a translation of Ghulam Husain's manuscript, it was hardly the book that Mustapha had wanted at all.

The whole process, which had begun as a way to salve Haji Mustapha's uneasy conscience, had, like the woman murdered as a result of his power to assign her to another man, come back to haunt him. The printers drained

---

228. Seid-Gholam-Hossein-Khan, *Seir Mutaqharin*, vol. 2, appendix pp. 5, 7.

229. Seid-Gholam-Hossein-Khan, *Seir Mutaqharin*, vol. 2, appendix p. 8.

230. Seid-Gholam-Hossein-Khan, *Seir Mutaqharin*, vol. 2, appendix p. 8. Gilchrist, *Dictionary*, p. v, described Hindustani "jargon" as "viler than English butchered by Negroes in the West, and mangled by Bungalees in the East Indies, and in fact is as remote from the proper Hindoostanee, as light is from darkness."

231. Seid-Gholam-Hossein-Khan, *Seir Mutaqharin*, vol. 2, appendix p. 28.

him of all his cash. The subscribers were on his back. Together they reduced him to "selling trinkets, plates, and books."[232] Finally, any hopes that the printed text would be able to make political capital by connecting England and Bengal were abruptly cut short when the ship carrying the consignment of volumes to London sank before reaching its destination. Instead of creating an authoritative and autonomous book that would excite the attention of English readers, effect the reforms in their government that he sought, and perhaps make him some money too, Mustapha found himself unable to escape from his text. His inability to control the processes of dissemination and the social and technical organization of the press meant that he had to live with the consequences. As he explained to the subscribers who wanted to return unwanted copies to him, "I shall contrive to return, and am firmly resolved to return, the money of all those, who having inadvertently subscribed several months ago, have thought themselves cheated out of their money"; nevertheless, the books "are already lumber upon my hands, and lumber that encroaches so much upon the dimensions of my habitation, and moreover requires so much care and solicitude, that by keeping those Books out of my view, they shall really confer a favour upon me, and render me a service." The whole affair had, he reported, "disgusted me with Printers and Books." He concluded that "[p]rinting in this country requires a young man and a rich one, and I am neither."[233]

Conclusion

The contrast drawn across this chapter between Halhed's *Grammar* and Mustapha's *Amusements of Posterity*—the first book published in British Bengal and a book that never came into existence in the way its publisher intended—demonstrates that what really mattered was all the work that needed to be undertaken to begin to realize the powerful promise of print. Of course, as Mustapha lamented, this was a matter of money and youthful energy. Yet when it came to the construction of an empire of print—to the work of empire in the age of mechanical reproduction—what really mattered were the politics of patronage and the conjoined social and technical organization of the printing press. In terms of patronage, Halhed had Hastings, and Mustapha was next to nobody. In terms of the press, Wilkins was able to craft an effective, if unequal, cross-cultural, cross-class, and cross-caste combination of forms of labor and knowledge that put the printing press to work

232. Seid-Gholam-Hossein-Khan, *Seir Mutaqharin*, vol. 2, appendix p. 30.
233. Seid-Gholam-Hossein-Khan, *Seir Mutaqharin*, vol. 2, appendix p. 31.

to produce legible, credible, and potentially authoritative texts. In contrast, Mustapha was undone by the empire's racialized social order. Identified and unmasked by printers and readers alike as a "Foreigner," a writer of "Bengalee English," and an untrustworthy translator, Mustapha was unable to control the press to produce a text that would have the political effects that he desired, or would even be taken seriously. It was, therefore, the management of printing—the hard work necessary for mechanical reproduction to appear to be mechanical—that made it an imperial technology. Printing's imperial promise could be realized only by labor in all sorts of cultural, political, and social registers. The ways in which printing might be made imperial, and empires made from print, could be technological matters, but they were never that alone.

There is, of course, the danger of suggesting that the Company's press did work successfully to produce an empire. As C. A. Bayly has demonstrated, Britain's imperial rule was more closely tied into existing Indian modes of information gathering and distribution than it was revolutionized by print.[234] As has been argued here, even where print could prove itself, it was only in the circumscribed arena of texts—such as forms and legal regulations—that could be subject to a prior standardization. These might then be resisted by those meant to use them, and they all came off the Company's press in a curious and unfamiliar "Englishman's Bengali" formed of strangely shaped characters. Moreover, the history of reading and the use of texts can always demonstrate the gulf between authorial intentions and readers' reception. Orientalist scholarship had its day, and only that; Halhed felt that his work on the *Grammar* was not appreciated by the Company; the printing of *pattās* never realized Wilkins's projections; and, as has been shown by others, both Halhed's *Code* and Jones's *Digest* were textual solutions to the problem of British rule in India through Indian law that were found to be unworkable in practice.[235] Yet printing as part of a political philosophy of imperial rule was more enduring. Jon Wilson demonstrates that the solutions adopted in the early nineteenth century relied even more upon the imperial promise of print. In the legal codes subsequently produced by English judges in India (and printed in Calcutta and Serampore) and later in Britain, the determinacy of the law was to be fixed by the clarity of print. As James Mill put it in his *History of British India* (1817), "Let the laws, what ever they...may be determined to be, receive what alone can bestow upon them a fixed or

234. Bayly, *Empire and Information*.
235. Rocher, *Orientalism, Poetry and the Millennium*; and Derrett, *Religion, Law and the State*.

real existence; let them all be expressed in a written form of words, words as precise and accurate as it is possible to make them, and let them be published in a book."[236]

This enthusiasm for imperial printing in the early nineteenth century can also be seen in Lord Wellesley's plans to make print one of the engines for the transformation of "the great machine of Government" in India. He argued that the empire could no longer be considered simply "a commercial concern" and that it had to be administered as "a sacred trust, and a permanent possession."[237] This produced a critique of the education and training of the Company's European civil servants that identified script as the problem:

> It would be superfluous to enter into any argument to demonstrate the absolute inefficiency of this class of young men to execute the duties of any station whatever in the civil service of the Company beyond the menial, laborious, unwholesome and unprofitable duty of a mere copying-clerk.

Spending all their time "[h]arassed with the ungrateful task of transcribing papers and accounts, or with other equally fatiguing and fruitless labours of a copying-clerk or index-maker," they were unable to learn anything useful, were "levelled with the native and Portuguese clerks," and were easily led into vice and dissipation.[238]

Wellesley's proposed solution was built around the printing press and the College of Fort William. The college would provide an appropriate education and moral training for newly arrived Company servants, fitting them to govern by teaching a combination of Oriental languages and the curriculum of an English public school. The "menial labour" of routine copying would be undertaken by a "Public Press."[239] This was to be a full-fledged department of the civil service employing sixty-four compositors and costing ninety thousand rupees a year, with its profits going to support the

236. Quoted in Wilson, *Governing Property, Making Law*, p. 224.

237. Richard Colley Wellesley (1836–37) *The Despatches, Minutes, and Correspondence of the Marquess Wellesley, K.G. During His Administration in India*, Montgomery Martin (ed.), 5 vols. (W. H. Allen and Co., London) vol. 2 p. 339; and Bayly, *Indian Society and the Making of the British Empire*.

238. *Despatches, Minutes, and Correspondence of the Marquess Wellesley*, vol. 2 p. 331–33, 348.

239. *Despatches, Minutes, and Correspondence of the Marquess Wellesley*, vol. 2 p. 348; and IOR P/5/20 *Bengal Public Consultations, 14th March to 9th April 1801*, 9 April 1801.

college. It would massively expand the amount of printing by extending it to the manuscript records and correspondence as well as forms and regulations. The intention was to produce "a degree of accuracy and expedition in the execution of the Public Business in every Department." Any remaining scribal copying could be undertaken by "the native or Portuguese writers" from whom the English civil servants would now be effectively separated.[240] There were other benefits too. The college would be a bulwark against the political and religious principles of the French Revolution, and a government press would "effectively silence" Calcutta printers judged both potentially seditious and "Useless to literature and to the Public."[241]

The plan was, however, unworkable. The Company was unimpressed by Wellesley's reasoning and objected to the cost of the college. The printing plan was even more fundamentally misconceived. It mistook the costs involved and expanded the use of printing far beyond its practicable and profitable uses into areas where only a few copies were required. As the superintendent of the Honourable Company's Press, admittedly a man who would stand to lose from the proposal, put it, the plan "offers nothing but delusive expectations founded on gratuitous assumptions, and a misconception of the proper use of printing."[242] Print's imperial promise had overstepped its limitations once again. As a result, there was no change in the organization of printing until 1815, when the government's printing was taken over by the Military Orphan Society. At that time over forty of the workers on the Honourable Company's thirteen presses had been with the Company for more than ten years. Only four of these were British, including Thomas Watley, a printer who had been Francis Gladwin's apprentice and had joined in 1786, along with the compositor William Rees. However, the longest-standing employees, who had started with Gladwin in 1784, were listed as Mooktaram Doss, a compositor (on a monthly salary of 22 rupees, compared with Rees's 100), Joogoo, the head pressman, and Mohorum, a pressman.[243] We know very little of these men, who had done so much of the work of mechanical reproduction. However, if they were like Panchānan Karmakār's son-in-law Manohar, they might also show the limitations of imperial printing. Manohar worked for the Serampore Mission Press for forty years. In that time he cut beautiful types to print their Scriptures but always worked "only under protection of his god," keeping "a little image in a niche

240. IOR P/5/21 *Bengal Public Consultations, 16th April to 28th May 1801*, 14 May 1801; and *Despatches, Minutes, and Correspondence of the Marquess Wellesley*, vol. 2 p. 334.
241. IOR P/5/20 9 April 1801.
242. IOR P/5/20 9 April 1801 (letter from Mr. Horsley, 16 March 1801).
243. IOR P/8/35 *Bengal Public Consultations, 17th to 23rd June 1814*, 23 June 1814.

above him." No matter how many bibles his types produced it was reported that "[t]he teaching of the missionaries had had no effect in weaning him from his ancestral idolatry."[244] The work of empire in the age of mechanical reproduction could never be taken for granted.

244. James Kennedy (1884) *Life and Work in Benares and Kumaon, 1839–1877* (T. Fisher Unwin, London) p. 7.

☙

POSTSCRIPT

In 1860, *The Printer*, a New York monthly journal devoted to "the art preser-
vative of all arts," published a curious and unattributed story of Indian ink:

> An extraordinary discovery has been made of a press in India. When War-
> ren Hastings was Governor-General of India, he observed that in the
> district of Benares, a little below the surface of the earth, is to be found a
> stratum of a kind of fibrous woody substance, of various thicknesses, in
> horizontal layers. Major Roebuck, informed of this, went out to a spot
> where an excavation had been made, displaying this singular phenomena.
> In digging somewhat deeper, for the purpose of farther research, they laid
> open a vault which, on examination, proved to be of some size, and, to
> their astonishment, they found a kind of printing press, set up in the
> vault, and on it movable types, placed as if ready for printing. Every
> inquiry was set on foot to ascertain the probable period at which such
> an instrument could have been placed there; for it was evidently not of
> modern origin, and from all the Major could collect, it appeared probable
> that the place had remained in the state in which it was found for at
> least one thousand years. We believe the worthy Major, on his return to
> England, presented one of the learned associations with a memoir con-
> taining many curious speculations on the subject.[1]

Mofakhkhar Hussain Khan, writing his doctorate in London in the 1970s
on printing in Bengali, could find no corroborating evidence, and could not

---

1. *The Printer*, 3:6 (15 November 1860) p. 130.

further identify Major Roebuck. He left it at that.[2] However, the tale of an ancient Indian printing press buried deep underground is an intriguing one, and it is tempting to speculate on what its telling might have meant in 1860.

By the early nineteenth century, prompted in part by the missionaries discussed in the previous chapter, the printing press was at the center of British attempts at reform in India that emphasized public instruction as the means of civic and religious transformation. This meant official support for a free press and a cheap postal service. Beyond the colonial state itself, the Calcutta branch of the Society for the Diffusion of Useful Knowledge was established in 1829, followed by various schoolbook societies. In the decades that followed, "the spread of the printed book in a subcontinent where it barely existed a generation before was a striking effect of colonial rule."[3] Yet that did not mean that the printing press simply reinforced colonial power. Indians may have come late to the printing press, but they took to it with great rapidity. The first indigenous vernacular presses of the 1810s and 1820s were run by men of the scribal high castes, such as Bābu Rām and Gangākīshor Bhattāchārya, who had already been teaching, writing, and printing didactic works in Sanskritized Bengali in association with the English. These scholar-printers, and the nouveau riche *abhijāta* aristocracy that sponsored their printing of the works of Brahminical high culture, formed the basis of attempts to purify Bengali language and literature in the early nineteenth century. By the 1850s they were being joined and superseded in this reformism by a new *bhadralok* professional, service, and rentier class. This educated middle class built its cultural claims to status and distinction on the dual basis of respectability and Hinduism. Both could be achieved through print and literary purification: a Bengal renaissance.[4] For these reformers the story of the buried press would have told of the ancient roots of Indian printing in a pre-Mughal and pre-British golden age of Hindu gods and kings and of its contemporary uses in bringing that past back to light.

However, the overturning of the sorts of constraints that had hindered Haji Mustapha's efforts at printing in Calcutta in the 1780s (see chapter 6) also occurred on the basis of already existing structures of social communication and information use in the "Indian ecumene" that had established

2. Mofakhkhar Hussain Khan (1976) *History of Printing in Bengali Characters up to 1866*, unpublished Ph.D. thesis, School of Oriental and African Studies, University of London.

3. C. A. Bayly (1996) *Empire and Information: Intelligence Gathering and Social Communication in India, 1780–1870* (Cambridge University Press, Cambridge) p. 216.

4. Anindita Ghosh (2006) *Power in Print: Popular Publishing and the Politics of Language and Culture in a Colonial Society* (Oxford University Press, New Delhi).

modes and means of discussion of religion, politics, and aesthetics. Many of the works that were produced were printed versions of the sorts of texts that were already circulating in manuscripts produced by scribes in the bazaar: religious texts, almanacs and works of astrology, medical books, and ancient ballads and novelettes.[5] In Bengal, the *bhadralok* reformers' cultural battles were waged against the cheap and popular print of the Battala area of north Calcutta, where hundreds of tiny printshops lined the narrow lanes.[6] The products of these presses run by low-caste artisans "were mostly remnants of pre-print literary traditions and a far cry from the Sanskritized, purified, and 'enlightened' literature of the vernacular reformists." The jobbing printers' worn types produced "Myths and fables spun in a syncretic and earthy colloquial language, racy and abusive dramas mocking the fallen bhadralok, uninhibited accounts of sexuality and belief in the supernatural" on cheap paper for a broad readership. In 1857 there were reported to be over half a million books printed for sale in Calcutta.[7] For these printers and their readers the tale of the buried press might well have signified the continuities of colonial Calcutta's print culture with its preprint, popular, oral and manuscript traditions, and the hubris of those—Indian or British—who would try to secure print and printing to themselves and deny it to others.

Newspaper publishers also soon turned the new media to old uses. Benares—where Major Roebuck's men were said to have discovered the ancient press—was the location in 1847 of the publication of the first newspaper in Devanāgarī and Sanskritized Hindi, *Sudhakar*, which "took up some of the literary and religious debates which had been characteristic of the Hindu segment of the ecumene."[8] This included the use of newspapers to develop a critique of the colonial government. Indeed, as C. A. Bayly argues, in their armed uprising against colonial rule the rebels of 1857 not only destroyed British printing presses as tools of oppression, but were willing to use the presses they themselves controlled to fight a modern propaganda war and broaden the struggle. In this they were supported by north Indian

5. Bayly, *Empire and Information*. This pattern is strikingly similar to the publication of popular religious texts, almanacs, and stories in early modern Europe and America; see David D. Hall (1996) "The world of print and collective mentality in seventeenth-century New England," in *Cultures of Print: Essays in the History of the Book* (University of Massachusetts Press, Amherst) pp. 79–96.

6. Nikhil Sarkar (1990) "Printing and the spirit of Calcutta," in Sukhanta Chaudhuri (ed.) *Calcutta: The Living City*, vol. 1: *The Past* (Oxford University Press, Calcutta) pp. 128–36; and Sukumar Sen (1968) "Early printers and publishers in Calcutta," *Bengal Past and Present*, 87, p. 64, who refers to the mid-nineteenth century as the "golden age" of cheap printing in Calcutta.

7. Ghosh, *Power in Print*, pp. 131, 139.

8. Bayly, *Empire and Information*, p. 241.

newspaper editors who continued to criticize the authorities throughout the fighting.

By 1860, then, the East India Company had been replaced as the rulers of the subcontinent and the British themselves were in two minds over the role of the press. It still held potential as the vehicle for spreading Western civilization, but was also feared as a tool of sedition.[9] The story of the discovery of an ancient, native, underground press just at the time when the British were beginning to build their empire in India undermined this hope and amplified these fears. A thousand-year-old Indian press with moveable type would deny European claims to this crucial invention, sending Major Roebuck back to the learned societies of London with evidence that if you dug deeper in India you discovered the primacy of Asian civilization. A press that was truly Indian, especially one that had remained hidden for so long, was an unsettling thing. Indeed, by the 1860s it has been said that "the Indian press and other internal lines of communication were running out of the government's control."[10] Most towns possessed a printing press or two, and, as in Battala, their products were increasingly original publications that tapped deep into Indian networks of communication rather than being translations of English newspapers and books. The story of an Indian press "ready for printing" for the last millennium and finally unearthed just as the British were making an empire on the ground in which it lay, might confirm what many Indians could see around them in the flurry of printed material that spread through cities, towns, and villages. The story certainly made the printing press Indian, and, where that was needed, justified its use by a curious fusion of tradition and modernity.[11] Perhaps, in its publication and republication in New York, the Americans who read the story were also sensing another British empire beginning to be challenged by the products of the press.

This tale of writing can, therefore, signal an ending: the role of printing and the press in the end of Company rule in India in 1858 and in the end of

---

9. The Reverend James Long's important 1859 report on the press in Bengal was prompted by fears of sedition, but concluded that the press should not be subject to heavy censorship; see Priya Joshi (2002) *In Another Country: Colonialism, Culture, and the English Novel in India* (Columbia University Press, New York) pp. 48–50.

10. Bayly, *Empire and Information*, p. 343.

11. Khan, *History of Printing*, pp. 410–11, notes the religious rituals that were followed in producing some religious texts in the nineteenth century so that they would be acceptable. For example, ink was prepared with Ganges water, and composition undertaken by Brahmins. Perhaps the story of the ancient press might also have made printing more acceptable to some Indian readers.

Britain's Indian empire in 1947.[12] It can also help to draw together the themes of this book. In its account of the displacement of the modern, European, and rational provenance of the press by resituating it within a different history that is ancient, Indian, and mysterious it signals both the promise and the uncertainty of the relationship between power and knowledge when the technologies of writing are brought to the fore.

Indian Ink seeks to investigate the European engagement with Asia through trade and empire using the lens offered by the history of writing, or the history of the book. In various ways its chapters have demonstrated different forms and technologies of writing, in script and print, at work in and around the English East India Company. Different ways of understanding writing have allowed specific conclusions to be drawn about the Company and its context in the two centuries between 1600 and 1800. In chapter 2, the royal letters from Elizabeth I and James I that were carried east on the Company's earliest voyages are understood as traveling objects in order to demonstrate how the nature of the relationship between Company and state shaped how the Company's generals and merchants encountered Asian polities. By treating these texts in terms of their production, carriage, and delivery, the chapter demonstrates the contingencies involved in the active construction of global trading connections on the basis of historically and geographically specific encounters and performances. In contrast, chapter 3 treats the routine administrative documentation and accounting of the late seventeenth-century expansion of Company trade on the Coromandel Coast and in the Bay of Bengal as Latourian "inscriptions" at work in making new economic forms. As the chapter shows, the "Regulating and new Methodiseing" pursued by Streynsham Master intended to use forms of writing—consultation books, account books, and letters between the factories and to London—to construct Company procedures and organizational forms. The ways in which that was done demonstrated attempts to determine what could be taken for decisions, modes of social order, and forms of authority. These transformations were to be secured by a microgeography of writing made evident in the factories' writing offices. Turning to print, chapter 4 shows that the Company in London had to engage with the press in order to defend itself and its privileges against rival English merchants, the Dutch East India Company, and the political machinations of king and Parliament.

12. See Margarita Barns (1940) *The Indian Press: A History of the Growth of Public Opinion in India* (George Allen & Unwin, London); Prem Narain (1970) *Press and Politics in India, 1885–1905* (Munshiram Manoharlal, Delhi); and Basanti Sinha (1994) *Press and National Movement in India (1911 to 1947)* (Manak Publications, Delhi), which, despite its title, is a local study of the press and nationalism in Bihar.

However, what the early seventeenth-century conflicts over *The Trades Increase* and the Amboina pamphlets demonstrate is that the Company's management of the economy of influence within which it operated meant that avoiding going into print was as important a political strategy as using the press. As the economic writers Gerard de Malynes, Edward Misselden, and Thomas Mun also found in the early 1620s, putting an argument out in print required a careful modulation of the relationships between private interests and public benefits, in a world where these were both being re-defined. By the late seventeenth century the explosion of printed material had changed this context and print needed to be met with more print. The Company, pursuing parliamentary influence, engaged vigorously in this new political print culture. This metropolitan world of print also provides the context for the discussion in chapter 5 of the first printing of stock prices for a broad audience in John Houghton's *Collection of Letters on Husbandry and Trade*. Situating Houghton's project within the culture of information, knowledge, and fact created by the Royal Society, and his own Tory politics, demonstrates the use of print in an attempt to make the workings of the stock market familiar within a changing economy. This chapter is concerned only with the representation of the East India Company as a single stock price within an increasingly extensive list. However, it argues that even this rather minimal form of writing was crucial to the Company's reputation and fortunes when understood in terms of the economics and politics of stock jobbing. Finally, chapter 6 investigates another printing project: the uses of printing as a technology of empire building in late eighteenth-century Bengal. The chapter shows the specificity of imperial printing, as one culture of print alongside Calcutta's commercial printers and the missionaries of Serampore, and argues that any hope of its effective use as an imperial technology depended upon the politics of patronage, the organization of the printing office, and the ways in which the printing press was put to work. The story of Haji Mustapha's problems printing the *Seir Mutaqharin* serves to demonstrate the social structures of printed communication in an age of empire.

These chapters, therefore, offer a series of historically and geographically differentiated perspectives on the English East India Company over two centuries. Each chapter takes a new or changing form or mode of writing or printing and situates its active role in making the Company and its work. The chapters also all operate together in pursuing the intention of this book to rethink questions of power and knowledge in the encounter between "East" and "West" by treating texts as objects, by understanding the social relations of writing technologies, and by situating those written objects and technologies within modes of social practice. In all of the chapters, albeit with different sorts and amounts of evidence, this means understanding the forms that texts take and then attending closely to their production, dissemination, and use. This treads a path that leads away from a sense that in this encounter and in the making of the very categories of "East" and "West," power resides simply in the representation of people and places. It leads towards combining analyses of the textual and the material, and towards notions of imperial power as "work in progress" which are signaled in chapter 1 as common concerns across a wide range of approaches to empire. Therefore, the different chapters demonstrate how the East India Company, and those who sought to profit from it, understood specific forms and ways of writing as something that could produce desirable effects in and on the world. The promise of the power of writing, and of its codifications and mobilizations of knowledge of various sorts, is evident in a variety of ways: the delivery of ornate manuscript letters from English monarchs as part of attempts to forge profitable and lasting connections between places that were oceans apart; the careful construction, compilation, and archiving of administrative documentation as an important part of the making of routine trading procedures in and between Indian factories and the Company's headquarters in London; the use or avoidance of the products of the press, in combination with speech and script, in the making of political arguments in favor of the Company; the familiarization of a broader public with the processes and prices of the new stock market; and the attempt to establish a centralized, routinized, and improved imperial administration based on the printing of Indian forms of documentation.

In terms of the East India Company itself, this certainly emphasizes the ways in which its purposes and practices changed dramatically over time. The book opens with tentative and speculative voyages and ends with the building of an eastern empire. It also emphasizes that at any one time there was a vast range of highly differentiated work going on to shape the Company and what it did. The Company's existence depended on many things, but it certainly depended upon the material work with texts that secured connections with distant places, managed the cultural politics of long-distance trade, fought for Company privileges in political arenas, familiarized the dangers of trading in stocks, and sought to render imperial administration routine. In each of these different contexts writing promised to secure the Company and its operations. Understanding the promise of writing in this way has meant attending to the role of the materiality of texts in producing its effects: the inks and parchment of royal letters, or the cutting of Bengali types. It has also emphasized how the geographies of texts—where they were produced, how they traveled, where they were stored, and where they were used—were crucial in these processes of making texts have effects. In the broadest sense, therefore, *Indian Ink* is an argument for understanding the history of knowledge geographically. Indeed, it is an argument for reconfiguring the social history of knowledge as the historical geography of knowledge.

Indian Ink also argues that understanding the historical geography of knowledge in this way entails recognizing that producing these effects with texts was not simply a matter of deploying the techniques and technologies of script and print in new contexts and reaping the benefits. Making those forms of writing and printing, and attempting to make them work, was always a matter of the social and political organization of people, spaces, and objects in particular historical contexts. Chapter 2 demonstrates how the making, transportation, and delivery of royal letters involved a range of different forms of work in heraldic scriptoria, on ships, and in Asian ports and palaces in order to make those pieces of parchment do what they were supposed to do. Chapter 3 shows Streynsham Master making and using the spaces of the factory—consultation rooms and writing offices (but also the new Anglican church at Fort St. George)—and reordering hierarchies of personnel within each factory and across the coast and the bay, in attempts to generate legitimate decisions, bring administrative order, and make authority work. Chapter 4 sets out the ways in which the political impact of texts was managed by working through a variety of regulatory mechanisms to restrict the circulation of texts. Where they were produced, they were met by attempts at manipulating political spaces, from the king's court to the coffeehouse, and the interpretations made by readers within them. Chapter 5 sets out the different, but connected, modes of practice of Exchange Alley and the Royal Society, showing how John Houghton used the practices of natural philosophy to shape the forms of knowledge that might be disseminated about stock trading. Doing so also meant drawing upon an extensive network of correspondents, and a constant battle to get his periodical to its readers. Perhaps most strikingly, chapter 6 argues that the imperial uses of print could not simply be secured by the technology of the press itself, but required—as the failed experiments of William Bolts, and the different disappointments experienced by James Augustus Hicky and Haji Mustapha show—the careful, and local, work of organizing patronage, making and managing a printing office, and composing a set of characters around the printing press.

In each case, the English East India Company undertook the organization of trade and empire through the social organization of writing and its geographies. Exploring how this was done, in script and in print, demonstrates

the active, difficult, and geographically specific construction of the relationships between power and knowledge—in both Europe and Asia—which
shaped and reshaped the ongoing encounter between East and West. These
social and material processes of organization and construction meant that
the promises of both power and knowledge that these forms of writing offered the Company were always uncertain. In part this was a matter of the
many contending interests that the Company brought together. This is evident in Streynsham Master's concern for both Company administration and
his own private trade (chapter 3), in debates over whether the future of the
English Company lay with the Dutch or in going it alone (chapter 4), and
in the different versions of empire that Warren Hastings and Philip Francis fought over in the Bengal Council (chapter 6). It was also a matter of
their opponents, both at home and abroad, who, in print and in Parliament
throughout the seventeenth and eighteenth centuries, challenged how the
Company conducted itself and even its right to exist. Finally, it was a matter
of the modes of encounter and interaction in Asia that were required for the
Company's business to prosper and, later, for its empire to function. Chapter
2 demonstrates both successes and failures in attempts to engage Asian polities with the Company's efforts to secure trading privileges. Chapter 3 shows
that the Company's Indian trade (and its servants' private trade) depended
upon the work of Indian merchants, leading to fears over their influence.
And chapter 6 demonstrates the necessary contribution of Indian teachers,
writers, blacksmiths, and pressmen to the making of the Honourable Company's Press and its imperial ambitions. In each case these relationships of
contradiction, conflict, interaction, and exclusion shaped the technologies
of writing and their uses by shaping the social and political organization of
the production, dissemination, and use of texts. In short, this book argues
for a version of writing as a political technology that not only is formed by
the social relations that made it work, but also acts to shape those social
relations. In this way the written word is understood as embedded in the
material and social worlds.

   This focus on materiality and the social organization of technologies of
script and print has produced an approach to writing and its geographies that
understands them as a vital part of practices that are actively involved in
shaping how the world works. The royal letters, ships' journals, accounting
practices, and commissions discussed in chapter 2 are shown to be, in their
making, transportation and use, actively and directly involved in the practices of kingship, captainship, and intercultural translation of which they
were a part. As chapter 3 argues, it was the organization of the documentation that established what it was to make a decision, to run an orderly

factory, and to engage in the proper forms of authority. Otherwise indeterminate forms of words, action, and interaction were given shape in writing, a shape that meant that they could be audited and evaluated. In chapter 4 the practices of politics, and the modes of political influence, are shown to be undertaken through different forms of print and the relationships they had with speech, performance, and script. Chapter 5 also shows politics being done through print, but in the stock market rather than in Parliament. Indeed, in Houghton's *Collection* we see a market in fictitious commodities being made for a particular audience, and made visible in specific ways, through the printing of stock prices alongside other forms of information. Finally, chapter 6 argues that it is necessary to write the histories of printing and empire together to demonstrate the ways that a print culture can be constructed of imperial materials, and how an empire comes out in print. In each case, this emphasis on writing and practice challenges distinctions between the discursive and the nondiscursive, between words and deeds. It shows writing as a vital part of the action rather than simply a reflection upon it. It foregrounds the active making of power and knowledge as a contested enterprise involving multiple agents and connects the realm of ideas firmly to the realm of practice. Understanding writing in this way means that this work with words was always both an intellectual and a practical matter. East India Company servants had specific ends in mind when they wrote, but each time they wrote, and each time they chose particular forms of words and forms of writing they were drawing on meanings and practices that were at once political, intellectual, and practical. This means that the engagement of the East India Company with the history of political ideas does not begin with its late eighteenth-century empire. The Company's operation was always, in both its overt pronouncements and buried deep in its practical routines, a matter of intellectual and political practice.

⁓

Overall, *Indian Ink* argues for an engagement between the histories of overseas trade and empire and the history of the book in order to understand a changing world. This confluence of ideas is based on the argument that consideration of the local spaces and places of knowledge making and use, and the wider networked geographies of its communication, can offer ways of understanding the relationship between knowledge and power in the making of a new world of trade and empire. I have argued, through a series of specific engagements with particular forms of script and print, that these historical geographies of writing and reading can demonstrate the complexities of the workings of power and knowledge as they are made. Considering texts as material objects, produced by political technologies, and embedded within practices that make up material, social, and symbolic worlds provides an account of the development of the relationship between Europe and Asia as a complex intellectual and practical process of making new economic and political forms and forging new relationships between places. *Indian Ink* demonstrates the active, material, and contested cultures of information and its technologies within which those relationships between distant places have been made in the past, and through which they continue to be made today.

BIBLIOGRAPHY

MANUSCRIPT SOURCES

Angus Library, Regent's Park College, Oxford
Baptist Missionary Society Archives

Bodleian Library, Oxford
Persian Manuscripts

British Library
Additional Charters
Additional Manuscripts
Sloane Manuscripts
Stowe Manuscripts

Huntington Library, Pasadena, California
Huntington Library Manuscripts

James Ford Bell Library, University of Minnesota, Minneapolis
Manuscript Collection

Oriental and India Office Collection, British Library
Additional Manuscripts
European Manuscripts
India Office Records

PRINTED RECORDS

Acts of the Privy Council of England, 1615–1616 (HMSO, London, 1925).
Arber, E. (ed.) (1875–94) *A Transcript of the Registers of the Company of Stationers of London,* 5 vols. (privately printed, London).
Birdwood, G., and Foster, W. (eds.) (1893) *The First Letter Book of the East India Company, 1600–1619* (Bernard Quaritch, London).
Danvers, F. C. (ed.) (1896) *Letters Received by the East India Company from its*

Servants in the East, vol. 1: *1602–1613* (Sampson Low, Marston & Company, London).

Fort William–India House Correspondence. Vol. IX Public Series, 1782–1785 (National Archives of India, Delhi, 1959).

Foster, W. (ed.) (1897) *Letters Received by the East India Company from its Servants in the East*, vol. 2 (Sampson Low, Marston & Company, London).

Jackson, W. A. (ed.) (1957) *Records of the Court of the Stationers' Company, 1602–1640* (Bibliographical Society, London).

Records of Fort St. George (1910) *Diary and Consultation Book, 1672–1678* (Government Press, Madras).

Records of Fort St. George (1911) *Despatches from England, 1670–1677* (Government Press, Madras).

Records of Fort St. George (1911) *Diary and Consultation Book, 1678–1679* (Government Press, Madras).

Records of Fort St. George (1911) *Diary and Consultation Book, 1679–1680* (Government Press, Madras).

Records of Fort St. George (1912) *Diary and Consultation Book, 1680–1681* (Government Press, Madras).

Records of Fort St. George (1913) *Diary and Consultation Book, 1681* (Government Press, Madras).

Records of Fort St. George (1914) *Despatches from England, 1680–1682* (Government Press, Madras).

Sainsbury, E. B. (ed.) (1925) *A Calendar of the Court Minutes of the East India Company, 1664–1667* (Clarendon Press, Oxford).

Sainsbury, E. B. (ed.) (1935) *A Calendar of the Court Minutes of the East India Company, 1674–1676* (Clarendon Press, Oxford).

Sainsbury, W. N. (ed.) (1862) *Calendar of State Papers. Colonial Series: East Indies, China and Japan, 1513–1616* (Longman, Green, Longman & Roberts, London).

Sainsbury, W. N. (ed.) (1878) *Calendar of State Papers. Colonial Series: East Indies, China and Japan, 1622–1624* (Longman & Co., London).

Sainsbury, W. N. (ed.) (1884) *Calendar of State Papers. Colonial Series: East Indies, China and Persia, 1625–1629* (Longman & Co., London).

Stevens, H. (ed.) (1886) *The Dawn of British Trade to the East Indies as Recorded in the Court Minutes of the East India Company, 1599–1603* (Henry Stevens, London).

Wellesley, R. C. (1836–37) *The Despatches, Minutes, and Correspondence of the Marquess Wellesley, K.G. During His Administration in India*, Montgomery Martin (ed.), 5 vols. (W. H. Allen and Co., London).

PUBLISHED SERIALS

The Asiatick Miscellany
Asiatick Researches

A Collection of Letters for the Improvement of Husbandry and Trade
The Course of the Exchange, &c.
Critical Review
Freke's Prices of Stocks, &c.
The Merchants Remembrancer
Monthly Review
The Prices of Merchandise in London
The Printer

WORKS PUBLISHED PRIOR TO 1820

A., W. (1690) *An Apology for the East-India Company* (Printed for the author, London).

The Answer of the East-India Company, to Two Printed Papers of Mr. Samuel White (n.d.).

Answer to a Late Tract Entituled, An Essay on the East-India Trade (Printed for Tho. Cockerill, London, 1697).

An Answer to Two Letters, Concerning the East-India Company (1676).

The Argument of a Learned Counsel, upon an Action of the Case Brought by the East-India-Company Against Mr. Thomas Sands, an Interloper (Printed for B. Aylmer, London, 1696).

The Argument of the Lord Chief Justice of the Court of King's Bench concerning the Great Case of Monopolies, between the East-India Company, Plaintiff, and Thomas Sandys, Defendant (Printed and sold by Randall Taylor, London [1689]).

An Arrest on the East India Privateer (1681).

Balfour, F. (1781) *The Forms of Herkern* (Printed at Calcutta).

The Bhagvat-Geeta, or Dialogues of Kreeshna and Arjoon. Translated by Charles Wilkins (Printed for C. Nourse, London, 1785).

Birch, T. (1756–57) *The History of the Royal Society of London*, 4 vols. (Printed for A. Millar, London).

Bolts, W. (1772) *Considerations on Indian Affairs; Particularly Respecting the Present State of Bengal and its Dependencies* (Printed for J. Almon in Piccadilly, P. Elmsley in the Strand, and Brotherton and Sewell in Cornhill, London).

A Brief Abstract of the Great Oppressions and Injuries which the late Managers of the East-India-Company Have Acted on the Lives, Liberties and Estates of their Fellow-Subjects (1698?).

Child, J.: *see also* Philopatris.

Child, J. (1668) *Brief Observations Concerning Trade and Interest of Money* (Printed for Elizabeth Calvert and Henry Mortlock, London).

[Child, J.] (1689) *A Supplement . . . to a former Treatise, concerning the East-India Trade* (London).

Child, J. (1693) *A New Discourse of Trade* (Printed and sold by John Everingham, London).

Coke, R. (1670) *A Discourse of Trade* (Printed by H. Brome and R. Home, London).

Colebrooke, H. T. (1798) *A Digest of Hindu Law On Contracts And Successions*, 4 vols. (printed at the Honourable Company's Press, Calcutta).

A Collection of Papers Relating to the East India Trade: Wherein are shewn the Disadvantages to a Nation, by conjoining any Trade to a Corporation with a Joint-Stock (Printed for J. Walthoe, London, 1730).

Coryate, T. (1611) *Coryates Crudities* (Printed by W.S., London).

Coverte, R. (1612) *A True and Almost Incredible Report of an Englishman that (Being Cast Away in the Good Ship Called the Assention in Cambaya the Furthest Part of the East Indies) Travelled by Land Through Many Unknowne Kingdomes, and Great Cities* (London)

Davis, J. (1595) *The Seamen's Secrets* (London).

[Defoe, D.] (1701) *The Free-Holders Plea Against Stock-Jobbing Elections of Parliament Men*, 2nd ed. (London).

[Defoe, D.] (1701) *The Villainy of Stock-Jobbers Detected, and the Causes of the Late Run Upon the Bank and Bankers Discovered and Considered* (London).

[Defoe, D.] (1719) *The Anatomy of Exchange-Alley. Or a System of Stock-Jobbing* (Printed by E. Smith, London).

A Dialogue Between a Director of the New East-India Company, and one of the Committee for preparing By-Laws for the said Company (Printed for Andrew Bell, London, 1699).

A Dialogue Between Jest, an East-India Stock-Jobber, and Earnest, an Honest Merchant (London, 1708).

[Digges, D.] (1612) *Of the Circumference of the Earth: Or, A Treatise of the North-east [sic] Passage* (Printed by W.W. for John Barnes, London).

Digges, D. (1615) *The Defence of Trade* (Printed by William Stansby for John Barnes, London).

Digges, T., and Digges, D. (1604) *Foure Paradoxes, or politique Discourses. 2 Concerning Militarie Discipline* [and] *2 of the Worthinesse of Warre and Warriors* (Printed by H. Lownes for Clement Knight, London).

A Discourse concerning the East-India Trade (1693) in Lord J. Somers (1751) *A Third Collection of Scarce and Valuable Tracts, on the most Interesting and Entertaining Subjects* (F. Cogan, London).

A Discourse Concerning Trade, and that in particular of the East-Indies (Printed and sold by Andrew Sowle, London, 1689).

Dow, A. (1768–72) *History of Hindostan*, 3 vols. (Printed for T. Becket and P. A. de Hondt, in the Strand, London).

An Elegy on the Death of the Old East-India Company (Printed for the Author, London, 1699).

The Emblem of Ingratitude (Printed for William Home, London, 1672).

An Essay Towards a Scheme or Model for Erecting a National East-India Joynt Stock or Company (Printed for the Author, London, 1691).

Free, Regulated Trade, Particularly to India, the Interest of England (1691?).

Gentleman, T. (1614) *Englands Way to Win Wealth, and to Employ Ships and Mariners* (Printed for Nathaniel Butter, London).

Gilchrist, J. (1787–90) *A Dictionary, English and Hindoostanee* (Stuart and Cooper, Calcutta).

Gladwin, F. (1780) *A Compendious Vocabulary English and Persian* (Malda).

Gladwin, F. (1783) *Ayeen Akbery: Or, The Institutes of the Emperor Akber*, 3 vols. (Calcutta).

Halhed, N. B. (1776) *A Code of Gentoo Laws; Or, Ordination of the Pundits* (London).

Halhed, N. B. (1778) *A Grammar of the Bengal Language* (Printed at Hoogly [sic] in Bengal).

Hall, R. (1781) *Help to Zion's Travellers* (Bristol).

Hill, J. (1689) *The Young Secretary's Guide* (H. Rhodes, London).

Hitchcock, R. (1580) *A Politique Platt for the honour of the Prince, the greate profite of the publique state, relief of the poore, preservation of the riche, reformation of Rogues and Idle persones, and the wealthe of thousands that knowes not how to live* (Printed by John Kyngston, London).

[Houghton, J.] (1677) *England's Great Happiness. Or, a Dialogue Between Content and Complaint* (Printed by J.M. for Edward Croft, London).

[Houghton, J.] [1683?] *An Account of the Bank of Credit in the City of London* (Printed by John Gain, London)

Houghton, J. (1691) *A Proposal for the Improvement of Husbandry and Trade*.

Houghton, J. (1693) *An Account of the Acres & Houses, with the Proportional Tax, &c. Of each County in England and Wales* (Printed for Randall Taylor, London).

Houghton, J. (1699) "A discourse of coffee," *Philosophical Transactions*, 21, pp. 311–17.

Houghton, J. (ed.) (1681–83) *A Collection of Letters for the Improvement of Husbandry and Trade* (Printed for John Lawrence, London).

An Impartial Vindication of the English East-India-Company (Printed by J. Richardson for Samuel Tidmarsh, London, 1688).

Insignia Bataviæ; or, the Dutch Trophies Display'd (Printed for Thomas Pyke, London, 1688).

Jones, W. (1771) *A Grammar of the Persian Language* (Printed by W. and J. Richardson, London).

A Journal of Several Remarkable Passages, before The Honourable House of Commons, and the ... Privy Council: Relating to the East India Trade [1693?].

A Letter from a Lawyer of the Inner Temple to His Friend in the Country, Concerning the East India Stock (London, 1698).

A Letter to a Friend Concerning the East-India Trade (Printed and sold by E. Whitlock, London, 1696).

A Letter to the Chairman of the East-India Company (Printed for J. Roberts, London, 1727).

Linschoten, J. H. von (1596) *Itinerario* (Amsterdam).

Linschoten, J. H. von (1598) *John Huigen Van Linschoten, His Discours of Voyages into Ye East & West Indies* (John Wolfe, London).

Malynes, G. de (1622) *Lex Mercatoria, or The Ancient Law-Merchant* (Printed by Adam Islip, London).

Malynes, G. de (1623) *The Center of the Circle of Commerce* (Printed by William Jones, London).

[Martyn, H.] (1701) *Considerations upon the East-India Trade* (Printed for A. and J. Churchill, London).

Misselden, E. (1622) *Free Trade. Or, the Meanes to Make Trade Flourish* (Printed by John Legatt, for Simon Waterson, London).

Misselden, E. (1623) *The Circle of Commerce. Or the Balance of Trade, in Defence of Free Trade* (Printed by John Dawson for Nicholas Bourne, London).

A Modest and Just Apology for; or Defence of the Present East-India-Company (London, 1690).

Mores, E. R. (1778) *A Dissertation Upon English Typographical Founders and Foundaries* (Privately Printed)

Moxon, J. (1683–84) *Mechanick Exercises on the Whole Art of Printing* (Printed for J. Moxon, London).

Mun, T. (1621) *A Discourse of Trade From England Unto the East Indies*, 2nd ed. (Printed by Nicholas Okes for John Piper, London).

Mun, T. (1664) *England's Treasure by Forraign Trade* (Printed by J.G. for Thomas Clark, London).

N., A. (1699) *England's Advocate, Europe's Monitor: Being an Intreaty for Help in Behalf of the English Silk-Weavers and Silk-Throwsters* (Printed for George Larkin, London).

A Narrative of the Insurrection Which Happened in the Zemeedary of Banaris (Printed by Order of the Governor General, Calcutta, 1782).

Newes out of East India (Printed for F. Coules, London, 1624).

Parker, S. (1681) *A Demonstration of the Divine Authority of the Law of Nature, and of the Christian Religion* (Printed by M. Flesher for R. Royston and R. Chiswell, London).

Philopatris [Josiah Child] (1681) *A Treatise Concerning the East India Trade* (Printed by J.R. for the Honourable the East-India Company, London).

Polloxfen, J. (1697) *England and East-India Inconsistent in their Manufactures* (London).

[Polloxfen, J.] (1697) *A Discourse of Trade, Coyn, and Paper Credit* (Printed for Brabazon Aylmer, London).

Proclamation: James II. East India Company (1 April 1685).

Proposals for Setling the East-India Trade (Printed and sold by E. Whitlock, London, 1696).

R., J. [Robert Kayll] (1615) *The Trades Increase* (Printed by Nicholas Okes, and sold by Walter Burre, London).

Reasons Humbly Offered Against Grafting or Splicing, and for Dissolving this Present East-India Company (London, 1690).

Reasons Proposed for the Encouragement of all People to Under-Write to the New Subscriptions, Appointed to be Made to the late East-India Companyes Stock (1693).

A Regulated Company more National than a Joint-Stock in the EAST-INDIA TRADE [London]

Regulations for the Administration of Justice in the Courts of Dewannee Adaulut . . . With a Bengal Translation (At the Honorable Company's Press, Calcutta, 1785).

Regulations for the Administration of Justice in the Courts of Mofussil Dewannee Adaulut, and in the Sudder Dewannee Adaulut (At the Hon'ble Company's Press, Calcutta, 1781).

Regulations Passed by the Governor General in Council of Bengal . . ., vol. 1: *1793, 1794, and 1795* (J. L. Cox, London, 1828).

Roberts, L. (1638) *Merchants Mappe of Commerce* (London).

Rules, Orders, and Directions Appointed and Established by the Governour and Committees of the East-India Company, for the Well Regulating and Managing their Affairs in the Parts of INDIA [London, 1680?].

S., T. (1697) *Reasons Humbly Offered for the Passing a Bill for the Hindering the Home Consumption of East-India Silks, Bengals &c.* (Printed by J. Bradford, London).

Sawyer, E. (ed.) (1725) *Memorials of the Affairs of State in the Reigns of Q. Elizabeth and K. James I. Collected (chiefly) from the original papers of . . . Sir Ralph Winwood, Kt,* 3 vols. (Printed by W.B. for T. Ward, London).

Segar, W. (1590) *The Booke of Honor and Armes* (London).

Segar, W. (1602) *Honor Military and Civill* (London).

Seid-Gholam-Hossein-Khan (1789) *A Translation of the Seir Mutaqharin; or, View of Modern Times,* 3 vols. (Printed for the Translator, Calcutta).

Some Considerations Offered touching the East-India Affairs [1698].

Some Idea of the Civil and Criminal Courts of Justice at Moorshoodabad, in a Letter to Capt. John Hawkshaw, At Behrampore, of the 30th May 1789 (Printed for the Author, Calcutta, 1789).

Some Reflections on a Pamphlet, Intituled England and East-India Inconsistent in their Manufactures (London, 1697).

Some Remarks upon the Present State of the East India Company's Affairs (London, 1690).

Spalding, A. (1614) *Dialogues in the English and Malay Languages* (London).

To Preserve the East-India Trade (Printed by Freeman Collins, London, 1695).

Translation of the Persian Abridgement of the Regulations . . . for the Sudder and Mofussul Dewanny Adaulets (At the Hon'ble Company's Press, Calcutta, 1783).

Two Letters Concerning the East India Company (London, 1676).

A True Relation of the Rise and Progress of the East India Company.
A True Relation of the Unjust, Cruell, and Barbarous Proceedings against the English at Amboyna in the East-Indies, by the Neatherlandish Governour and Councel there (Printed by H. Lownes for Nathanael Newberry, London, 1624).

Verelst, H. (1772) *A View of the Rise, Progress and Present State of the English Government in Bengal* (Printed for J. Nourse; Brotherton and Sewell; and T. Evans, London).

Ward, W. (1811) *Account of the Writings, Religion, and Manners, of The Hindoos,* 4 vols. (Printed at the Mission Press, Serampore).

White, G. (1689) *A Letter to Mr Nathaniel Tenche* (London).

White, G. (1691) *An Account of the Trade to the East-Indies* (London).

Wilkins, C. (1781) *A Translation of a Royal Grant of Land by One of the Ancient Rajaas of Hindostan* (Printed at Calcutta by the Translator).

Wilkins, C. (1808) *A Grammar of the Sanskrita Language* (Printed for the Author, London).

WORKS PUBLISHED AFTER 1820

Adams, S. (1983) "Spain or Netherlands? the dilemmas of early Stuart foreign policy," in H. Tomlinson (ed.) *Before the English Civil War: Essays on Early Stuart Politics and Government* (Macmillan Press, London) pp. 79–101.

Adams, S. (1985) "Foreign policy and the parliaments of 1621 and 1624," in K. Sharpe (ed.) *Faction and Parliament: Essays on Early Stuart History* (Methuen, London) pp. 139–71.

Alam, M., and S. Subrahmanyam (eds.) (1998) *The Mughal State, 1526–1750* (Oxford University Press, New Delhi).

Alavi, S. (ed.) (2002) *The Eighteenth Century in India* (Oxford University Press, New Delhi).

Allaudin, S., and R. K. Rout (1996) *Libraries and Librarianship during Muslim Rule in India* (Reliance Publishing House, Delhi).

Alvi, S. S. (ed.) (1989) *Mauīzah-i Jahāngīrī of Muhammad Bāqir Najm-i Sānī: Advice on the Art of Government—An Indo-Islamic Mirror for Princes* (State University of New York Press, Albany).

Amory, H., and D. D. Hall (eds.) (2000) *A History of the Book in America,* vol. 1: *The Colonial Book in the Atlantic World* (Cambridge University Press, Cambridge).

Anderson, B. (1983) *Imagined Communities: Reflections on the Origin and Spread of Nationalism* (Verso, London).

Andrews, K. R. (1964) *Elizabethan Privateering: English Privateering during the Spanish War, 1585–1603* (Cambridge University Press, Cambridge).

Andrews, K. R. (1984) *Trade, Plunder and Settlement: Maritime Enterprise and the Genesis of the British Empire, 1480–1630* (Cambridge University Press, Cambridge).

Appadurai, A. (1996) "Number in the colonial imagination," in *Modernity at Large: Cultural Dimensions of Globalization* (University of Minnesota Press, Minneapolis) pp. 114–35.

Appleby, J. C. (1998) "War, politics, and colonization, 1558–1625," in N. Canny (ed.) *The Oxford History of the British Empire*, vol. 1: *The Origins of Empire* (Oxford University Press, Oxford) pp. 55–78.

Appleby, J. O. (1978) *Economic Thought and Ideology in Seventeenth-Century England* (Princeton University Press, Princeton, NJ).

Arasaratnam, S. (1986) *Merchants, Companies and Commerce on the Coromandel Coast, 1650–1740* (Oxford University Press, Delhi)

Arasaratnam, S. (1986) "Society, power, factionalism and corruption in early Madras, 1640–1746," *Indica* 23, pp. 113–34.

Arasaratnam, S. (1987) "India and the Indian Ocean in the seventeenth century," in A. Das Gupta and M. N. Pearson (eds.) *India and the Indian Ocean, 1500–1800* (Oxford University Press, Calcutta) pp. 94–130.

Arasaratnam, S. (1995) *Maritime Trade, Society and European Influence in Southern Asia, 1600–1800* (Variorum, Aldershot).

Aravamudan, S. (1999) *Tropicopolitans: Colonialism and Agency, 1688–1804* (Duke University Press, Durham).

Armitage, D. (2000) *The Ideological Origins of the British Empire* (Cambridge University Press, Cambridge).

Azizuddin Husain, S. M. (ed.) (1990) *Raqaim-i-Karaim (Epistles of Aurangzeb)* (Idarah-i Adabiyat-i Delli, Delhi).

Balen, M. (2002) *A Very English Deceit: The South Sea Bubble and the World's First Great Financial Scandal* (Fourth Estate, London).

Ballantyne, T. (2002) *Orientalism and Race: Aryanism in the British Empire* (Palgrave, Basingstoke).

Barber, G. (1982) "Book imports and exports in the eighteenth century," in R. Myers and M. Harris (eds.) *Sale and Distribution of Books from 1700* (Oxford Polytechnic Press, Oxford) pp. 77–105.

Barbour, R. (2000) "'There is our commission': writing and authority in *Measure for Measure* and the London East India Company," *Journal of English and German Philology*, 99:2, pp. 193–214.

Barbour, R. (2003) *Before Orientalism: London's Theatre of the East, 1576–1626* (Cambridge University Press, Cambridge).

Barns, M. (1940) *The Indian Press: A History of the Growth of Public Opinion in India* (George Allen & Unwin, London).

Barrow, I. J., and D. E. Haynes (2004) "The colonial transition: South Asia, 1780–1840," *Modern Asian Studies*, 38:3, pp. 469–78.

Bayly, C. A. (1983) *Rulers, Townsmen and Bazaars: North Indian Society in the Age of British Expansion, 1770–1870* (Cambridge University Press, Cambridge).

Bayly, C. A. (1988) *Indian Society and the Making of the British Empire* (Cambridge University Press, Cambridge).

Bayly, C. A. (1996) *Empire and Information: Intelligence Gathering and Social Communication in India, 1780–1870* (Cambridge University Press, Cambridge).

Beal, P. (1998) *In Praise of Scribes: Manuscripts and Their Makers in Seventeenth-Century England* (Clarendon Press, Oxford).

Benjamin, W. (1992) "The work of art in the age of mechanical reproduction" in Walter Benjamin, *Illuminations* (Fontana, London) pp. 211–44.

Berry, H. (2003) *Gender, Society and Print Culture in Late-Stuart England: The Cultural World of the Athenian Mercury* (Ashgate, Aldershot).

Beveridge, H. (ed.) (1978 [1909–14]) *The Tūzuk-i-Jahāngīrī, or Memoirs of Jahāngīr.* Translated by Alexander Rogers, 2 vols. (Munshiram Manoharlal, New Delhi).

Bhabha, H. (1994) *The Location of Culture* (Routledge, London).

Blair, A. (2004) "Focus: scientific readers—an early modernist's perspective," *Isis*, 95, pp. 420–30.

Blussé, L., and F. Gaastra (eds.) (1981) *Companies and Trade: Essays on Overseas Trading Companies during the Ancien Régime* (Leiden University Press, Leiden).

Boehmer, E. (1995) *Colonial and Postcolonial Literature: Migrant Metaphors* (Oxford University Press, Oxford).

Bond, D. F. (ed.) (1965) *The Spectator*, 5 vols. (Clarendon Press, Oxford).

Boon, J. A. (1982) *Other Tribes, Other Scribes: Symbolic Anthropology in the Comparative Study of Cultures, Histories, Religions and Texts* (Cambridge University Press, Cambridge).

Bowen, H. V. (1991) *Revenue and Reform: The Indian Problem in British Politics, 1757–1773* (Cambridge University Press, Cambridge).

Bowen, H. V. (1993) "'The pests of human society': stockbrokers, jobbers and speculators in mid-eighteenth-century Britain," *History*, 78, pp. 38–53.

Bowen, H. V. (2002) "'No longer mere traders': continuities and change in the metropolitan development of the East India Company, 1600–1834," in H. V. Bowen, M. Lincoln, and N. Rigby (eds.) *The Worlds of the East India Company* (Boydell Press, Woodbridge) pp. 19–32.

Bowen, H. V. (2005) *The Business of Empire: The East India Company and Imperial Britain, 1756–1833* (Cambridge University Press, Cambridge).

Boyle, R. (2001) *The Correspondence of Robert Boyle*, 6 vols., M. Hunter, A. Clericuzio, and L. M. Principe (eds.) (Pickering and Chatto, London).

Braddick, M. J. (1998) "The English government, war, trade, and settlement, 1625–1688," in N. Canny (ed.) *The Oxford History of the British Empire*, vol. 1: *The Origins of Empire* (Oxford University Press, Oxford) pp. 286–308.

Brakel, L. F. (1975) "State and statecraft in seventeenth-century Aceh," *Monographs of the Malaysian Branch of the Royal Asiatic Society 6: Pre-colonial State Systems in Southeast Asia*, pp. 56–66.

Breckenridge, C. A., and P. van der Veer (eds.) (1993) *Orientalism and the Postcolonial Predicament: Perspectives on South Asia* (University of Pennsylvania Press, Philadelphia).

Brenner, R. (1993) *Merchants and Revolution: Commercial Change, Political Conflict and London's Overseas Traders, 1550–1653* (Cambridge University Press, Cambridge).

Brewer, J. (1989) *The Sinews of Power: War, Money and the English State, 1688–1783* (Unwin Hyman, London).

Brewer, J. (1997) *The Pleasures of the Imagination: English Culture in the Eighteenth Century* (HarperCollins, London).

Brockington, J. L. (1989) "Warren Hastings and Orientalism," in G. Carnall and C. Nicholson (eds.) *The Impeachment of Warren Hastings: Papers from a Bicentenary Commemoration* (Edinburgh University Press, Edinburgh) pp. 91–108.

Buck, P. (1977) "Seventeenth-century political arithmetic: civil strife and vital statistics," *Isis*, 68, pp. 67–84.

Buckland, C. E. (1999) *Dictionary of Indian Biography*, 2 vols. (Cosmo Publications, New Delhi).

Burke, P. (2000) *A Social History of Knowledge: From Gutenberg to Diderot* (Polity Press, Cambridge).

Burnett, D. G. (2000) *Masters of All They Surveyed: Exploration, Geography, and a British El Dorado* (University of Chicago Press, Chicago).

Busteed, H. E. (1908) *Echoes from Old Calcutta: Being Chiefly Reminiscences of the Days of Warren Hastings, Francis, and Impey*, 4th ed. (W. Thacker & Co, London).

Callon, M. (ed.) (1998) *The Laws of the Markets* (Blackwell, Oxford and The Sociological Review).

Cañizares-Esguerra, J. (2001) *How to Write the History of the New World: Histories, Epistemologies, and Identities in the Eighteenth-Century Atlantic World* (Stanford University Press, Stanford).

Cannon, G. (1990) *The Life and Mind of Oriental Jones* (Cambridge University Press, Cambridge).

Carlos, A. M. (1994) "Bonding and the agency problem: evidence from the Royal African Company, 1672–1691," *Explorations in Economic History*, 31, pp. 313–35.

Carlos, A. M., and S. Hejeebu (2006) "Specific information and the English chartered companies," paper presented at the 14th International Economic History Congress, Helsinki, Finland.

Carlos, A. M., and S. Nicholas (1996) "Theory and history: seventeenth-century joint-stock chartered trading companies," *Journal of Economic History*, 56:4, pp. 916–24.

Carruthers, B. G. (1996) *City of Capital: Politics and Markets in the English Financial Revolution* (Princeton University Press, Princeton, NJ).

Carswell, J. (1960) *The South Sea Bubble* (Cresset Press, London).

Carter, T. F. (1925) *The Invention of Printing in China and Its Spread Westward* (Columbia University Press, New York).

Casid, J. H. (2005) *Sowing Empire: Landscape and Colonization* (University of Minnesota Press, Minneapolis).

Chakrabarty, D. (2000) *Provincializing Europe: Postcolonial Thought and Historical Difference* (Princeton University Press, Princeton, NJ).

Chakrabarty, P. (1983) *Anglo-Mughal Commercial Relations, 1583–1717* (O.P.S. Publishers, Calcutta).

Chan, M. (ed.) (1995) *The Life of the Lord Keeper North by Roger North* (Edwin Mellen Press, Lampeter).

Chartier, R. (1988) *Cultural History: Between Practices and Representations* (Polity Press, Cambridge).

Chartier, R. (1994) *The Order of Books: Readers, Authors, and Libraries in Europe between the Fourteenth and Eighteenth Centuries* (Stanford University Press, Stanford).

Chartier, R. (1995) *Forms and Meanings: Texts, Performances, and Audiences from Codex to Computer* (University of Pennsylvania Press, Philadelphia).

Chartier, R. (1997) *On the Edge of the Cliff: History, Language, and Practices* (Johns Hopkins University Press, Baltimore).

Chartier, R. (ed.) (1989) *The Culture of Print: Power and the Uses of Print in Early Modern Europe* (Polity Press, Cambridge)

Chatterjee, K. (1998) "History as self-representation: the recasting of a political tradition in late eighteenth-century eastern India," *Modern Asian Studies*, 32:4, pp. 913–48.

Chaudhuri, K. N. (1965) *The English East India Company: The Study of an Early Joint-Stock Company, 1600–1640* (Frank Cass, London).

Chaudhuri, K. N. (1978) *The Trading World of Asia and the English East India Company, 1660–1760* (Cambridge University Press, Cambridge).

Chaudhuri, K. N. (1981) "The East India Company in the seventeenth and eighteenth centuries: a pre-modern multinational organization," in L. Blussé and F. Gaastra (eds.) *Companies and Trade: Essays on Overseas Trading Companies during the Ancien Régime* (Leiden University Press, Leiden) pp. 29–46.

Chaudhuri, K. N. (1985) *Trade and Civilization in the Indian Ocean: An Economic History from the Rise of Islam to 1750* (Cambridge University Press, Cambridge).

Chaudhuri, K. N. (1991) "Reflections on the organizing principle of premodern trade," in J. D. Tracy (ed.) *The Political Economy of Merchant Empires: State Power and World Trade, 1350–1750* (Cambridge University Press, Cambridge) pp. 421–42.

Chaudhuri, S. (1975) *Trade and Commercial Organization in Bengal, 1650–1720: With Special Reference to the English East India Company* (Firma K. L. Mukhopadhyay, Calcutta).

Chaudhuri, S. B. (1968) "Early English printers and publishers in Calcutta," *Bengal Past and Present*, 87, pp. 67–77

Clark, G. N., and W. J. M. Eysinga (1951) *The Colonial Conferences between England and the Netherlands in 1613 and 1615*. Part 2. (E. J. Brill, Leiden).

Clayton, D. W. (2000) *Islands of Truth: The Imperial Fashioning of Vancouver Island* (University of British Columbia Press, Vancouver).

Cohn, B. S. (1996) *Colonialism and Its Forms of Knowledge: The British in India* (Princeton University Press, Princeton, NJ).

Colclough, S. (2000) *Reading Experience 1700–1840: An Annotated Register of Sources for the History of Reading in the British Isles* (University of Reading, Reading).

Cooper, F. (2005) *Colonialism in Question: Theory, Knowledge, History* (University of California Press, Berkeley).

Cowan, B. (2005) *The Social Life of Coffee: The Emergence of the British Coffeehouse* (Yale University Press, New Haven).

Crill, R. (2004) "Visual responses: depicting Europeans in south Asia," in A. Jackson and A. Jaffer (eds.) *Encounters: The Meeting of Asia and Europe, 1500–1800* (V&A Publications, London) pp. 188–99.

Cunningham, B., and M. Kennedy (eds.) (1999) *The Experience of Reading: Irish Historical Perspectives* (Rare Books Group of the Library Association of Ireland, Dublin).

Darnton, R. (1979) *The Business of Enlightenment: A Publishing History of the Encyclopédie, 1775–1800* (Belknap Press of Harvard University Press, Cambridge, Mass.).

Darnton, R. (1982) "What is the history of books?" *Daedalus*, 111:3, pp. 65–83.

Darnton, R. (1984) *The Great Cat Massacre and Other Episodes in French Cultural History* (Penguin Books, Harmondsworth).

Darnton, R. (1986) "First steps towards a history of reading," *Australian Journal of French Studies*, 23:1, pp. 5–30.

Das Gupta, A. K. (1962) "Acheh in the seventeenth-century Asian trade," *Bengal Past and Present*, 81, pp. 37–49.

Das Gupta, A. K. (1998) "Trade and politics in eighteenth-century India," in M. Alam and S. Subrahmanyam (eds.) *The Mughal State, 1526–1750* (Oxford University Press, New Delhi) pp. 361–97.

Daston, L. (1991) "Baconian facts, academic civility, and the prehistory of objectivity," *Annals of Scholarship*, 8:3-4, pp. 337–64.

Daston, L., and K. Park (1998) *Wonders and the Order of Nature, 1150–1750* (Zone Books, New York).

Datta, R. (2000) *Society, Economy and the Market: Commercialisation in Rural Bengal, c. 1760–1800* (Manohar, New Delhi).

Davies, K. G. (1952) "Joint-stock investment in the later seventeenth century," *Economic History Review*, 4, pp. 283–301.

Dening, G. (1980) *Islands and Beaches: Discourse on a Silent Land, Marquesas, 1774–1880* (University Press of Hawaii, Honolulu).

Dening, G. (1992) *Mr Bligh's Bad Language: Passion, Power and Theatre on the Bounty* (Cambridge University Press, Cambridge).

Derrett, J. D. M. (1960) "Nandakumar's forgery," *English Historical Review*, 75, pp. 223–38.

Derrett, J. D. M. (1968) *Religion, Law and the State in India* (Faber & Faber, London).

Dickson, P. G. M. (1967) *The Financial Revolution in England: A Study in the Development of Public Credit, 1688–1756* (Macmillan, London).

Diehl, K. S. (1990) *Printers and Printing in the East Indies to 1850*, vol. 1: *Batavia 1600–1850* (Aristide D. Caratzas, New Rochelle, New York).

Drace-Francis, A. (2006) "A provincial imperialist and a *Curious Account of Wallachia*: Ignaz von Born," *European History Quarterly*, 36:1, pp. 61–89.

Drayton, R. (2000) *Nature's Government: Science, Imperial Britain, and the "Improvement" of the World* (Yale University Press, New Haven).

Drewes, G. W. J., and P. Voorhoeve (eds.) (1958) *Adat Atjèh: Reproduced in Facsimile from a Manuscript in the India Office Library* (Martinus Nijhoff, The Hague).

Driver, F. (2001) *Geography Militant: Cultures of Exploration and Empire* (Blackwell, Oxford).

Driver, F., and D. Gilbert (eds.) (1999) *Imperial Cities: Landscape, Display and Identity* (Manchester University Press, Manchester).

Driver, F., and L. Martins (eds.) (2005) *Tropical Visions in an Age of Empire* (University of Chicago Press, Chicago).

Dubois, L. (2004) *Avengers of the New World: The Story of the Haitian Revolution* (Belknap Press of Harvard University Press, Cambridge, Mass.).

Dubois, L. (2006) "An enslaved Enlightenment: rethinking the intellectual history of the French Atlantic," *Social History*, 31:1, pp. 1–14.

Edney, M. H. (1997) *Mapping an Empire: The Geographical Construction of British India, 1765–1843* (University of Chicago Press, Chicago).

Eisenstein, E. L. (1979) *The Printing Press as an Agent of Change: Communications and Cultural Transformations in Early Modern Europe* (Cambridge University Press, Cambridge).

Eisenstein, E. L. (2002) "An unacknowledged revolution revisited," *American Historical Review*, 107:1, pp. 87–105.

Ellis, G. W. J. A. (ed.) (1829) *The Ellis Correspondence*, 2 vols. (Henry Colburn, London).

Ellis, M. (2004) *The Coffee-House: A Cultural History* (Wiedenfeld & Nicholson, London).

Erikson, E., and P. Bearman (2004) "Routes into networks: the structure of English trade in the East Indies, 1601–1833," *Institute for Social and Economic Research and Policy Working Paper 04-07*, Columbia University, New York.

Evelyn, J. (1955) *The Diary of John Evelyn*, 6 vols., E. S. de Beer (ed.) (Clarendon Press, Oxford).

Fawcett, C. G. H. (1952) *The English Factories in India*, vol. 2 (n.s.): *The Eastern Coast and Bengal, 1670–1677* (Clarendon Press, Oxford).

Feather, J. (1985) *The Provincial Book Trade in Eighteenth-Century England* (Cambridge University Press, Cambridge).

Febvre, L., and H.-J. Martin (1976 [1958]) *The Coming of the Book: The Impact of Printing, 1450–1800* (Verso, London).

Finkelstein, A. (2000) *Harmony and Balance: An Intellectual History of Seventeenth-Century English Economic Thought* (University of Michigan Press, Ann Arbor).

Forbes, E. G., and (for M. Forbes) L. Murdin and F. Willmoth (eds.) (1995–2002) *The Correspondence of John Flamsteed, The First Astronomer Royal.* 3 vols. (Institute of Physics Publishing, Bristol and Philadelphia).

Foster, W. (ed.) (1899) *The Embassy of Sir Thomas Roe to the Court of the Great Mogul, 1615–1619* (Oxford University Press, Oxford).

Foster, W. (ed.) (1905) *The Journal of John Jourdain, 1608–1617* (Hakluyt Society, Cambridge).

Foster, W. (ed.) (1926) *The Embassy of Sir Thomas Roe to India, 1615–1619*, revised ed. (Oxford University Press, London).

Foster, W. (ed.) (1934) *The Voyage of Thomas Best to the East Indies, 1612–14* (Hakluyt Society, London).

Foster, W. (ed.) (1940) *The Voyages of Sir James Lancaster to Brazil and the East Indies, 1591–1603* (Hakluyt Society, London).

Foster, W. (ed.) (1943) *The Voyage of Sir Henry Middleton to the Moluccas, 1604–1606* (Hakluyt Society, London).

Franklin, M. J. (2002) "'The Hastings Circle': writers and writing in Calcutta in the last quarter of the eighteenth century," in E. J. Clery, C. Franklin, and P. Garside (eds.) *Authorship, Commerce and the Public: Scenes of Writing, 1750–1850* (Palgrave, London) pp. 186–202.

Franklin, M. J. (ed.) (2000) *Representing India: Indian Culture and Imperial Control in Eighteenth-Century British Orientalist Discourse*, 9 vols. (Routledge, London).

Frasca-Spada, M., and N. Jardine (eds.) (2000) *Books and the Sciences in History* (Cambridge University Press, Cambridge).

Furber, H. (1976) *Rival Empires of Trade in the Orient, 1600–1800* (University of Minnesota Press, Minneapolis).

Gauci, P. (2001) *The Politics of Trade: The Overseas Merchant in State and Society, 1660–1720* (Oxford University Press, Oxford).

Ghosh, A. (2003) "An uncertain 'coming of the book': early print cultures in colonial India," *Book History*, 6, pp. 23–55.

Ghosh, A. (2006) *Power in Print: Popular Publishing and the Politics of Language and Culture in a Colonial Society* (Oxford University Press, New Delhi).

Ginzburg, C. (1980 [1976]) *The Cheese and the Worms: The Cosmos of a Sixteenth-Century Miller* (Johns Hopkins University Press, Baltimore).

Glaisyer, N. (2000) "Readers, correspondents and communities: John Houghton's A Collection for Improvement of Husbandry and Trade (1692–1703)," in A. Shepard and P. Withington (eds.) *Communities in Early Modern England: Networks, Place, Rhetoric* (Manchester University Press, Manchester) pp. 235–51.

Goody, J. (1986) *The Logic of Writing and the Organization of Society* (Cambridge University Press, Cambridge).

Grant, J. (1856) "Warren Hastings in slippers," *Calcutta Review*, 26, pp. 59–141.

Gray, A. (1887) *The Voyage of François Pyrard of Laval to the East Indies, the Maldives, the Moluccas and Brazil*, 2 vols. (Hakluyt Society, London).

Greenblatt, S. (1991) *Marvelous Possessions: The Wonder of the New World* (Clarendon Press, Oxford).

Grewal, J. S. (1970) *Muslim Rule in India: The Assessments of British Historians* (Oxford University Press, Calcutta).

Grotius, H. (1916) *The Freedom of the Seas, or the Right Which Belongs to the Dutch to Take Part in the East Indian Trade*. Translated by R. van Deman Magoffin and edited by J. B. Scott (Oxford University Press, New York).

Guha, R. (1983) *Elementary Aspects of Peasant Insurgency in Colonial India* (Oxford University Press, Delhi).

Guha, R. (1996) *A Rule of Property for Bengal: An Essay on the Idea of Permanent Settlement* (Duke University Press, Durham).

Guha, R., and G. C. Spivak (eds.) (1988) *Selected Subaltern Studies* (Oxford University Press, Oxford).

Habermas, J. (1989) *The Structural Transformation of the Public Sphere: An Inquiry into a Category of Bourgeois Society* (Polity Press, Cambridge).

Hackel, H. B., and P. C. Mancall (2004) "Richard Hakluyt the younger's notes for the East India Company in 1601: a transcription of Huntington Library manuscript EL 2360," *Huntington Library Quarterly*, 67:3, pp. 423–36.

Haellquist, K. R. (ed.) (1991) *Asian Trade Routes* (Curzon Press, London).

Hakluyt, R. (1907 [1598–1600]) *The Principal Navigations, Voyages, Traffiques & Discoveries of the English Nation*, 8 vols. (J. M. Dent and Sons, London).

Halasz, A. (1997) *The Marketplace of Print: Pamphlets and the Public Sphere in Early Modern England* (Cambridge University Press, Cambridge).

Hall, C. (2002) *Civilising Subjects: Metropole and Colony in the English Imagination, 1830–1867* (Polity Press, Cambridge).

Hall, D. D. (1996) *Cultures of Print: Essays in the History of the Book* (University of Massachusetts Press, Amherst).

Hall, M. B. (1965) "Oldenburg and the art of scientific communication," *British Journal for the History of Science*, 2:8, pp. 277–90.

Hall, M. B. (1983) "Oldenburg, the *Philosophical Transactions*, and technology," in J. G. Burke (ed.) *The Uses of Science in the Age of Newton* (University of California Press, Berkeley) pp. 21–47.

Hallward, N. L. (1920) *William Bolts: A Dutch Adventurer under John Company* (Cambridge University Press, Cambridge).

Hancock, D. (1995) *Citizens of the World: London Merchants and the Integration of the British Atlantic Community, 1735–1785* (Cambridge University Press, Cambridge).

Hancock, D. (2000) "'A world of business to do': William Freeman and the foundations of England's commercial empire, 1645–1707," *William and Mary Quarterly*, 3rd ser., 57:1, pp. 3–34.

Hancock, D. (ed.) (2002) *The Letters of William Freeman, London Merchant, 1678–1685*, London Record Society, vol. 36.

Harris, R. C. (2002) *Making Native Space: Colonialism, Resistance, and Reserves in British Columbia* (University of British Columbia Press, Vancouver).

Harris, M. (1997) "Exchanging information: print and business at the Royal Exchange in the late seventeenth century," in A. Saunders (ed.) *The Royal Exchange* (London Topographical Society, London) pp. 188–97.

Harris, S. J. (1998) "Long-distance corporations, big sciences, and the geography of knowledge," *Configurations*, 6, pp. 269–304.

Hejeebu, S. (2005) "Contract enforcement in the English East India Company," *Journal of Economic History*, 65:2, pp. 496–523.

Helgerson, R. (1992) *Forms of Nationhood: The Elizabethan Writing of England* (University of Chicago Press, Chicago).

Hooke, R. (1935) *The Diary of Robert Hooke, 1672–1680*, H. W. Robinson and W. Adams (eds.) (Taylor & Francis, London).

Hoppitt, J. (1993) "Reforming Britain's weights and measures, 1660–1824," *English Historical Review*, 108, pp. 516–40.

Hopwood, A. G., and P. Miller (eds.) (1994) *Accounting as Social and Institutional Practice* (Cambridge University Press, Cambridge).

Houghton, W. E., Jr. (1941) "The history of trades: its relation to seventeenth-century thought," *Journal of the History of Ideas*, 2, pp. 33–60.

Hunt, M. R. (1996) *The Middling Sort: Commerce, Gender, and the Family in England, 1680–1780* (University of California Press, Berkeley).

Hunter, M. (1982) *The Royal Society and Its Fellows, 1660–1700: The Morphology of an Early Scientific Institution*, British Society for the History of Science Monograph 4 (British Society for the History of Science, Chalfont St. Giles).

Hunter, M. (ed.) (1994) *Robert Boyle by Himself and His Friends with a Fragment of William Wotton's Lost Life of Boyle* (William Pickering, London).

Iliffe, R. C. (1992) "'In the warehouse': privacy, property and propriety in the early Royal Society," *History of Science*, 30, pp. 29–68.

Iliffe, R. C. (1995) "Author-mongering: the 'editor' between producer and consumer," in A. Bermingham and J. Brewer (eds.) *The Consumption of Culture, 1600–1800: Image, Object, Text* (Routledge, London) pp. 166–92.

Innis, H. A. (1986 [1950]) *Empire and Communications* (Press Porcépic, Victoria).

Islam, R. (1970) *Indo-Persian Relations: A Study of the Political and Diplomatic Relations between the Mughal Empire and Iran* (Iranian Culture Foundation, Teheran).

Israel, J. I. (1989) *Dutch Primacy in World Trade, 1585–1740* (Clarendon Press, Oxford).

Ito, T. (1984) *The World of the Adat Aceh: A Historical Study of the Sultanate of Aceh*. Ph.D. thesis, Australian National University.

Ito, T., and A. Reid (1985) "From harbour autocracies to 'feudal' diffusion in seventeenth-century Indonesia: the case of Aceh," in E. Leach, S. N. Mukherjee, and J. Ward (eds.) *Feudalism: Comparative Studies* (Sydney Association for Studies in Society and Culture, Sydney) pp. 197–213.

Jackson, A., and A. Jaffer (eds.) (2004) *Encounters: The Meeting of Asia and Europe, 1500–1800* (V&A Publications, London).

Jackson, I. (2004) "Approaches to the history of readers and reading in eighteenth-century Britain," *Historical Journal*, 47:4, pp. 1041–54.

Jacob, J. R. (1980) "Restoration ideologies and the Royal Society," *History of Science*, 18, pp. 25–38.

Jardine, L. (1990) "Mastering the uncouth: Gabriel Harvey, Edmund Spenser and the English experience in Ireland," in J. Henry and S. Sutton (eds.) *New Perspectives on Renaissance Thought: Essays in the History of Science, Education and Philosophy in Memory of Charles B. Schmitt* (Duckworth, London) pp. 68–82.

Jardine, L., and A. Grafton (1990) "'Studied for action': How Gabriel Harvey read his Livy," *Past and Present*, 129, pp. 30–78.

Jardine, N. (2000) "Books, texts, and the making of knowledge," in M. Frasca-Spada and N. Jardine (eds.) *Books and the Sciences in History* (Cambridge University Press, Cambridge) pp. 393–407.

Jasanoff, M. (2005) *Edge of Empire: Lives, Culture, and Conquest in the East, 1750–1850* (Alfred A. Knopf, New York).

Johns, A. (1998) "Science and the book in modern cultural historiography," *Studies in History and Philosophy of Science*, 29A:2, pp. 167–94.

Johns, A. (1998) *The Nature of the Book: Print and Knowledge in the Making* (University of Chicago Press, Chicago).

Johns, A. (2002) "How to acknowledge a revolution," *American Historical Review*, 107:1, pp. 106–25.

Jones, S. R. H., and S. P. Ville (1996) "Efficient transactors or rent-seeking monopolists? The rationale for early chartered trading companies," *Journal of Economic History*, 56:4, pp. 898–915.

Jones, S. R. H., and S. P. Ville (1996) "Theory and evidence: understanding chartered trading companies," *Journal of Economic History*, 56:4, pp. 925–26.

Joseph, B. (2004) *Reading the East India Company, 1720–1840: Colonial Currencies of Gender* (University of Chicago Press, Chicago).

Joshi, P. (2002) *In Another Country: Colonialism, Culture, and the English Novel in India* (Columbia University Press, New York).

Keay, J. (1991) *The Honourable Company: A History of the English East India Company* (HarperCollins, London).

Kennedy, J. (1884) *Life and Work in Benares and Kumaon, 1839–1877* (T. Fisher Unwin, London).

Khan, G. (1998) *Indian Muslim Perceptions of the West during the Eighteenth Century* (Oxford University Press, Karachi).

Khan, M. H. (1976) *History of Printing in Bengali Characters up to 1866*. Ph.D. thesis, School of Oriental and African Studies, University of London.

Khan, M. S. (1962) "The early history of Bengali printing," *Library Quarterly*, 32, pp. 51–61.

Khan, M. S. (1966) "Early Bengali printed books," *Gutenberg Jahrbuch*, pp. 200–208.

Khatchikian, L. (1996) "The ledger of the merchant Hovhannes Joughayetsi," in S. Subrahmanyam (ed.) *Merchant Networks in an Early Modern World* (Variorum, Aldershot) pp. 125–58.

Kishlansky, M. (1996) *A Monarchy Transformed: Britain 1603–1714* (Penguin, Harmondsworth).

Klein, H. S. (1999) *The Atlantic Slave Trade* (Cambridge University Press, Cambridge).

Klein, P. W. (1981) "The origins of trading companies," in L. Blussé and F. Gaastra (eds.) *Companies and Trade: Essays on Overseas Trading Companies during the Ancien Régime* (Leiden University Press, Leiden) pp. 17–28.

Lambert, D., and A. Lester (2006) "Introduction: imperial spaces, imperial subjects," in D. Lambert and A. Lester (eds.) *Colonial Lives across the British Empire: Imperial Careering in the Long Nineteenth Century* (Cambridge University Press, Cambridge) pp. 1–31.

Latour, B. (1987) *Science in Action: How to Follow Scientists and Engineers through Society* (Harvard University Press, Cambridge, Mass.).

Latour, B. (1993) *We Have Never Been Modern* (Harvester Wheatsheaf, Hemel Hempstead).

Latour, B. (1999) *Pandora's Hope: Essays on the Reality of Science Studies* (Harvard University Press, Cambridge, Mass.).

Law, J. (1986) "On the methods of long-distance control: vessels, navigation and the Portuguese route to India," in J. Law (ed.) *Power, Action and Belief: A New Sociology of Knowledge* (Routledge and Kegan Paul, London) pp. 234–63.

Lawson, P. (1987) *The East India Company: A History* (Longman, London).

Lepore, J. (1999) *The Name of War: King Philip's War and the Origins of American Identity* (Vintage Books, New York).

Linebaugh, P., and M. Rediker (2000) *The Many-Headed Hydra: Sailors, Slaves, Commoners, and the Hidden History of the Revolutionary Atlantic* (Beacon Press, Boston).

Livingstone, D. N. (2003) *Putting Science in Its Place: Geographies of Scientific Knowledge* (University of Chicago Press, Chicago).

Livingstone, D. N. (2005) "Science, text and space: thoughts on the geography of reading," *Transactions of the Institute of British Geographers*, 30:4, pp. 391–401.

Lloyd, M. (1979) "Sir Charles Wilkins, 1749–1836," *India Office Library and Records: Report for the Year 1978*, pp. 9–39.

Love, H. (1993) *Scribal Publication in Seventeenth-Century England* (Clarendon Press, Oxford).

Love, H. (2002) "Oral and scribal texts in early modern England," in J. Barnard and
　　D. F. McKenzie (eds.) *The Cambridge History of the Book*, vol. 4: *1557–1695*
　　(Cambridge University Press, Cambridge) pp. 97–121.
Lovejoy, P. E. (2000) *Transformations in Slavery: A History of Slavery in Africa*,
　　2nd ed. (Cambridge University Press, Cambridge).
Lux, D. S., and H. J. Cook (1998) "Closed circles or open networks? communicating
　　at a distance during the scientific revolution," *History of Science*, 36, pp. 179–
　　211.
Mackenzie, J. (1995) *Orientalism: History, Theory and the Arts* (Manchester Univer-
　　sity Press, Manchester).
Majeed, J. (1995) "'The Jargon of Indostan': an exploration of jargon in Urdu and East
　　India Company English," in P. Burke and R. Porter (eds.) *Languages and Jargons:
　　Contributions to a Social History of Language* (Polity Press, Cambridge) pp. 182–
　　205.
Markham, C. R. (ed.) (1877) *The Voyages of Sir James Lancaster to the East Indies*
　　(Hakluyt Society, London).
Markham, C. R. (ed.) (1878) *The Hawkins' Voyages during the Reigns of Henry VIII,
　　Queen Elizabeth and James I* (Hakluyt Society, London).
Marshall, P. J. (1973) "Warren Hastings as scholar and patron," in A. Whiterman,
　　J. S. Bromley, and P. G. M. Dickson (eds.) *Statesmen, Scholars, and Merchants*
　　(Oxford University Press, Oxford) pp. 242–62.
Marshall, P. J. (1987) "Private British trade in the Indian Ocean before 1800," in A.
　　Das Gupta and M. N. Pearson (eds.) *India and the Indian Ocean, 1500–1800*
　　(Oxford University Press, Calcutta) pp. 276–300.
Marshall, P. J. (1987) *Bengal: The British Bridgehead, Eastern India 1740–1828* (Cam-
　　bridge University Press, Cambridge).
Marshall, P. J. (1998) "The English in Asia to 1700," in N. Canny (ed.) *The Oxford
　　History of the British Empire*, vol. 1: *The Origins of Empire* (Oxford University
　　Press, Oxford) pp. 264–85.
Marshman, J. C. (1859) *The Life and Times of Carey, Marshman, and Ward*, 2 vols.
　　(Longman & Co, London).
Martin, J. B. (1892) *"The Grasshopper" in Lombard Street* (Leadenhall Press, London).
Massarella, D. (1990) *A World Elsewhere: Europe's Encounters with Japan in the
　　Sixteenth and Seventeenth Centuries* (Yale University Press, New Haven).
McKenzie, D. F. (1985) *Oral Culture, Literacy and Print in Early New Zealand: The
　　Treaty of Waitangi* (Victoria University Press with the Alexander Turnbull Li-
　　brary Endowment Trust, Wellington).
McKenzie, D. F. (1986) *Bibliography and the Sociology of Texts* (British Library, Lon-
　　don).
McKenzie, D. F. (2002) *Making Meaning: "Printers of the Mind" and Other Essays*,
　　edited by P. D. McDonald and M. F. Suarez (University of Massachusetts Press,
　　Amherst).

McKerrow, R. B. (ed.) (1910) *A Dictionary of Printers and Booksellers in England, Scotland and Ireland, and of Foreign Printers of English Books, 1557–1640* (Bibliographical Society, London).

McLuhan, M. (1969 [1962]) *The Gutenberg Galaxy: The Making of Typographic Man* (Signet Books, New York).

Meilink-Roelofsz, M. A. P. (1962) *Asian Trade and European Influence in the Indonesian Archipelago between 1500 and about 1630* (Martinus Nijhoff, The Hague).

Mentz, S. (1996) "English private trade on the Coromandel Coast, 1660–1690: diamonds and country trade," *Indian Economic and Social History Review*, 33:2, pp. 155–73.

Mentz, S. (2005) *The English Gentleman Merchant at Work: Madras and the City of London, 1660–1740* (Museum Tusculanum Press, Copenhagen).

Merwick, D. (1999) *Death of a Notary: Conquest and Change in Colonial New York* (Cornell University Press, Ithaca).

Mignolo, W. D. (2003) *The Darker Side of the Renaissance: Literacy, Territoriality, and Colonization*, 2nd ed. (University of Michigan Press, Ann Arbor).

Miller, P. (1992) "Accounting and objectivity: the invention of calculating selves and calculable spaces," *Annals of Scholarship*, 9, pp. 61–86.

Miller, P. (2001) "Governing by numbers: why calculative practices matter," *Social Research*, 68, pp. 379–96.

Mirowski, P. (1981) "The rise (and retreat) of a market: English joint-stock shares in the eighteenth century," *Journal of Economic History*, 41, pp. 559–77.

Misra, B. B. (1959) *The Central Administration of the East India Company* (Manchester University Press, Manchester).

Mitchell, C. P. (1997) "Safavid imperial *tarassul* and the Persian *inshā'* tradition," *Studia Iranica*, 26, pp. 173–209.

Mitchell, C. P. (2000) *Sir Thomas Roe and the Mughal Empire* (Area Study Centre for Europe, University of Karachi, Karachi).

Mitchell, T. (2002) *Rule of Experts: Egypt, Techno-politics, Modernity* (University of California Press, Berkeley).

Mohiuddin, M. (1971) *The Chancellery and Persian Epistolography under the Mughals, From Babur to Shah Jahan (1526–1628)* (Iran Society, Calcutta).

Monaghan, E. J. (1993) "'Able and willing to read': the meaning of literacy to the Indians of colonial Martha's Vineyard," in G. Brooks, A. K. Pugh, and N. Hall (eds.) *Further Studies in the History of Reading* (United Kingdom Reading Association, Widnes) pp. 43–59.

Monteyne, J. (2000) *The Space of Print and Printed Spaces in Restoration London, 1669–1685*. Ph.D. thesis, University of British Columbia.

Montrose, L. A. (1999) "Idols of the Queen: policy, gender, and the picturing of Elizabeth I," *Representations*, 68, pp. 108–61.

Moreland, W. H. (ed.) (1934) *Peter Floris, His Voyage to the East Indies in the Globe, 1611–1615* (Hakluyt Society, London).

Murdur, S. P., L. S. Wakankar, and P. M. Ghosh (1986) "Text composition in Devana-gari," *Sesame Bulletin*, 1:1, pp. 18–27.

Myers, R., and M. Harris (eds.) (1990) *Spreading the Word: The Distribution Net-works of Print, 1550–1850* (St Paul's Bibliographies, Winchester).

Narain, P. (1970) *Press and Politics in India, 1885–1905* (Munshiram Manoharlal, Delhi).

Neal, L. (1988) "The rise of a financial press: London and Amsterdam, 1681–1810," *Business History*, 30, pp. 163–78.

Neal, L. (1990) *The Rise of Financial Capitalism: International Capital Markets in the Age of Reason* (Cambridge University Press, Cambridge).

Neville-Sington, P. (1997) "The primary Purchas bibliography," in L. E. Pennington (ed.) *The Purchas Handbook: Studies of the Life, Times and Writings of Samuel Purchas, 1577–1626*, 2 vols. (Hakluyt Society, London) 2, pp. 465–573.

Ngugi wa Thiong'o (1986) *Decolonising the Mind: The Politics of Language in African Literature* (James Currey, London).

Nicols, J. (1823) *The Progresses and Public Processions of Queen Elizabeth*, 3 vols. (John Nicols and Son, London).

Nijman, J. (1994) "The VOC and the expansion of the world-system 1602–1799," *Political Geography*, 13:3, pp. 211–27.

Nussbaum, F. A. (ed.) (2003) *The Global Eighteenth Century* (Johns Hopkins University Press, Baltimore).

Ogborn, M. (1998) "The capacities of the state: Charles Davenant and the manage-ment of the Excise, 1683–1698," *Journal of Historical Geography*, 24, pp. 289–312.

Oldenburg, Henry (1965–86) *The Correspondence of Henry Oldenburg*, 13 vols., A. R. Hall and M. B. Hall (eds.) (University of Wisconsin Press, Madison).

Ong, W. J. (1988 [1982]) *Orality and Literacy: The Technologizing of the Word* (Rout-ledge, London).

Osborn, J. (2002) "India and the East India Company in the public sphere of eighteenth-century Britain," in H. V. Bowen, M. Lincoln, and N. Rigby (eds.) *The Worlds of the East India Company* (Boydell Press, Woodbridge) pp. 201–21.

Pagden, A. (1995) *Lords of All the World: Ideologies of Empire in Spain, Britain and France, c.1500–c.1800* (Yale University Press, New Haven).

Parry, J. (1985) "The Brahmanical tradition and the technology of the intellect," in J. Overing (ed.) *Reason and Morality* (Tavistock, London) pp. 200–225.

Pearson, M. N. (1991) "Merchants and states," in J. D. Tracy (ed.) *The Political Econ-omy of Merchant Empires: State Power and World Trade, 1350–1750* (Cambridge University Press, Cambridge) pp. 41–116.

Pinch, W. R. (1997) "Same difference in India and Europe," *History and Theory*, 38:3, pp. 389–407.

Pincus, S. (1995) "'Coffee politicians does create': coffeehouses and restoration politi-cal culture," *Journal of Modern History*, 67, pp. 807–34.

Pocock, J. G. A. (1957) *The Ancient Constitution and the Feudal Law: A Study of English Historical Thought in the Seventeenth Century* (Cambridge University Press, Cambridge).

Poovey, M. (1998) *A History of the Modern Fact: Problems of Knowledge in the Sciences of Wealth and Society* (University of Chicago Press, Chicago).

Porter, T. M. (1995) *Trust in Numbers: The Pursuit of Objectivity in Science and Public Life* (Princeton University Press, Princeton, NJ).

Prakash, O. (1981) "European trade and south Asian economies: some regional contrasts, 1600–1800," in L. Blussé and F. Gaastra (eds.) *Companies and Trade: Essays on Overseas Trading Companies during the Ancien Régime* (Leiden University Press, Leiden) pp. 189–205.

Price, J. M. (1954) "Notes on some London price-currents, 1667–1715," *Economic History Review*, 7:2, pp. 240–50.

Priolkar, A. K. (1958) *The Printing Press in India* (Marathi Samshodhana Mandala, Bombay).

Purchas, S. (1905 [1625]) *Hakluytus Posthumus, or Purchas His Pilgrimes*, 20 vols. (James MacLehose & Sons, Glasgow).

Qayyum, M. A. (1982) *A Critical Study of the Early Bengali Grammars: Halhed to Haughton* (Asiatic Society of Bangladesh, Dhaka).

Raj, K. (2000) "Colonial encounters and the forging of new knowledge and national identities: Great Britain and India, 1760–1850," *Osiris*, 15, pp. 119–34.

Raj, K. (2003) "Circulation and the emergence of modern mapping: Great Britain and early colonial India, 1764–1820," in C. Markovits, J. Pouchepadass, and S. Subrahmanyam (eds.) *Society and Circulation: Mobile People and Itinerant Cultures in South Asia, 1750–1950* (Permanent Black, Delhi) pp. 23–54.

Rama, A. (1996) *The Lettered City* (Duke University Press, Durham).

Rappaport, J. (1994) "Object and alphabet: Andean Indians and documents in the colonial period," in E. H. Boone and W. D. Mignolo (eds.) *Writing without Words: Alternative Literacies in Mesoamerica and the Andes* (Duke University Press, Durham) pp. 271–92.

Raven, J., H. Small, and N. Tadmor (eds.) (1996) *The Practice and Representation of Reading in England* (Cambridge University Press, Cambridge).

Ray, R. K. (1998) "Indian society and the establishment of British supremacy, 1765–1818," in P. J. Marshall (ed.) *The Oxford History of the British Empire*, vol. 2: *The Eighteenth Century* (Oxford University Press, Oxford) pp. 508–29.

Raymond, J. (2003) *Pamphlets and Pamphleteering in Early Modern Britain* (Cambridge University Press, Cambridge).

Reid, A. (1975) "Trade and the problem of royal power in Aceh: three stages—c. 1550–1700," *Monographs of the Malaysian Branch of the Royal Asiatic Society 6: Precolonial State Systems in Southeast Asia*, pp. 45–55.

Rhodes, D. E. (1969) *The Spread of Printing. Eastern Hemisphere: India, Pakistan, Ceylon, Burma and Thailand* (Vangendt & Co, Amsterdam).

Richards, J. F. (1984) "Norms of comportment among imperial Mughal officers," in B. D. Metcalf (ed.) *Moral Conduct and Authority: The Place of Adab in South Asian Islam* (University of California Press, Berkeley) pp. 255–89.

Richards, J. F. (1986) *Document Forms for Official Orders of Appointment in the Mughal Empire* (E. J. W. Gibb Memorial Trust, Cambridge).

Richards, J. F. (1995) *The Mughal Empire* (Cambridge University Press, Cambridge).

Richards, J. F. (1998) "The formulation of imperial authority under Akbar and Jahangir," in M. Alam and S. Subrahmanyam (eds.) *The Mughal State, 1526–1750* (Oxford University Press, New Delhi) pp. 126–67.

Roach, J. (1996) *Cities of the Dead: Circum-Atlantic Performance* (Columbia University Press, New York).

Robinson, F. (1993) "Technology and religious change: Islam and the impact of print," *Modern Asian Studies*, 27:1, pp. 229–51.

Rocher, R. (1983) *Orientalism, Poetry, and the Millennium: The Checkered Life of Nathaniel Brassey Halhed, 1751–1830* (Motilal Banarsidass, Delhi).

Rocher, R. (1989) "The career of Rādhākānta Tarkavāgīśa: an eighteenth-century pandit in British employ," *Journal of the American Oriental Society*, 109:4, pp. 627–33.

Rogers, J. M. (1993) *Mughal Miniatures* (British Museum, London).

Rogers, K. (1935) *Old London: Cornhill, Threadneedle Street and Lombard Street, Old Houses and Signs* (Whitefriars Press, London and Tonbridge).

Ross, F. G. E. (1999) *The Printed Bengali Character and Its Evolution* (Curzon, London).

Said, E. W. (1978) *Orientalism* (Routledge and Kegan Paul, London).

Sarkar, N. (1990) "Printing and the spirit of Calcutta," in S. Chaudhuri (ed.) *Calcutta: The Living City*, vol. 1: *The Past* (Oxford University Press, Calcutta) pp. 128–36.

Satow, Sir E. M. (ed.) (1900) *The Voyage of Captain John Saris to Japan, 1613* (Hakluyt Society, London).

Saxe, E. L. (1979) *Fortune's Tangled Web: Trading Networks of English Entrepreneurs in Eastern India. 1657–1717.* Ph.D. thesis, Yale University.

Schiebinger, L., and C. Swan (eds.) (2005) *Colonial Botany: Science, Commerce, and Politics in the Early Modern World* (University of Pennsylvania Press, Philadelphia).

Schimmel, A. (1990) *Calligraphy and Islamic Culture* (I. B. Tauris, London).

Schimmel, A. (2004) *The Empire of the Great Mughals: History, Art and Culture* (Reaktion Books, London).

Scott, W. R. (1910–12) *The Constitution and Finance of English, Scottish and Irish Joint-Stock Companies to 1720*, 3 vols. (Cambridge University Press, Cambridge).

Secord, J. A. (2000) *Victorian Sensation: The Extraordinary Publication, Reception, and Secret Authorship of Vestiges of the Natural History of Creation* (University of Chicago Press, Chicago).

Secord, J. A. (2004) "Knowledge in transit," *Isis*, 95, pp. 654–72.

Seed, P. (1995) *Ceremonies of Possession in Europe's Conquest of the New World, 1492–1640* (Cambridge University Press, Cambridge).

Sen, S. (1968) "Early printers and publishers in Calcutta," *Bengal Past and Present*, 87, pp. 59–66.

Sen, S. (1998) *Empire of Free Trade: The East India Company and the Making of the Colonial Marketplace* (University of Pennsylvania Press, Philadelphia).

Sen, S. (2002) *Distant Sovereignty: National Imperialism and the Origins of British India* (Routledge, London).

Sen, S. (2004) "Liberal empire and illiberal trade: the political economy of 'responsible government' in early British India," in K. Wilson (ed.) *A New Imperial History: Culture, Identity and Modernity in Britain and the Empire, 1660–1840* (Cambridge University Press, Cambridge) pp. 136–54.

Shapin, S. (1994) *A Social History of Truth: Civility and Science in Seventeenth-Century England* (University of Chicago Press, Chicago).

Shapin, S., and Schaffer, S. (1985) *Leviathan and the Air-Pump: Hobbes, Boyle, and the Experimental Life* (Princeton University Press, Princeton, NJ).

Shapiro, B. J. (2000) *A Culture of Fact: England, 1550–1720* (Cornell University Press, Ithaca).

Sharpe, K. (2000) *Reading Revolutions: The Politics of Reading in Early Modern England* (Yale University Press, New Haven).

Sharpe, K. (2000) *Remapping Early Modern England: The Culture of Seventeenth-Century Politics* (Cambridge University Press, Cambridge).

Shaw, G. (1981) *Printing in Calcutta to 1800* (Bibliographic Society, London).

Sherman, W. H. (1995) *John Dee: The Politics of Reading and Writing in the English Renaissance* (University of Massachusetts Press, Amherst).

Singh, J. G. (1996) *Colonial Narratives/Cultural Dialogues: "Discoveries" of India in the Language of Colonialism* (Routledge, London).

Sinha, B. (1994) *Press and National Movement in India (1911 to 1947)* (Manak Publications, Delhi).

Sivasundaram, S. (2005) "Trading knowledge: the East India Company's elephants in India and Britain," *Historical Journal*, 48:1, pp. 27–63.

Sorrenson, R. (1996) "The ship as a scientific instrument in the eighteenth century," *Osiris*, 11, pp. 221–36.

Spivak, G. C. (1988) "Subaltern studies: deconstructing historiography," in R. Guha and G. C. Spivak (eds.) (1988) *Selected Subaltern Studies* (Oxford University Press, Oxford) pp. 3–32.

Spufford, M. (1981) *Small Books and Pleasant Histories: Popular Fiction and Its Readership in Seventeenth-Century England* (Methuen, London).

Steensgaard, N. (1974) *The Asian Trade Revolution of the Seventeenth Century: The East India Companies and the Decline of the Caravan Trade* (University of Chicago Press, Chicago).

Steensgaard, N. (1981) "The companies as a specific institution in the history of European expansion," in L. Blussé and F. Gaastra (eds.) *Companies and Trade: Essays on Overseas Trading Companies during the Ancien Régime* (Leiden University Press, Leiden) pp. 245–64.

Stokes, E. (1959) *The English Utilitarians and India* (Clarendon Press, Oxford).

Stoler, A. L., and F. Cooper (1997) "Between metropole and colony: rethinking a re-
    search agenda," in F. Cooper and A. L. Stoler (eds.) *Tensions of Empire: Colonial
    Cultures in a Bourgeois World* (University of California Press, Berkeley) pp. 1–56.

Stubbs, M. (1982) "John Beale, philosophical gardener of Herefordshire. Part I. Prelude
    to the Royal Society (1608–1663)," *Annals of Science*, 39, pp. 463–89.

Stubbs, M. (1989) "John Beale, philosophical gardener of Herefordshire. Part II. The
    improvement of agriculture and trade in the Royal Society (1663–1683)," *Annals
    of Science*, 46, pp. 323–63.

Styles, J. (2000) "Product innovation in early modern London," *Past and Present*, 168,
    pp. 124–69.

Subrahmanyam, S., and C. A. Bayly (1988) "Portfolio capitalists and the political
    economy of early modern India," *Indian Economic and Social History Review*
    25:4, pp. 401–24.

Subrahmanyam, S. (2002) "Frank submissions: the Company and the Mughals be-
    tween Sir Thomas Roe and Sir William Norris," in H. V. Bowen, M. Lincoln, and
    N. Rigby (eds.) *The Worlds of the East India Company* (Boydell Press, Wood-
    bridge) pp. 69–96.

Supple, B. E. (1964) *Commercial Crisis and Change in England, 1600–1642: A Study
    in the Instability of a Mercantile Economy* (Cambridge University Press, Cam-
    bridge).

Sutherland, L. S. (1952) *The East India Company in Eighteenth-Century Politics*
    (Clarendon Press, Oxford).

Teltscher, K. (1997) *India Inscribed: European and British Writing on India, 1600–
    1800* (Oxford University Press, Delhi).

Temple, Sir R. C. (1911) *The Diaries of Streynsham Master, 1675–1680*, 2 vols. (In-
    dian Record Series, John Murray, London).

Thomas, N. (1991) *Entangled Objects: Exchange, Material Culture, and Colonialism
    in the Pacific* (Harvard University Press, Cambridge, Mass.).

Thompson, J. E. (1994) *Mercenaries, Pirates, and Sovereigns: State-Building and
    Extraterritorial Violence in Early Modern Europe* (Princeton University Press,
    Princeton, NJ).

Tiffin, C., and A. Lawson (eds.) (1994) *De-Scribing Empire: Post-colonialism and
    Textuality* (Routledge, London).

Todorov, T. (1984) *The Conquest of America* (Harper and Row, New York).

Tracy, J. D. (ed.) (1991) *The Political Economy of Merchant Empires: State Power and
    World Trade, 1350–1750* (Cambridge University Press, Cambridge).

Travers, R. (2001) *Contested Notions of Sovereignty in Bengal under British Rule.*
    Ph.D. thesis, University of Cambridge.

Travers, R. (2004) "'The real value of the lands': the *Nawabs*, the British and the land
    tax in eighteenth-century Bengal," *Modern Asian Studies*, 38:3, pp. 517–58.

Travers, R. (2005) "Ideology and British expansion in Bengal," *Journal of Imperial and
    Commonwealth History*, 33:1, pp. 7–27.

Tribble, E. B. (1993) *Margins and Marginality: The Printed Page in Early Modern England* (University Press of Virginia, Charlottesville).

Waldstreicher, D. (1999) "Reading the runaways: self-fashioning, print culture, and confidence in slavery in the eighteenth-century mid-Atlantic," *William and Mary Quarterly*, 3rd ser., 56: 2, pp. 243–72.

Wallerstein, I. (1980) *The Modern World System II: Mercantilism and the Consolidation of the European World-Economy, 1600–1750* (Academic Press, London).

Warner, M. (1990) *The Letters of the Republic: Publication and the Public Sphere in Eighteenth-Century America* (Harvard University Press, Cambridge, Mass.).

Watson, I. B. (1980) *Foundation for Empire: English Private Trade in India, 1659–1760* (Vikas, New Delhi).

Watson, I. B. (1987) "Indian merchants and English private interests: 1659–1760," in A. Das Gupta and M. N. Pearson (eds.) *India and the Indian Ocean, 1500–1800* (Oxford University Press, Calcutta) pp. 301–16.

Weitzman, S. (1929) *Warren Hastings and Philip Francis* (Manchester University Press, Manchester).

Welch, R. (2002) "The book in Ireland from the Tudor re-conquest to the battle of the Boyne," in J. Barnard and D. F. McKenzie (eds.) *The Cambridge History of the Book*, vol. 4: *1557–1695* (Cambridge University Press, Cambridge) pp. 701–18.

Williams, G. (1997) *The Great South Sea: English Voyages and Encounters 1570–1750* (Yale University Press, New Haven).

Wilson, H. H. (1855) *A Glossary of Judicial and Revenue Terms* (William H. Allen and Co., London).

Wilson, J. E. (2000) *Governing Property, Making Law: Land, Local Society and Colonial Discourse in Agrarian Bengal, c. 1785–1830*. Ph.D. thesis, University of Oxford.

Wilson, J. E. (2005) "'A thousand countries to go to': peasants and rulers in late eighteenth-century Bengal," *Past and Present*, 189, pp. 81–109.

Wilson, K. (1995) *The Sense of the People: Politics, Culture and Imperialism in England, 1715–1785* (Cambridge University Press, Cambridge).

Wilson, K. (2003) *The Island Race: Englishness, Empire and Gender in the Eighteenth Century* (Routledge, London).

Wilson, K. (ed.) (2004) *A New Imperial History: Culture, Identity and Modernity in Britain and the Empire, 1660–1840* (Cambridge University Press, Cambridge).

Wolf, E. R. (1982) *Europe and the People Without History* (University of California Press, Berkeley).

Wolff, J. (1993) "On the road again: metaphors of travel in cultural criticism," *Cultural Studies*, 7, pp. 224–39.

Wootton, D. (ed.) (2003) *Divine Right and Democracy: An Anthology of Political Writing in Stuart England* (Hackett, Cambridge).

Young, R. J. C. (2001) *Postcolonialism: An Historical Introduction* (Blackwell, Oxford).

Yule, Colonel Sir H. (ed.) (1887) *The Diary of William Hedges*, 3 vols. (Hakluyt Society, London).

Zaret, D. (1992) "Religion, science, and printing in the public spheres of seventeenth-century England," in Craig Calhoun (ed.) *Habermas and the Public Sphere* (MIT Press, Cambridge, Mass.) pp. 221–34.

Zaret, D. (2000) *Origins of Democratic Culture: Printing, Petitions, and the Public Sphere in Early-Modern England* (Princeton University Press, Princeton, NJ).

INDEX

writing (*cont.*)

251–53; Indian civil society and, 18–19, 217; Islamic diplomacy and, 36–37, 62–64, 65–66; management of, 99–103; material forms of, xxiii, 5–6, 21, 42–45, 56, 62–64, 70, 76; navigation and, 49–50; non-alphabetic forms, 13; parliamentary practices of, 106; plainness as a valued style of, 96, 97, 134–35, 137, 153, 192; power and, 12–22, 57, 252–53, 272–73; in precolonial Bengal, 17–18, 217; separation of public and private and, 72, 81–82, 92,

100, 101; ships' cargoes and, 52; speech and, 10; state power and, 19; style of, 118, 134–35; as a technology of preservation and reproduction, 34; trade and, 34–35, 69–103. *See also* manuscripts; print; texts

Young Secretary's Guide (Hill), 96

zamīndars: imperial administration and, 214–15, 228, 251; imperial authority and, 233–34; writing and, 18